Birkhäuser

Applied and Numerical Harmonic Analysis

Dmitriy Bilyk • Emil Iacob •
Andrei Martinez-Finkelshtein •
Alexander M. Stokolos
Editors

Recent Advances in Approximation and Potential Theory

In Honor of Edward Saff

Birkhäuser

Editors
Dmitriy Bilyk
School of Mathematics
University of Minnesota
Minneapolis, MN, USA

Emil Iacob
Department of Mathematical Sciences
Georgia Southern University
Statesboro, GA, USA

Andrei Martinez-Finkelshtein
Department of Mathematics
Baylor University
Waco, TX, USA

Alexander M. Stokolos
Department of Mathematical Sciences
Georgia Southern University
Statesboro, GA, USA

ISSN 2296-5009 ISSN 2296-5017 (electronic)
Applied and Numerical Harmonic Analysis
ISBN 978-3-031-95550-1 ISBN 978-3-031-95551-8 (eBook)
https://doi.org/10.1007/978-3-031-95551-8

Mathematics Subject Classification: 33C45, 42-XX, 41-06, 31-06

This book is published under the imprint Birkhäuser, www.birkhauser-science.com by the registered company Springer Nature Switzerland AG
The registered company address is: Gewerbestrasse 11, 6330 Cham, Switzerland

This volume is dedicated to Edward Saff's 80th birthday.

Ed Saff, November 2004

ANHA Series Preface

The *Applied and Numerical Harmonic Analysis* (ANHA) book series aims to provide the engineering, mathematical, and scientific communities with significant developments in harmonic analysis, ranging from abstract harmonic analysis to basic applications. The title of the series reflects the importance of applications and numerical implementation, but richness and relevance of applications and implementation depend fundamentally on the structure and depth of theoretical underpinnings. Thus, from our point of view, the interleaving of theory and applications and their creative symbiotic evolution is axiomatic.

Harmonic analysis is a wellspring of ideas and applicability that has flourished, developed, and deepened over time within many disciplines and by means of creative cross-fertilization with diverse areas. The intricate and fundamental relationship between harmonic analysis and fields such as signal processing, partial differential equations (PDEs), and image processing is reflected in our state-of-the-art ANHA series.

Our vision of modern harmonic analysis includes a broad array of mathematical areas, e.g., wavelet theory, Banach algebras, classical Fourier analysis, time-frequency analysis, deep learning, and fractal geometry, as well as the diverse topics that impinge on them.

For example, wavelet theory can be considered an appropriate tool to deal with some basic problems in digital signal processing, speech and image processing, geophysics, pattern recognition, biomedical engineering, and turbulence. These areas implement the latest technology from sampling methods on surfaces to fast algorithms and computer vision methods. The underlying mathematics of wavelet theory depends not only on classical Fourier analysis, but also on ideas from abstract harmonic analysis, including von Neumann algebras and the affine group. This leads to a study of the Heisenberg group and its relationship to Gabor systems, and of the metaplectic group for a meaningful interaction of signal decomposition methods.

The unifying influence of wavelet theory in the aforementioned topics illustrates the justification for providing a means for centralizing and disseminating information from the broader, but still focused, area of harmonic analysis. This will be a key role of ANHA. We intend to publish with the scope and interaction that such a host of issues demands.

Along with our commitment to publish mathematically significant works at the frontiers of harmonic analysis, we have a comparably strong commitment to publish major advances in the following applicable topics in which harmonic analysis plays a substantial role:

*Analytic Number theory * Antenna Theory * Artificial Intelligence * Biomedical Signal Processing * Classical Fourier Analysis * Coding Theory * Communications Theory * Compressed Sensing * Crystallography and Quasi-Crystals * Data Mining * Data Science * Deep Learning * Digital Signal Processing * Dimension Reduction and Classification * Fast Algorithms * Frame Theory and Applications * Gabor Theory and Applications * Geophysics * Image Processing * Machine Learning * Manifold Learning * Numerical Partial Differential Equations * Neural Networks * Phaseless Reconstruction * Prediction Theory * Quantum Information Theory * Radar Applications * Sampling Theory (Uniform and Non-uniform) and Applications * Spectral Estimation * Speech Processing * Statistical Signal Processing * Super-resolution * Time Series * Time-Frequency and Time-Scale Analysis * Tomography * Turbulence * Uncertainty Principles *Waveform design * Wavelet Theory and Applications

The above point of view for the ANHA book series is inspired by the history of Fourier analysis itself, whose tentacles reach into so many fields.

In the last two centuries Fourier analysis has had a major impact on the development of mathematics, on the understanding of many engineering and scientific phenomena, and on the solution of some of the most important problems in mathematics and the sciences. Historically, Fourier series were developed in the analysis of some of the classical PDEs of mathematical physics; these series were used to solve such equations. In order to understand Fourier series and the kinds of solutions they could represent, some of the most basic notions of analysis were defined, e.g., the concept of "function." Since the coefficients of Fourier series are integrals, it is no surprise that Riemann integrals were conceived to deal with uniqueness properties of trigonometric series. Cantor's set theory was also developed because of such uniqueness questions.

A basic problem in Fourier analysis is to show how complicated phenomena, such as sound waves, can be described in terms of elementary harmonics. There are two aspects of this problem: first, to find, or even define properly, the harmonics or spectrum of a given phenomenon, e.g., the spectroscopy problem in optics; second, to determine which phenomena can be constructed from given classes of harmonics, as done, for example, by the mechanical synthesizers in tidal analysis.

Fourier analysis is also the natural setting for many other problems in engineering, mathematics, and the sciences. For example, Wiener's Tauberian theorem in Fourier analysis not only characterizes the behavior of the prime numbers, but is a fundamental tool for analyzing the ideal structures of Banach algebras. It also provides the proper notion of spectrum for phenomena such as white light. This latter process leads to the Fourier analysis associated with correlation functions in filtering and prediction problems. These problems, in turn, deal naturally with Hardy spaces in complex analysis, as well as inspiring Wiener to consider communications engineering in terms of feedback and stability, his cybernetics. This latter theory develops concepts to understand complex systems such as learning and cognition and neural networks; and it is arguably a precursor of deep learning and its spectacular interactions with data science and AI.

Nowadays, some of the theory of PDEs has given way to the study of Fourier integral operators. Problems in antenna theory are studied in terms of unimodular trigonometric polynomials. Applications of Fourier analysis abound in signal processing, whether with the fast Fourier transform (FFT), or filter design, or the adaptive modeling inherent in time-frequency-scale methods such as wavelet theory.

The coherent states of mathematical physics are translated and modulated Fourier transforms, and these are used, in conjunction with the uncertainty principle, for dealing with signal reconstruction in communications theory. We are back to the raison d'etre of the ANHA series!

College Park, MD *John Benedetto*
Wojciech Czaja
Boston, MA *Kasso Okoudjou*

Foreword

Edward Saff has played a leading role in the fields of approximation theory, potential theory, and complex analysis for many decades. Dr. Saff was born in New York in 1944 to a family with two older brothers. He is married to Loretta Singer; they have three daughters and seven grandchildren.

Edward Saff graduated from the Georgia Institute of Technology in 1964 earning the bachelor's degree with the highest honors. In 1968, he received his Ph.D. degree in Mathematics from University of Maryland, College Park, under the guidance of a renowned mathematician, Joseph Walsh. He joined the University of South Florida (USF) in 1969 after spending a year at Imperial College London as a Fulbright Fellow. In 1986, Ed Saff became a Distinguished Research Professor at USF. In 2001, he moved to Vanderbilt University, Nashville, TN, to become a Professor and the Director of the Center for Constructive Approximation. In 2023, Ed Saff was named a Distinguished Professor of Mathematics there. Edward Saff was elected an AMS Inaugural Fellow in 2012 and a SIAM Fellow in 2023.

There are many highly valued research contributions of Ed Saff. Probably the highest valued among them is the seminal paper published with H. Mhaskar in 1984 and their subsequent papers together with the works of E. Rakhmanov. This sequel of papers created a foundation for a comprehensive study of orthogonal and extremal polynomials with respect to exponential weights. In particular, this paper introduced the Mhaskar–Rakhmanov–Saff number. In 1985, together with H. Mhaskar, Ed Saff wrote another groundbreaking work on the support of the sup-norm of a weighted orthogonal polynomial making a significant contribution to potential theory with external fields and laying the groundwork for the subsequent very highly rated research monograph "Logarithmic Potentials with External Fields" (joint with V. Totik).

In 1990s, Ed Saff proved his first results on distribution of points on the sphere (jointly with E. Rakhmanov, M. Zhou, and A. Kuijlaars). This work is related to the famous Thomson's problem on ground state depositions of electrons in an atom, to the discovery of the Carbon-60 molecule (this discovery was awarded the Nobel Prize), to the 7th Smale's problem for the twenty-first century, and to the theory of spherical designs and spherical cubatures. The Math Intelligencer article

"Distributing many points on the sphere" has over 1500 citations on Google Scholar and its impact expands well beyond fields in Mathematics.

One of the most notable research results of Ed Saff in this direction is the "Poppy-seed bagel" theorem where he (jointly with D. Hardin and later also with S. Borodachov) established the asymptotic behavior of the minimal Riesz energy value and of optimal configurations on rectifiable sets in the hyper-singular case. This work was featured on the cover of the November 2004 issue of *Notices of the AMS*, in a very popular French journal *Science & Vie* (July 2005), and in an interview of Ed Saff to National Public Radio (in December 2004). It also inspired the subsequent research monograph "Discrete Energy on Rectifiable Sets" showcasing Ed Saff's more recent research contributions.

Towson University
Purdue University-Fort Wayne
January 2025

Sergiy Borodachov
Peter D. Dragnev

Preface

The volume arose as a follow-up to the International Conference on Approximation and Potential Theory, held September 22–24, 2023, at Georgia Southern University, Savannah campus, supported by the NSF grant No. 1955202. The conference brought together many experts in the approximation and potential theory, postdoctorates, and graduate students in an effort to promote advances in the area, in particular, to young mathematicians. It was a great networking opportunity for junior faculty and graduate students, who had a first-hand opportunity to further their field knowledge by interacting with established researchers.

In total, there were 45 participants, 17 invited talks, and 16 contributed talks. A center of gravity of the conference was Distinguished Professor, Director of Center for Constructive Approximation at Vanderbilt University, Edward B. Saff. The group photo below confirms this statement.

First row: Elena Berdysheva, Tetiana Stepaniuk, Azita Mayeli, Erik Lundberg, Ryan Matzke, Iryna Petrenko, Diana Son.
Second row: Nathaniel Tenpas, Emma D'Aniella, Minh Quan Hoang Vu, Brian Simanek.
Third row: Chandler Haight, Dmitriy Bilyk.
Fourth row: Liudmyla Kryvonos, Edward Saff, Andrei Martinez-Finkelshtein.
Fifth row: Laurent Moonens, Alexey Solyanik, Luis Rodriguez, Alex Iosevich, Nicolas Nagel.
Six row: Sergiy Borodachov, Laura De Carli, Ivy Collins, David Roach, Md Mahmud Hasan.
Seventh row: Oleg Asipchuk, Doug Hardin, Nurgissa Yessirkegenov.

Waco, TX, USA — Andrei Martínez-Finkelshtein
Statesboro, GA, USA — Alexander M. Stokolos
Minneapolis, MN, USA — Dmitriy Bilyk
Statesboro, GA, USA — Emil Iacob

Acknowledgments

This volume would not have been possible without the contributions of all of our authors. The editors are grateful for the time and effort these colleagues put into writing their manuscripts.

All articles were peer-reviewed, so we thank our referees who took the time to help make this volume a reality.

We are also indebted to our editor at Springer, Christopher Tominich, who was accommodating and patient with us as we put together the volume.

And last but not least, we should mention the National Science Foundation and the Department of Mathematical Sciences at Georgia Southern University who generously supported the conference. Without their support, the project would never have come to fruition.

Contents

Ed Saff at 80: Career and Distinguished Scientific Contributions

Sergiy Borodachov and Peter D. Dragnev

Edward Barry Saff, born on January 2, 1944, in Brooklyn, New York, to Irving and Rose (Koslow) Saff, comes from a family with a diverse set of professions and experiences. His father worked as a sales manager at a large department store. Ed's two older brothers include Harvey Saff, a podiatrist who founded the Pompano Podiatry Group, and Donald Saff, a well-established artist, art historian, and educator who founded Graphic Studio. Ed's maternal grandfather emigrated from the Russian Empire in 1903, fleeing anti-Jewish violence, specifically the pogroms occurring in his village of Rossava in the Kyiv region. Ed married Loretta Singer on July 3, 1966. Together, they have three daughters—Lisa Jill, Tracy Karen, and Alison Michelle—and seven grandchildren.

Ed's distinguished academic career began with his graduation from the Georgia Institute of Technology in 1964, where he earned his Bachelor's degree with the highest honors. Notably, as a senior (at the age of 19), he was asked to teach classes to his fellow students, showcasing his early aptitude for education. He continued his studies at the University of Maryland, College Park, where he pursued a Doctor of Philosophy degree in Mathematics under the guidance of the renowned analyst Joseph Walsh. His thesis, titled "Interpolation and Functions of Class $H(k, a, 2)$", was defended in 1968.

As a graduate student, Ed received a NASA Fellowship and, after earning his Ph.D., he was awarded a prestigious Fulbright Fellowship at Imperial College London, England, spending a year there under the mentorship of Walter Hayman.

S. Borodachov (✉)
Department of Mathematics, Towson University, Towson, MD, USA
e-mail: sborodachov@towson.edu

P. D. Dragnev
Department of Mathematical Sciences, Purdue University, Fort Wayne, IN, USA
e-mail: dragnevp@pfw.edu

D. Bilyk et al. (eds.), *Recent Advances in Approximation and Potential Theory*, Applied and Numerical Harmonic Analysis,
https://doi.org/10.1007/978-3-031-95551-8_1

Fig. 1 Edward B. Saff. Courtesy of Loretta Saff

Ed joined the University of South Florida in Tampa, Florida in 1969, where he became a Distinguished Research Professor in 1986. For the academic year 1978–1979, Ed was recognized with a Guggenheim Fellowship, which he held while at the University of Oxford.

In 1985, Ed founded and directed the Institute for Constructive Mathematics at the University of South Florida. He also served as an Associate Director of the Florida Regional Center for Excellence in Mathematics, Science, Computers and Technology between 1983 and 1990, and served as a member of the Florida House of Representatives Special Task Force on Science, Mathematics and Computer Science. During his tenure at the University of South Florida, Ed received the Distinguished Scholar Award, the College of Natural Sciences Outstanding Scholar Award, the Sigma Xi Outstanding Researcher Award, and the Professional Excellence Program Award.

At an early stage of his career, Ed co-organized jointly with a renowned Russian complex analyst, Andrei Gonchar, an ad hoc meeting among researchers from both the East and West at the International Congress of Mathematics in Helsinki in 1978. Its purpose was to discuss the state of knowledge at that time on Padé approximation. Gonchar's input at that meeting had a significant influence on Ed's further research direction. Later, they organized two major U.S.-U.S.S.R. approximation theory conferences, the first one in Tampa, in 1990 and the second one in Leningrad in 1991. These meetings had a substantial impact since they helped facilitate a number of lasting collaborations that built and strengthened communications among researchers from different geopolitical backgrounds.

In summer 2006, Ed Saff visited the Ukraine searching for information about his ancestors who lived in Rossava village not far from Kyiv. One startling fact that he was able to confirm during that visit was that 'Tsap' not 'Saff' was his original family name. After presenting his talk at a workshop at the Taras Shevchenko National University Kyiv, Ed together with Igor Shevchuk and his recent Ph.D.

student (one of the current authors) traveled to Rossava village. There, he had a chance to talk to several residents who told him what they knew about the Tsap family who once lived in the village. During the trip, Ed was invited to visit a home in the village which, according to local people, was on the site of the home where his grandparents once lived. The grandmotherly lady ('babushka') who lived in the house met the visitors at the door enthusiastically waving the deed of ownership. She was afraid that Ed came to claim the ownership rights, but later she was reassured that this was not the case. The babushka invited Ed and his colleagues to stay at her home over night, but they rejected that gracious proposal, a decision they later regretted. Indeed, after extensively searching for a comfortable hotel, the closest they came was to one in nearby Myronivka. It turned out to be a hotel for train engine operators and had a suspiciously low price. It had spartan furnishings with beds having iron meshes instead of mattresses and a showering room with a rusty look. But a night in that uncomfortable hotel did not derail the travelers. The next day they returned to Rossava and were invited by the babushka to have a picnic lunch in her backyard. Reportedly, during that lunch while enjoying some delicious Borsch, Ed jokingly said "I hope you won't poison me to prevent from me claiming the ownership". Of course, everyone was healthy after the lunch and Ed returned from that trip with many great memories and impressions. He recorded them in an amusing self-published book (available on Amazon).

In 2001, Ed moved to Vanderbilt University, taking on the role of a Professor and the Director of the Center for Constructive Approximation. During the period from 2004 to 2007, he served as Executive Dean of the College of Arts and Science. In 2005, he received, together with his close collaborator Doug Hardin, Vanderbilt University's Chancellor's Research Award, and in 2023, he was named Distinguished Professor of Mathematics.

Ed Saff's research contributions span the topics of Approximation Theory (polynomial and rational interpolation and approximation, Padé approximation, and orthogonal polynomials), Potential Theory (minimal energy and polarization problems, universal optimums and bounds), Complex Analysis (Hardy spaces, special functions, and conformal mappings), as well as Numerical Analysis. He has published nearly 300 mathematical research articles, co-authored 10 research monographs and textbooks, and co-edited 12 research volumes. In particular, he was the co-editor, with his mathematical brother Theodore J. Rivlin, of a volume of Joseph L. Walsh's "Selected Works" that was published in 2000.

The research works of Ed's have been cited more than 4200 times and by more than 2000 different authors (according to MathSciNet) and more than 14,600 times (according to Google Scholar). He collaborated with more than 140 different researchers. The most cited research monographs co-authored by Ed are "Logarithmic Potentials with External Fields" with V. Totik, which received 1,000+ citations (MatSciNet) and 2500+ citations (Google Scholar), which has recently appeared in second edition, and "Discrete Energy on Rectifiable Sets" with S. Borodachov and D. Hardin, both published by Springer. Among textbooks that Ed co-authored for students, we mention "Fundamentals of Differential Equations and Boundary Value Problems" (with R.K. Nagle and A.D. Snider), "Fundamentals of Complex

Fig. 2 Ed Saff receiving the diploma of a Foreign Member of the Bulgarian Academy of Sciences, Sozopol, 2013. Courtesy of P. Dragnev

Analysis", and "Fundamentals of Matrix Analysis with Applications" (both with A.D. Snider).

Dr. Saff's contributions to the field of mathematics have been widely recognized. In addition to the previously mentioned Guggenheim Fellowship, he was elected an Inaugural Fellow of the American Mathematical Society in 2012 and a SIAM Fellow (Society for Industrial and Applied Mathematics) in 2023. Additionally, he was elected as a Foreign Member of the Bulgarian Academy of Sciences in 2013 and, in 2024, he received an Honorary Doctorate from Taras Schevchenko National University Kyiv, Ukraine. During the ceremony for the latter award, his extensive contributions to mentoring and nurturing future generations of scientists were highlighted.

Over the course of his career, Dr. Saff has mentored 19 Ph.D. students and 13 postdoctoral researchers, demonstrating his profound influence on the academic community. For a list of his Ph.D. students, see the Appendix.

Dr. Saff's research has been supported by numerous prestigious grants from organizations such as NSF and NATO. He has also made significant contributions to academic publishing. Ed is a Founding Editor (together with Ronald DeVore) and has served as co-Editor-in-Chief and Managing Editor of *Constructive Approximation* for nearly 40 years. He was also co-founding editor (with S. Ruscheweyh)

Fig. 3 On behalf of the selection committee, Ed Saff (left) presents the Popov Prize to Arno Kuijlaars, 2001. Courtesy of E. Saff

and co-Editor-in-Chief of *Computational Methods and Function Theory* and an Editor for *Journal of Approximation Theory*. Dr. Saff's international engagement is especially noteworthy. He was a Pichorides Distinguished Lecturer at the University of Crete in Heraklion and an Erskine Fellow at the University of Canterbury in New Zealand. Ed has also served as an advisor and NATO collaborator with a French research team at INRIA Sophia-Antipolis, co-directed an Australian Research Council Discovery Award, and maintained an annual visiting research collaboration at the University of Cyprus in Nicosia. Additionally, he has organized a series of international CMFT (Computational Methods and Function Theory) research conferences, including outreach efforts to mathematicians from developing countries. His significant contributions in bridging Western and Eastern Block mathematics since the 1980s were recognized by the Bulgarian, Ukrainian and Russian Academies of Sciences.

It is impossible to mention all significant scientific contributions by Ed Saff even in several articles. An excellent story about Ed's biography with personal memories by his colleagues as well as his path as a researcher and mentor was told by Doron

Fig. 4 Ed Saff and Vili Totik with D. Lubinsky, H. Mhaskar and J. Geronimo at a conference, in 1980s. Courtesy of E. Saff

Lubinsky in paper [53] dedicated to Ed Saff's 70-th birthday. Our emphasis here will be mainly on some of Ed's deep and fascinating work that was done relatively recently.

Among Ed's most cited research works is his paper [55] written in 1985 with H. Mhaskar. In this paper, they characterized the set supporting the uniform norms of weighted polynomials $(w(x))^n P_n(x)$, $n \in \mathbb{N}$, where $P_n(x)$ is any polynomial of degree at most n and the weight w satisfies certain (not very restrictive) assumptions. More precisely, they proved that there exists a set S in the support Σ of w of positive inner logarithmic capacity such that the uniform norm of $(w(x))^n P_n(x)$ on S equals that on Σ. This work extended earlier results by Kemperman and Lorentz [45], Saff and Varga [61], and others. Saff and Mhaskar characterize the set S as an extremal set of a certain functional. One of applications of their result is the study of incomplete polynomials in the sense of Lorentz.

In early 1990s, Ed became interested in minimal energy problems and discretization of manifolds and became one of the main leaders in this field. Ed Saff's interest in energy minimization on the sphere was, in particular, inspired by the discovery in 1985 of fullerenes [48], new forms of the element carbon in which the atoms are arranged in closed shells.[1] The most known of these is the soccer-ball shaped C_{60}

[1] R. Curl, H. Kroto, and R. Smalley were awarded a Nobel Prize in 1996 for this discovery.

molecule with a pattern of 12 pentagons and 20 hexagons located on a sphere called buckminsterfullerene or buckyball.

One of the first papers in this direction was published by Dr. Saff in 1994 with E.A. Rakhmanov and Y.M. Zhou [57]. It contains several significant results which set directions for further research. The first one is the theorem on area-regular partition of the unit sphere $\mathbb{S}^2$ in $\mathbb{R}^3$. A collection $\{D_i\}_{i=1}^N$ of N closed subsets of $\mathbb{S}^2$ is called an *area-regular partition of* $\mathbb{S}^2$ *into N parts* if it covers $\mathbb{S}^2$, for each pair of indices $i \neq j$, sets D_i and D_j have disjoint interiors with respect to $\mathbb{S}^2$, and the interior of each set D_i has area $\frac{4\pi}{N}$. The existence of an absolute constant C such that, for all $N \geq 2$, the sphere can be partitioned into N subsets of equal area with the diameter of each subset at most $C/\sqrt{N}$ is well known together with the fact that the C cannot be less than 4, since among all subsets of $\mathbb{S}^2$ with fixed area $\frac{4\pi}{N}$, the spherical cap has minimal diameter, which asymptotically equals $\frac{4}{\sqrt{N}}$. Ed and his co-authors E. Rakhmanov and Y. Zhou were able to find an explicit constant close to the optimal one.

Theorem 1 *For any* $N \geq 2$*, there exists an area-regular partition of* $\mathbb{S}^2$ *into N parts with the diameter of each part* $\leq 7/\sqrt{N}$.

They also showed that the constant 7 can be improved to 6 for sufficiently large N. Results of this type were used by Bondarenko, Radchenko, and Viazovska [3] to solve the famous N^2-conjecture by Korevaar and Myers [47] about the order of growth of the smallest cardinality of a spherical N-design on the Euclidean sphere of a given dimension.

The logarithmic energy minimization problem on the sphere $\mathbb{S}^2$ requires finding the quantity

$$\mathcal{E}_{\log}(\mathbb{S}^2, N) := \inf_{x_1,\ldots,x_N \in \mathbb{S}^2} \sum_{1 \leq i \neq j \leq N} \ln \frac{1}{|x_i - x_j|} \tag{1}$$

and point configurations $\{x_1, \ldots, x_N\} \subset \mathbb{S}^2$ attaining the infimum in (1). Optimal points are known as elliptic Fekete points as this problem is equivalent to that of finding N points on $\mathbb{S}^2$ with the largest possible product of pairwise distances. Ed Saff and his co-authors showed in [57] that for C_N defined by

$$\mathcal{E}_{\log}(\mathbb{S}^2, N) = -\frac{1}{4} \ln\left(\frac{4}{e}\right) N^2 - \frac{1}{4} N \ln N + C_N N,$$

there holds

$$\liminf_{N\to\infty} C_N \geq -\frac{1}{4}\left(\ln\frac{\pi}{2} + b\ln\left(1 - e^{-a}\right)\right) = -0.112768770\ldots,$$

$$\limsup_{N\to\infty} C_N \leq -\frac{1}{4}\ln\frac{\pi\sqrt{3}}{2} - \frac{\pi}{8\sqrt{3}} = -0.0234973\ldots,$$

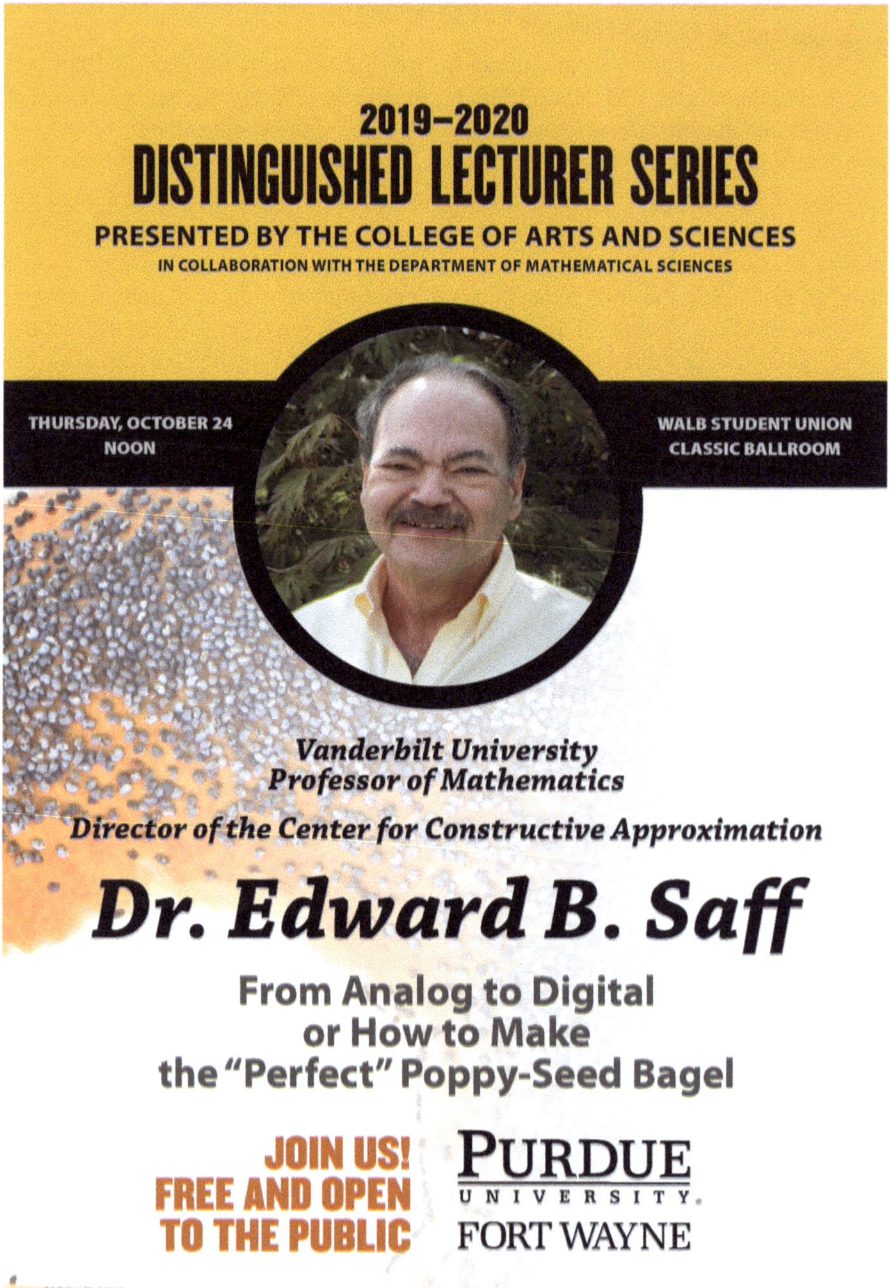

Fig. 5 Poster presenting the Distinguished Lecture Series given by Ed Saff at Purdue University Fort Wayne in October 2019. Poster provided by Purdue University Fort Wayne

Fig. 6 Ed Saff and his colleagues at the 2007 Oberwolfach workshop. Courtesy of P. Dragnev

$$a := \frac{2\sqrt{2\pi}}{\sqrt{27}}\left(\sqrt{2\pi} + \sqrt{2\pi + \sqrt{27}}\right), \quad \text{and} \quad b := \frac{\sqrt{2\pi + \sqrt{27}} - \sqrt{2\pi}}{\sqrt{2\pi + \sqrt{27}} + \sqrt{2\pi}}.$$

They also conjectured that $\lim_{N\to\infty} C_N$ exists. In [23, Conjecture 4] Brauchart, Hardin and Saff conjectured the limit value to be

$$\lim_{N\to\infty} C_N = 2\log 2 + \frac{1}{2}\log\frac{2}{3} + 3\log\frac{\sqrt{\pi}}{\Gamma(1/3)}.$$

Recently, Bétermin and Sandier [2] proved that $\lim_{N\to\infty} C_N$ exists and is bounded by the quantity on the right-hand side above.

Ed and his co-authors [58] were first to show that point configurations minimizing the logarithmic energy are well-separated. More precisely, the following result holds.

Theorem 2 *If a subset* $\{x_1^*, \ldots, x_N^*\} \subset \mathbb{S}^2$ *attains the infimum in* (1), *then*

$$\min_{1\le i\ne j\le N} \left|x_i^* - x_j^*\right| \ge \frac{3/5}{\sqrt{N}}.$$

Fig. 7 Ed Saff during the Workshop at BIRS, Banff, Canada, November 2007. Courtesy of P. Dragnev

Later, the constant in this lower bound was improved by Dubickas [39] to 7/4 and by Dragnev [30] to 2.

In connection with these results, we mention the Steven Smale's seventh problem for the twenty-first century, asking for a polynomial-time algorithm (in N) that would produce N points $x_1, \dots, x_N$ on the sphere such that

$$\sum_{1 \le i \ne j \le N} \ln \frac{1}{|x_i - x_j|} - \mathcal{E}_{\log}(\mathbb{S}^2, N) \le c \ln N,$$

where c is a universal constant. This problem, that has been an inspiration for Ed and his collaborators, is still unsolved.

The most cited paper written by Ed was his next paper on energy minimization [49] published jointly with A. Kuijlaars in 1997. It received about 200 citations according to MatSciNet and about 1500 according to Google Scholar. This paper provided algorithms for distributing points on the sphere, namely equal area points and generalized spiral points. These algorithms have been used for testing radar systems used in aircraft. Numerical evidence suggesting that the general pattern for optimal configurations on the sphere $\mathbb{S}^2$ of large cardinalities N is a (almost)

Fig. 8 Ed Saff with Peter Boyvalenkov, Peter Dragnev, Doug Hardin, and Maya Stoyanova, ICERM, Providence, RI, 2018. Courtesy of P. Dragnev

regular hexagonal arrangement perturbed "slightly" to fit on the sphere was also presented. They considered the logarithmic minimal energy (1) and the Coulomb minimal energy, which is a special case of the minimal Riesz s-energy,

$$\mathcal{E}_s(\mathbb{S}^d, N) := \inf_{x_1,\ldots,x_N \in \mathbb{S}^d} \sum_{1\le i\ne j\le N} \frac{1}{|x_i - x_j|^s} \tag{2}$$

with $s = 1$ and $d = 2$.

A year later, Saff and Kuijlaars showed in another very highly cited paper [50] that

$$\mathcal{E}_d(\mathbb{S}^d, N) = \frac{\mathrm{Vol}(B_d)N^2}{\mathcal{H}_d(\mathbb{S}^d)} \ln N(1 + o(1)), \quad N \to \infty,$$

where $\mathrm{Vol}(B_d)$ is the volume of the unit ball in $\mathbb{R}^d$ and $\mathcal{H}_d(\mathbb{S}^d)$ is the surface area of the unit sphere in $\mathbb{R}^{d+1}$. For $s > d$, they also proved that $\mathcal{E}_s(\mathbb{S}^d, N) \asymp N^{1+s/d}$, $N \to \infty$, with optimal N-point configurations being well-separated as $N \to \infty$ and conjectured the existence of the limit $\lim_{N\to\infty} \frac{\mathcal{E}_s(\mathbb{S}^d,N)}{N^{1+s/d}}$.

Fig. 9 Ed Saff and his colleagues at the International Conference "Construction Theory of Functions" in Sozopol, Bulgaria, June 9–15, 2013. Courtesy of P. Dragnev

This conjecture was proved in one of Ed Saff's further results with Doug Hardin, which became one of their very notable contributions to the fields of energy minimization and discretizing manifolds, the so-called Poppy-Seed Bagel Theorem [9, 42, 43]. He and his co-authors proved it in mid 2000s. The first version of this result appeared in a cover article "Discretizing Manifolds via Minimum Energy Points" in the November 2004 issue of *Notices of the American Mathematical Society*. Edward Saff was interviewed by National Public Radio (December 11, 2004) about the impact of this scientific result. An article about this theorem also appeared in a very popular French journal *Science & Vie* in July 2005 and in the *Online Research Journal of Vanderbilt University* in November 2004.

We now state rigorously the Poppy-Seed Bagel Theorem. For $s > 0$ and $N \geq 2$, the minimal Riesz s-energy of an infinite compact set $A \subset \mathbb{R}^{d'}$ is defined as

$$\mathcal{E}_s(A, N) := \inf_{x_1,\ldots,x_N \in A} \sum_{1 \leq i \neq j \leq N} \frac{1}{|x_i - x_j|^s}.$$

In the case $d' = 3$, $s = 1$, and A being a sphere, the problem of energy minimization is known as the Thomson's problem [66]. J.J. Thomson, a Nobel Prize Laureate, used it in the beginning of the twentieth century to explain the plum pudding model of an atom. This model described an atom as a positively charged sphere within which electrostatic forces determined the positions of negatively charged electrons.

A sphere is a representative of the class of rectifiable sets. We say that a subset $A \subset \mathbb{R}^{d'}$ is called d-rectifiable, $d \leq d'$, if A is an image of a bounded set from $\mathbb{R}^d$ under a Lipschitz mapping. We use $\mathcal{H}_\alpha$, $\alpha > 0$, to denote the α-dimensional Hausdorff measure in $\mathbb{R}^{d'}$. Ed jointly with Doug Hardin [42, 43] and later with S. Borodachov [9] proved the following.

Theorem 3 (Poppy-Seed Bagel Theorem) *Suppose $A \subset \mathbb{R}^{d'}$ is a closed d-rectifiable set ($d \leq d'$). Then for $s > d$,*

$$\lim_{N\to\infty} \frac{\mathcal{E}_s(A, N)}{N^{1+s/d}} = \frac{C_{s,d}}{\mathcal{H}_d(A)},$$

where $C_{s,d}$ is a positive constant independent of A.

If, in addition, $\mathcal{H}_d(A) > 0$, then any asymptotically s-energy minimizing sequence of N-point configurations on A is uniformly distributed on A (in the weak limit sense) with respect to $\mathcal{H}_d$-measure.*

This result contrasts with the potential-theoretic case $0 < s < d$ when the order of growth of the minimal s-energy does not depend on the Riesz parameter s (it remains N^2 throughout the range) and the limit distribution of energy minimizing configurations as N grows is non-uniform in general and is given by the corresponding equilibrium measure on A.

Of a special interest in Theorem 3 is the constant $C_{s,d}$. When $d = 1$, we have $C_{s,1} = 2\zeta(s)$, $s > 1$, where $\zeta(s)$ is the Riemann zeta-function (see the earlier paper

Fig. 10 Ed Saff in front of the Bulgarian Academy of Sciences, Sofia, Bulgaria, 2013. Courtesy of P. Dragnev

[54] by Ed and his collaborators). The values of constants $C_{s,8}$ and $C_{s,24}$ follow from the universal optimality results by H. Cohn, A. Kumar, S. Miller, D. Radchenko, and M. Viazovska in work [28] which extended the results by M. Viazovska [67][2] and this group [27] on the best-packing property of E_8 and Leech lattices in $\mathbb{R}^8$ and $\mathbb{R}^{24}$, respectively.

Often, it is important that points in sequences of optimal configurations do not get "too close" to each other. More precisely, it is desirable to have closest points in the configuration at a distance that has the same order of decay (as cardinality N grows) as the best N-point packing radius of A. In this direction, Ed co-authored the following separation result for Riesz energy minimizers first on the sphere (with A. Kuijlaars [50]), next on rectifiable manifolds (with D. Hardin [43]), and then on general compact sets [9].

Theorem 4 *Let $0 < \alpha \le d'$. Suppose $A \subset \mathbb{R}^{d'}$ is a compact set with $\mathcal{H}_\alpha(A) > 0$. Then for any $s > \alpha$, there is a constant $c_s = c_s(A, \alpha) > 0$ such that any Riesz s-energy minimizing configuration $\omega_N^* := \{x_1, \dots, x_N\} \subset A$ satisfies*

$$\min_{1 \le i \ne j \le N} |x_i - x_j| \ge \frac{c_s}{N^{1/\alpha}}, \quad s > \alpha, \quad N \ge 2.$$

Extensions of these results to the case of weighted energy,

$$E_s^w(\omega_N) = \sum_{1 \le i \ne j \le N} \frac{w(x_i, x_j)}{|x_i - x_j|^s},$$

where $w : A \times A \to [0, \infty]$ is what is called in [9] a CPD-weight function (here, CPD stands for "(almost) continuous and positive on the diagonal"), were also established by Ed and his co-authors in [9]. These results, in particular, provided an algorithm for placing a large number of points in a region or on a surface having a distribution close to a prescribed one.

As the cardinality of a point configuration on a manifold grows, the computations of energy minimizers become more and more complex. This restricts the size of the point set a machine can handle thus affecting the scope of applications. To lower the complexity, Ed proposed to truncate the energy sum keeping in it only pairs of points that are at most C_N times further apart than the best-packing distance, where C_N grows "slowly" to infinity as $N \to \infty$. This allowed, for example, to calculate near-optimal configurations on $\mathbb{S}^2$ of $N = 30{,}000$ points and on a 3D spherical shell of as many as $N = 500,000$ points [10] for geoscience applications. The following general result that makes this possible was proved by Ed and his collaborators in [10].

[2] M. Vizaovska was awarded a fields medal in 2022 for proving that the densest sphere packing in $\mathbb{R}^8$ is given by E_8 lattice.

Fig. 11 Ed Saff with Doug Hardin and Sergiy Borodachov during a Research in Pairs stay at Oberwolfach in December 2013 to work on the book "Discrete Energy on Rectifiable Sets". Courtesy of S. Borodachov

Theorem 5 *Let $A \subset \mathbb{R}^p$ be a compact d-rectifiable set with $\mathcal{H}_d(A) > 0$ and let w be a CPD-weight function on $A \times A$. Suppose Φ is a non-negative, bounded function on $(0, \infty)$ such that $\lim_{t\to 0^+} \Phi(t) = 1$ and $\{r_N\}_{N\in\mathbb{N}}$ is a sequence of positive numbers such that $\lim_{N\to\infty} r_N N = \infty$.*

Let $\{v_N\}_{N\in\mathbb{N}}$ be the sequence of weights

$$v_N(x, y) := \Phi\left(\frac{|x-y|}{r_N}\right) w(x, y), \quad x, y \in A, \quad x \neq y.$$

If $s > d$, then

$$\lim_{N\to\infty} \frac{\mathcal{E}_s^{v_N}(A, N)}{N^{1+s/d}} = \frac{C_{s,d}}{\left[\mathcal{H}_d^{s,w}(A)\right]^{s/d}},$$

where the constant $C_{s,d}$ is the same as in Theorem 3 and

$$\mathcal{H}_d^{s,w}(B) = \int_B w(x, x)^{-d/s}\, d\mathcal{H}_d(x), \quad B \subset A, \quad B \text{ is Borel measurable.}$$

Furthermore, any sequence $\{\omega_N\}_{N\in\mathbb{N}}$ *of asymptotically energy minimizing* N*-point configurations on* A*; i.e., such that*

$$\lim_{N\to\infty} \frac{E_s^{v_N}(\omega_N)}{\mathcal{E}_s^{v_N}(A,N)} = 1,$$

is uniformly distributed on A *with respect to the probability measure* $h_d^{s,w} := \frac{\mathcal{H}_d^{s,w}|_A(\cdot)}{\mathcal{H}_d^{s,w}(A)}$ *as* $N \to \infty$.

Theorem 5 provides a faster algorithm for obtaining large point configurations on a given region or a manifold that are approximately distributed with a prescribed density. In the case $s = d$, results similar to Theorems 3 and 5 were also established by Ed and his co-authors in [9, 10, 42, 43].

A second important result established in [10] is the quasi-uniformity of sequences of optimal N-point configurations on A under additional assumptions on A and on the sequence $\{v_N\}_{N\in\mathbb{N}}$ of the corresponding weights. We recall that a sequence $\{\omega_N\}_{N=2}^{\infty}$ of N-point configurations on A is called quasi-uniform if the ratio of its covering radius relative to the set A to the separation distance of the configurations stays bounded as $N \to \infty$.

Fig. 12 Ed Saff with Nikos Stylianopoulos and Peter Dragnev during a conference at Institut Mittag-Leffler in Stockholm, Sweeden. Courtesy of P. Dragnev

Another interesting direction of Ed Saff's research is on the behavior of the support of the equilibrium measure on the surface of a torus (circle of revolution) in $\mathbb{R}^3$ with respect to the logarithmic potential. He and his co-authors [22, 44] discovered a surprising fact that the support does not occupy the whole surface of the torus, but rather leaves off the inner (negative curvature) part of its surface. Computational experiments conducted by Ed and Doug Hardin (see the cover of the November 2004 issue of *Notices of the AMS*) suggest a similar phenomenon for Riesz s-energy on the torus with s sufficiently close to 0.

The limiting case as $s \to \infty$ of the minimal Riesz s-energy problem is the best-packing problem requiring to find N-point configurations on a compact set A maximizing the smallest pairwise distance between their points; i.e., configurations attaining the supremum below

$$\delta_N(A) := \sup_{\{x_1,\ldots,x_N\}\subset A} \min_{1\leq i\neq j\leq N} \left|x_i - x_j\right|.$$

This problem has been extensively studied, in particular, on the sphere (see, e.g., [5] for a review). However, best-packing configurations on the sphere $\mathbb{S}^2 \subset \mathbb{R}^3$ are only known for $N = 2, \ldots, 14$, and 24. Even fewer exact solutions have been found for minimum Riesz s-energy, namely, for $N = 2, \ldots, 6$, and 12 with the case $N = 5$ only partially settled by R. Schwartz with a computer-assisted proof for s up to $s_+ := 15 + \frac{25}{512}$ [62, 63] (see also [35] for a non-computer-assisted proof in the case of logarithmic energy). R. Schwartz observed that the triangular bi-pyramid minimizes the Riesz 5-point s-energy for s less than some computable constant $s^* \approx 15.05$ and has a higher s-energy than a certain square-based pyramid for $s > s^*$. It is still an open question whether a certain square-based pyramid minimizes the 5-point s-energy for $s > s_+$. Ed jointly with A. Bondarenko and D. Hardin showed [4] that, as $s \to \infty$, any family of s-energy minimizing 5-point configuration on $\mathbb{S}^2$ must converge to the square-based pyramid with its base on an equator.

Concerning the behavior of best-packing distance and best-packing N-point configurations as N gets large, Ed and his co-authors generalized known results for flat domains [46] and smooth manifolds to a more general class of d-rectifiable sets and characterized $(\mathcal{H}_d, d)$-rectifiable compact sets to which these known results can be extended [8]. We recall that a set $A \subset \mathbb{R}^{d'}$ of a finite $\mathcal{H}_d$-measure is $(\mathcal{H}_d, d)$-rectifiable if it is representable as a union of at most countably many d-rectifiable sets and of a set of $\mathcal{H}_d$-measure zero. Let also Δ_d denote the highest density of packing non-overlapping balls of equal radii in $\mathbb{R}^d$.

Theorem 6 (See [8]) *Let $d \leq d'$, where d, d' are integers, and $A \subset \mathbb{R}^{d'}$ be an infinite compact $(\mathcal{H}_d, d)$-rectifiable set. Then the equality*

$$\lim_{N\to\infty} \delta_N(A)N^{1/d} = 2(\Delta_d/\beta_d)^{1/d} \cdot \mathcal{H}_d(A)^{1/d}$$

Fig. 13 Ed Saff's incredible sense of humor comes across in many ways. Courtesy of P. Dragnev

holds if and only if $\mathcal{M}_d(A) = \mathcal{H}_d(A)$, *where* $\mathcal{M}_d(A)$ *is the* d*-dimensional Minkowski content of* A.

If A *is* d*-rectifiable with* $\mathcal{H}_d(A) > 0$, *then every sequence* $\{\omega_N\}_{N=2}^{\infty}$ *of best-packing configurations on* A *such that* $\#\omega_N = N$ *is asymptotically uniformly distributed on* A *with respect to* $\mathcal{H}_d$.

Motivated by the separation result for configurations on $\mathbb{S}^2$ minimizing logarithmic energy, Ed, along with Peter Dragnev extended the separation property to minimizing configurations of (2) for suitable s. Namely, given a minimizer $\omega_N^* = \{x_1^*, \ldots, x_N^*\} \subset \mathbb{S}^d$, let $\Delta_{N,s,d} := \min_{i \neq j} |x_i^* - x_j^*|$. Fixing a point in ω_N^*, the authors considered the external field $Q_N(x) = 1/(N-2)|x - x_N^*|^s$ it exerts on the other $N-1$ points. Proving that discrete s-energy minimizers are contained in the equilibrium support with external field Q_N and finding the equilibrium support explicitly, the authors derived the following separation result [38].

Theorem 7 *For* $d-2 \leq s < d$, *any sequence* $\{\omega_N^*\}_{N=2}^{\infty}$ *of minimal* s*-energy point configurations is well-separated. More precisely,*

$$\Delta_{N,s,d} \geq \frac{K_{s,d}}{N^{1/d}}, \quad K_{s,d} := \left(\frac{2\mathcal{B}(d/2, 1/2)}{\mathcal{B}(d/2, (d-s)/2)}\right)^{1/d},$$

where $\mathcal{B}(x, y)$ *is the Beta function. In particular,*

$$\Delta_{N,d-1,d} \geq \frac{2^{1/d}}{n^{1/d}}, \qquad \Delta_{N,s,2} \geq \frac{2\sqrt{1-s/2}}{\sqrt{N}}.$$

Riesz external field problems were further investigated by Brauchart, Dragnev, and Saff for axis-supported external fields [20, 21]. In particular, they answered the following question of A. A. Gonchar: given a unit positive charge in equilibrium on the unit sphere and a positively charged particle with unit charge approaching the unit sphere, at what distance will a negative charge start to appear. In the particular case of $\mathbb{S}^2$ and Coulomb interaction ($s = 1$), the distance is given by the golden ratio $(1 + \sqrt{5})/2$. A fascinating polynomial sequence of polynomials arising from Gonchar's problem was considered in [24].

In this regard the second author was challenged by Ed Saff at a conference in Oberwolfach, Germany in 2007 to join for a jog. Being 20 years younger, he naively thought he could keep up with Ed. Being an avid jogger and marathon runner, Ed kept a very fast pace and insisted on discussing the recent project, which left the younger mathematician out of breath, but with a new theorem, the point-mass question turned into an axis-supported external field.

Another interesting result of Ed Saff deals with a discrete Riesz external fields in the particular case when $s = d - 2$. He and his co-authors (Brauchart, Dragnev, and Womersley) showed that for external field exerted by finitely many particles $\{x_i\}$ on the sphere of positive charges $\{q_i\}$, when the magnitudes q_i of the charges are relatively small, the extremal support is the sphere with perfect spherical caps around x_i and radii depending on the q_i removed (see Fig. 14, published in [25]). Inspired by this work a young mathematician Alan Legg, along with second author characterized the equilibrium support for any magnitudes q_i (rightmost picture on Fig. 14) via quadrature domains for the special case of logarithmic interaction on $\mathbb{S}^2$, see [51]. Ed joined them in further investigation of the external field problem in a subsequent article [34].

In a sequence of papers [31–33] Ed, along with Peter Dragnev, Bent Fuglede, Doug Hardin and Natalia Zorii considered external field and constrained energy problems for Green energy and touching or intersecting condensers, a research that stemmed from a Research in Pairs stay at MFO.

Relatively recently, Ed Saff became interested in a problem somewhat dual to energy minimization known as the *polarization problem*. It requires placing N points on a given compact set A so that the absolute minimum of their total Riesz s-potential over A is the largest possible among all N-point configurations. This problem (sometimes referred to as max-min polarization) appears to be more difficult in general than the minimal energy problem. It is still not clear the extent to which linear programming techniques can be used for max-min polarization (F. Vallentin has had some recent success in this direction). Also, the exact solution even in the case of A being a unit circle $\mathbb{S}^1$ was known only for Riesz s-potentials with $s = 2$ [1] and for any $s \neq 0$ for $N = 3$ [56, 64]. Using techniques from approximation theory, Ed jointly with T. Erdélyi [40] showed that N equally spaced points on $\mathbb{S}^1$ remain optimal for $s = 4$. After that, jointly with D. Hardin and A. Kendall, he established the optimality of N equally spaced points on $\mathbb{S}^1$ [41] for any $s > -1, s \neq 0$. Together with the optimality of a regular simplex on a d-dimensional sphere among arbitrary configurations of the same cardinality [6, 65] and of a regular cross-polytope among antipodal ones [18], these results remain the only cases when

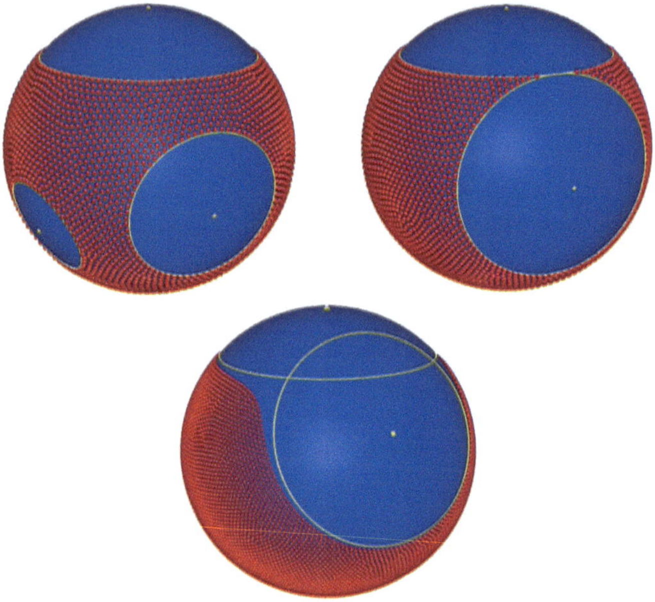

Fig. 14 Support of the equilibrium measure in the presence of a discrete potential external field (source: [25])

an exact solution to the max-min polarization problem on the sphere is known. For the case of a ball, Ed and T. Erdélyi proved that for $0 < s \leq d-2$, $d \geq 3$, the optimal max-min s-polarization configuration on a ball consists of N points that lie at the center.

Concerning the asymptotics for the max-min polarization problem, Ed and his collaborators [7] studied the behavior of the weighted max-min polarization

$$\mathcal{P}_s^w(A,N) = \sup_{x_1,\ldots,x_N \subset A} \inf_{x \in A} \sum_{i=1}^{N} \frac{w(x,x_i)}{|x-x_i|^s}$$

on compact subsets A of d-dimensional C^1-manifolds embedded in $\mathbb{R}^p$ with the boundary relative to the manifold having $\mathcal{H}_d$-measure zero, where $d \leq p$ and w is a CPD weight function on $A \times A$. It was proved that

$$\lim_{N\to\infty} \frac{\mathcal{P}_s^w(A,N)}{N^{s/d}} = \frac{\sigma_{s,d}}{\left[\mathcal{H}_d^{s,w}(A)\right]^{s/d}}, \qquad \text{for} \quad s > d,$$

with every asymptotically optimal sequence of N-point configurations on A being uniformly distributed with respect to the $\mathcal{H}_d^{s,w}$-measure. The limiting case of the max-min polarization problem as $s \to \infty$ is the optimal covering problem asking for a given N for a configuration ω_N on A with the smallest covering radius $\eta(\omega_N, A)$ relative to A, which is defined as follows

$$\eta(\omega_N, A) := \sup_{x\in A} \min_{y\in\omega_N} |x-y| .$$

This problem has been studied extensively, in particular, when A is a sphere, see, e.g., [5] for a review.

In view of the above mentioned result for a ball from [40], optimal points for max-min polarization need not be well-separated. Ed jointly with A. Reznikov and A. Volberg [60] established a week separation property[3] for sequences $\{\omega_N\}_{N\in\mathbb{N}}$ of N-point configurations optimal for max-min s-polarization, $s > d$, on strongly convex bodies[4] A of $\mathbb{R}^d$ as well as on $A = \mathbb{S}^d \subset \mathbb{R}^{d+1}$ or A being a spherical cap on $\mathbb{S}^d$.

Ed jointly with A. Reznikov found the leading term of the expected covering radius $\eta(\omega_N, \mathbb{S}^d)$ of N random points with independent uniform distributions over $\mathbb{S}^d$ and over a certain class of smooth manifolds. For $\mathbb{S}^d$, they proved that [59]

$$\lim_{N\to\infty} \mathbb{E}\left[\eta(\omega_N, \mathbb{S}^d)\right]\left(\frac{N}{\ln N}\right)^{1/d} = \left(\frac{(d+1)\beta_{d+1}}{\beta_d}\right)^{1/d} .$$

This generalized the result previously known only for the circle $\mathbb{S}^1$ from order statistics.

Ed has also made extensive contributions in the area of universal bounds for energy of spherical designs. These bounds are valid for potentials given by any absolutely monotone function of the dot product of points on the sphere and are determined by the strength of the design, its cardinality, and the dimension of the sphere. He published 8 papers jointly with P. Boyvalenkov, P. Dragnev, D. Hardin, and M. Stoyanova.

A point configuration $\omega_N = \{x_1, \ldots, x_N\} \subset \mathbb{S}^d \subset \mathbb{R}^{d+1}$ is called a spherical m-design, $m \geq 0$, if for every polynomial p in $d+1$ variables,

$$\frac{1}{N}\sum_{i=1}^{N} p(x_i) = \frac{1}{\mathcal{H}_d(\mathbb{S}^d)} \int_{\mathbb{S}^d} p(x)dx.$$

It was proved by H. Cohn and A. Kumar (see [26] and references therein) that every sharp code ω_N on $\mathbb{S}^d$; i.e., an m-distance set which is a spherical $(2m-1)$-design for some $m \in \mathbb{N}$, minimizes the energy

---

[3] Weak separatedness of the sequence $\{\omega_N\}_{N\in\mathbb{N}}$ means that there are constants $C, M > 0$ such that any ball in the space of radius $CN^{-1/d}$ contains at most M points from ω_N.

[4] A body $A \subset \mathbb{R}^d$ is strongly convex if the boundary of A is a $(d-1)$-dimensional C^2-smooth manifold with a non-degenerate Gaussian curvature.

Fig. 15 Ed Saff together with his scientific son Peter Dragnev and scientific grandson Alan Legg in Castro Urdiales, Spain, 2019. Courtesy of P. Dragnev

$$E_g(\omega_N) := \sum_{\substack{x,y\in\omega_N \\ x\neq y}} g(x\cdot y) \tag{3}$$

among all N-point configurations on $\mathbb{S}^d$, where g is any absolutely monotone potential function on $[-1, 1)$. They also established a similar optimality property for the 600-cell on $\mathbb{S}^3$.

Inspired by these results and by the earlier works of V. Levenshtein (see [52] and references therein), Ed's collaborative group [12] established the following bound. A finite sequence of ordered pairs $\{(\alpha_i, \rho_i)\}_{i=1}^k$, where $-1 \leq \alpha_1 < \ldots < \alpha_k < 1$ and $\rho_i > 0$, $i = 1, \ldots, k$, is called a $1/N$-quadrature rule exact for polynomials of degree up to τ if

$$f_0 := \gamma_n \int_{-1}^{1} f(t)(1-t^2)^{(n-3)/2} dt = \frac{f(1)}{N} + \sum_{i=1}^{k} \rho_i f(\alpha_i)$$

for all polynomials f of degree up to τ. Here the constant γ_n is selected such that the Gegenbauer weight $\gamma_n(1-t^2)^{(n-3)/2}$ is a probability density on $[-1, 1]$. Denote by $L(n, s)$, $s \in [-1, 1)$, the Levenshtein upper bound [52, Equation (6.12)] for the cardinality N of a point configuration on $\mathbb{S}^{n-1}$ with dot products between distinct points not exceeding s. Let $D(n, \tau)$ be the Delsarte-Goethals-Seidel lower bound for cardinality of τ-designs on $\mathbb{S}^{n-1}$ [29]. For a given cardinality N, denote by $\tau := \tau(n, N)$ the unique positive integer such that $D(n, \tau) < N \leq D(n, \tau + 1)$ and let $k := \lceil \frac{\tau+1}{2} \rceil$. Denote by $s = \alpha_k$ the unique solution to the equation $L(n, s) = N$. Then there exiist a unique $1/N$ quadrature rule $\{(\alpha_i, \rho_i)\}_{i=1}^k$ exact for polynomials of degree up to τ (known as Radau or Lobatto quadrature), see [52, Section 5].

Consider the linear program:

$$\mathcal{W}(n, N, \tau; h) := \sup \left(N^2 f_0 - N f(1) \right),$$

where the supremum is taken over all positive definite polynomials f of degree $\leq \tau$ such that $f \leq h$ on $[-1, 1)$. The following theorem was established in [12].

Theorem 8 *Let $n, N \geq 2$ and h be an absolutely monotone function on $[-1, 1)$. Suppose τ is the unique positive integer such that $D(n, \tau) < N \leq D(n, \tau + 1)$ and choose $k = \lceil \frac{\tau+1}{2} \rceil$. If $\{(\alpha_i, \rho_i)\}_{i=1}^k$ is the unique $1/N$ quadrature rule exact for polynomials of degree up to τ, then*

$$E_h(\omega_N) \geq N^2 \sum_{i=1}^{k} \rho_i h(\alpha_i)$$

holds for any N-point configuration $\omega_N \subset \mathbb{S}^{n-1}$. Furthermore,

$$\mathcal{W}(n, N, \tau; h) = N^2 \sum_{i=1}^{k} \rho_i h(\alpha_i).$$

In [12], Ed Saff and his co-authors established similar bounds corresponding to spherical designs with non-trivial indexes. This helped them [16] simplify the original proof from [26] of the fact that the 600-cell minimizes energy (3) among all 120-point configurations on $\mathbb{S}^3$ for any absolutely monotone potential g. They also

Fig. 16 Ed Saff with Sergiy Borodachov, Peter Boyvalenkov, Peter Dragnev, Doug Hardin, and Maya Stoyanova during the UP24 Workshop, Sofia, Bulgaria, December 2024. Courtesy of M. Stoyanova

extended Theorem 8 to Hamming spaces [13], to more general polynomial metric spaces [14], and to the case of weighted energy [11] also proving similar upper bounds in [15, 17].

It is interesting to note that Ed Saff and Peter Dragnev came to a full circle dealing with the counterpart of the Gegenbauer polynomials for Hamming spaces, namely the Krawtchouck polynomials, whose zero asymptotics they considered in [36, 37].

Recently, Ed Saff and his collaborators [18] established sharp lower and upper bounds for the potential

$$P_g(x, \omega_N) = \sum_{i=1}^{N} g(x \cdot x_i), \quad x \in \mathbb{S}^{n-1},$$

with an absolutely monotone potential function g and $\omega_N = \{x_1, \ldots, x_N\}$ being a spherical design on $\mathbb{S}^{n-1}$ of a given strength. These bounds allowed them to prove that the 600-cell is a solution to the min-max polarization problem (dual to the max-

Fig. 17 Ed and Loretta Saff at the CMFT'09 in Ankara, Turkey. Courtesy of E. Saff

min polarization). They also showed that these bounds are attained by many sharp configurations [19].

We wish Ed Saff to be in great health, to live many years, and to prove many more fascinating mathematical theorems!

Appendix: Ph.D. Students of Ed Saff

Michael Lachance, Ph.D. 1982
Mehrdad Simkani, Ph.D. 1987
Xin Li, Ph.D. 1989
K. Pan, Ph.D. 1990
X. (Matthew) He, Ph.D. 1991
Xiaoyan Liu, Ph.D. 1992
Igor Pritsker, Ph.D. 1995
Yanmu Zhou, Ph.D. 1995
Peter Dragnev, Ph.D. 1997
Plamen Simeonov, Ph.D. 1997
Viktor Maymeskul, Ph.D. 2001
Sergiy Borodachov, Ph.D. 2006

Erwin Mina-Diaz, Ph.D. 2006
Maxym Yattselev, Ph.D. 2007
Abey Lopez Garcia, Ph.D. 2010
Nattapong Bosuwan, Ph.D. 2013
Timothy Michaels, Ph.D. 2017
Oleksandr Vlasiuk, Ph.D. 2018
Minh Quan H. Vu, Ph.D. 2024

References

1. G. Ambrus, K. Ball, T. Erdélyi, Chebyshev constants for the unit circle, *Bull. Lond. Math. Soc.* **45** (2013), no. 2, 236–248.
2. L. Bétermin and E. Sandier, Renormalized energy and asymptotic expansion of optimal logarithmic energy on the sphere, *Constr. Approx.* **47** (2018), 39–74.
3. A. Bondarenko, D. Radchenko, M. Viazovska, Optimal asymptotic bounds for spherical designs, *Annals of Mathematics*, **178** (2013), 443–452.
4. A. V. Bondarenko, D. P. Hardin, E. B. Saff, Mesh ratios for best-packing and limits of minimal energy configurations, *Acta Math. Hungar.* **142** (2014), no. 1, 118–131.
5. K. Böröczky Jr., *Finite packing and covering*. Cambridge Tracts in Mathematics, **154** (Cambridge University Press, Cambridge, 2004).
6. S.V. Borodachov, Polarization problem on a higher-dimensional sphere for a simplex, *Discrete and Computational Geometry*, **67** (2022), no. 2, 525–542.
7. S. V. Borodachov, D. P. Hardin, A. Reznikov, E. B. Saff, Optimal discrete measures for Riesz potentials, *Trans. Amer. Math. Soc.* **370** (2018), no. 10, 6973–6993.
8. S.V. Borodachov, D.P. Hardin, E.B. Saff, Asymptotics of best-packing on rectifiable sets, *Proc. Am. Math. Soc.* **135** (2007), no. 8, 2369–2380.
9. S.V. Borodachov, D.P. Hardin, and E.B. Saff, Asymptotics for discrete weighted minimal Riesz energy problems on rectifiable sets, *Trans. Amer. Math. Soc.*, **360** (2008), 1559–1580.
10. S. V. Borodachov, D. P. Hardin, E. B. Saff, Low complexity methods for discretizing manifolds via Riesz energy minimization, *Found. Comput. Math.*, **14** (2014), no. 6, 1173–1208.
11. S.V. Borodachov, P.G. Boyvalenkov, P.D. Dragnev, D.P. Hardin, E.B. Saff, M.M. Stoyanova, Energy bounds for weighted spherical codes and designs via linear programming, *Analysis and Mathematical Physics*, 15 (2025), no. 1, art. 19.
12. P. G. Boyvalenkov, P. D. Dragnev, D. P. Hardin, E. B. Saff, M. M. Stoyanova, Universal Lower Bounds for Potential Energy of Spherical Codes, *Constructive Approximation*, **44** (2016), 385–415.
13. P. G. Boyvalenkov, P. D. Dragnev, D. P. Hardin, E. B. Saff, M. M. Stoyanova, Energy bounds for codes and designs in Hamming spaces, *Des. Codes Cryptogr.*, **82** (2017), no. 1–2, 411–433.
14. P. G. Boyvalenkov, P. D. Dragnev, D. P. Hardin, E. B. Saff, M. M. Stoyanova, Energy bounds for codes in polynomial metric spaces, *Anal. Math. Phys.* **9** (2019), no. 2, 781–808.
15. P. G. Boyvalenkov, P. D. Dragnev, D. P. Hardin, E. B. Saff, M. M. Stoyanova, Upper bounds for energies of spherical codes of given cardinality and separation, *Des. Codes Cryptogr.* **88** (2020), no. 9, 1811–1826.
16. P. G. Boyvalenkov, P. D. Dragnev, D. P. Hardin, E. B. Saff, M. M. Stoyanova, Bounds for spherical codes: the Levenshtein framework lifted, *Math. Comp.* **90** (2021), no. 329, 1323–1356.
17. P.G. Boyvalenkov, P.D. Dragnev, D.P. Hardin, E.B. Saff, M.M. Stoyanova, Universal bounds for size and energy of codes of given minimum and maximum distances, *IEEE Trans. Inform. Theory*, **67** (2021), no. 6, 3569–3584.
18. P. G. Boyvalenkov, P. D. Dragnev, D. P. Hardin, E. B. Saff, M. M. Stoyanova, On polarization of spherical codes and designs, *J. Math. Anal. Appl.* **524** (2023), no. 1, 127065.

19. P.G. Boyvalenkov, P.D. Dragnev, D.P. Hardin, E.B. Saff, M.M. Stoyanova, Universal minima of discrete potentials for sharp spherical codes, to appear in *Rev. Mat. Iberoam.*, arXiv:2211.00092.
20. Brauchart, J.S., Dragnev, P.D., Saff, E.B.: Riesz extremal measures on the sphere for axis-supported external fields.*J. Math. Anal. Appl.* **356** (2009), no. 2, 769–792.
21. Brauchart, J.S., Dragnev, P.D., Saff, E.B., Riesz External Field Problems on the Hypersphere and Optimal Point Separation, *Potential Anal.* **41** (2014), 647–678.
22. J. S. Brauchart, D. P. Hardin, E. B. Saff, Riesz energy and sets of revolution in $\mathbb{R}^3$, *Contemporary Math. Functional Analysis and Complex Analysis*, **481** (2009), 47–57.
23. J. S. Brauchart, D. P. Hardin, E. B. Saff, The next-order term for optimal Riesz and logarithmic energy asymptotics on the sphere, *Contemp. Math.*, **578** (2012), 31–61
24. Brauchart, J.S., Dragnev, P.D., Saff, E.B., van de Woestijne, C.E.: A fascinating polynomial sequence arising from an electrostatics problem on the sphere. *Acta Math. Hungar.* **137** (2012), no. 1–2, 10–26.
25. J. S. Brauchart, P. D. Dragnev, E. B. Saff, and R. Womersley, Logarithmic and Riesz Equilibrium for Multiple Sources on the Sphere: the Exceptional Case, in *Contemporary Computational Mathematics - a celebration of the 80th birthday of Ian Sloan*, Eds. Josef Dick, Frances Y. Kuo, Henryk Wozniakowski, Springer, (2018).
26. H. Cohn, A. Kumar, Universally optimal distribution of points on spheres, *J. of Amer. Math. Soc.* **20**, 99–148, (2007).
27. H. Cohn, A. Kumar, S. Miller, D. Radchenko, M. Viazovska, The sphere packing problem in dimension 24, *Annals of Mathematics*, **185** (2017), no. 3, 1017–1033.
28. H. Cohn, A. Kumar, S. Miller, D. Radchenko, M. Viazovska, Universal optimality of the E_8 and Leech lattices and interpolation formulas, *Annals of Mathematics*, **196** (2022), no. 3, 983–1082.
29. P. Delsarte, J.-M. Goethals, J. J. Seidel, Spherical codes and designs, *Geom. Dedicata*, **6** (1977), 363–388.
30. P.D. Dragnev, On the separation of logarithmic points on the sphere, *Approximation Theory X: Abstract and Classical Analysis*, **39** (2002), 137–144.
31. P. D. Dragnev, B. Fuglede, D. P. Hardin, E. B. Saff, and N. Zorii, Minimum Riesz energy problems for a condenser with "touching plates", Potential Analysis 44, v. 3 (2016), 543–577.
32. P. D. Dragnev, B. Fuglede, D. P. Hardin, E. B. Saff, and N. Zorii, Condensers with touching plates and constrained minimum Riesz and Green energy problems, Constr. Approx. 50, 3 (2019), 369–401.
33. P. D. Dragnev, B. Fuglede, D. P. Hardin, E. B. Saff, and N. Zorii, Constrained minimum Riesz energy problems for a condenser with intersecting plates, J. D'Analyse Math., 140 (2020), 117-159.
34. P.D. Dragnev, A.R. Legg, E.B. Saff, Point source equilibrium problems with connections to weighted quadrature domains, *Indagationes Mathematicae*, **34** (2023), no. 2, 200–218.
35. P. D. Dragnev, D. A. Legg, and D. W. Townsend, Discrete logarithmic energy on the sphere, *Pacific J. Math.* **207** (2002), 345–357.
36. P. D. Dragnev and E. B. Saff, Constrained energy problems with applications to orthogonal polynomials of a discrete variable, *J. Anal. Math.* **72**, (1997), 223–259.
37. P.D. Dragnev, E.B. Saff, A Problem in Potential Theory and Zero Asymptotics of Krawtchouk Polynomials Authors, *Journal of Approximation Theory*, **102** (2000), no. 1, 120–140.
38. P. D. Dragnev and E. B. Saff, *Riesz spherical potentials with external fields and minimal energy points separation* , Potential Anal. **26** (2007), 139–162.
39. A. Dubickas, On the maximal product of distances between points on the sphere, *Lithuanian Math. J.* **36**, no. 3 (1996), 241–248.
40. T. Erdélyi, E. Saff, Riesz polarization inequalities in higher dimensions, *J. Approx. Theory* **171** (2013), 128–147.
41. D.P. Hardin, A.P. Kendall, E.B. Saff, Polarization optimality of equally spaced points on the circle for discrete potentials, *Discrete Comput. Geom.* **50** (2013), no. 1, 236–243.
42. D.P. Hardin and E.B. Saff, Discretizing manifolds via minimum energy points, *Notices of the American Math. Soc.*, November 2004, 1186–1194.

43. D. Hardin and E.B. Saff, Minimal Riesz energy point configurations for rectifiable d-dimensional manifolds, *Advances in Mathematics*, **193** (2005), 174–204.
44. D. P. Hardin, E. B. Saff, H. Stahl, Support of the logarithmic equilibrium measure on sets of revolution in $\mathbb{R}^3$, *Journal of Mathematical Physics*, **48** (2007), 022901.
45. J.H.B. Kemperman, G.G. Lorentz, Bounds for polynomials with applications, *Nederl. Akad. Wetensch. Proc. Set. A*, **82** (1979), 13–26.
46. A. N. Kolmogorov, V. M. Tihomirov, ϵ-entropy and ϵ-capacity of sets in function spaces, *Uspehi Mat. Nauk*, **14** (1959), no. 2, 3–86.
47. J. Korevaar and J. L. H. Meyers, Spherical Faraday cage for the case of equal point charges and Chebyshev-type quadrature on the sphere, *Integral Transforms Spec. Funct.* **1** (1993), 105–117.
48. H. W. Kroto, J. R. Health, S. C. O'Brien, R. F. Curl, R. E. Smalley, "C_{60}: buckminsterfullerene", *Nature*, **318** (1985), 162–163.
49. A.B.J. Kuijlaars, E.B. Saff, Distributing many points on a sphere, *The Mathematical Intelligencer*, **19** (1997), 5–11,
50. A.B.J. Kuijlaars and E.B. Saff, Asymptotics for minimal discrete energy on the sphere, *Trans. Amer. Math. Soc.* **350** (1998), no. 2, 523–538.
51. A.R. Legg, P.D. Dragnev, Logarithmic equilibrium on the sphere in the presence of multiple point charges, *Constructive Approximation*, **54** (2021), 237–257.
52. V. I. Levenshtein, *Universal bounds for codes and designs*, Handbook of Coding Theory, V. S. Pless and W. C. Huffman, Eds., Elsevier, Amsterdam, Ch. 6, 499–648, (1998).
53. D.S. Lubinsky, Ed Saff at three score and ten, *Contemporary Mathematics*, **661** (2016), 1–27.
54. A. Martinez-Finkelshtein, V. Maymeskul, E. Rakhmanov, E.B. Saff, Asymptotics for minimal discrete Riesz energy on curves in Rd, Canad. J. Math. 56 (2004), 529–552.
55. H.N. Mhaskar, E.B. Saff, Where does the sup norm of a weighted polynomial live? *Constructive Approximation*, **1** (1985), 71–91.
56. N. Nikolov, R. Rafailov, On the sum of powered distances to certain sets of points on the circle, *Pacific J. Math.* **253** (2011), no. 1, 157–168.
57. E.A. Rakhmanov, E.B. Saff, and Y.M. Zhou, Minimal discrete energy on the sphere, *Mathematical Research Letters*, **1** (1994), 647–662.
58. E.A. Rakhmanov, E. B. Saff, and Y. M. Zhou, Electrons on the sphere, *Computational Methods and Functional Theory*, R. M. Ali, St. Ruscheweyh, and E. B. Saff, (eds.), Singapore: World Scientific (1995), 111–127.
59. A. Reznikov, E. B. Saff, The covering radius of randomly distributed points on a manifold, *Int. Math. Res. Not. IMRN*, (2016), no.19, 6065–6094.
60. A. Reznikov, E. Saff, A. Volberg, Covering and separation of Chebyshev points for non-integrable Riesz potentials, *J. Complexity*, **46** (2018), 19–44.
61. E.B. Saff, R.S. Varga, On incomplete polynomials, in: *Numerische Methoden der Approximationstheorie* (L. Collatz, G. Meinardus, H. Werner, eds.) Basel: Birkhäiuser Verlag, 1978, pp. 281–298. (ISNM, vol. 42).
62. R. Schwartz, The five-electron case of Thomson's problem, *Exp. Math.* **22** (2013), no. 2, 157–186.
63. R. Schwartz, Five point energy minimization: a synopsis, *Constr. Approx*, **51** (2020), no.3, 537–564.
64. K. Stolarsky, The sum of the distances to certain pointsets on the unit circle, *Pacific J. Math.* **59** (1975), no.1, 241–251.
65. Y. Su, Discrete minimal energy on flat tori and four-point maximal polarization on $\mathbb{S}^2$. 2015. Thesis (Ph.D.), Vanderbilt University.
66. J.J. Thompson, On the structure of the atom: an investigation of the stability and periods of oscillation of a number of corpuscles arranged at equal intervals around the circumference of a circle; with application of the results to the theory of atomic structure, *Philos. Mag.*, **7** (1904), no. 6, 237–265.
67. M. Viazovska, The sphere packing problem in dimension 8, *Annals of Mathematics*, **185** (2017), no. 3, 991–1015.

Exponential Bases on Modified Domains

Oleg Asipchuk

1 Introduction

The main purpose of this paper is to explore the stability of exponential bases under small modifications of the domain.

An exponential basis on a domain $D \subset \mathbb{R}^d$ is an unconditional Schäuder basis for $L^2(D)$ in the form of $\{e^{2\pi i \lambda_n \cdot x}\}_{n \in \mathbb{Z}^d}$ with $\lambda_n \in \mathbb{R}^d$. A canonical example of exponential basis is the Fourier basis $\mathcal{E} = \{e^{2\pi i n x}\}_{n \in \mathbb{Z}}$ for $L^2([0, 1))$. In this case, the domain is an interval on the real line, and the corresponding set of exponent forms a complete and orthonormal basis for $L^2(0, 1)$.

The study of exponential bases on domains in $\mathbb{R}^d$ is a topic of significant interest in harmonic analysis and approximation theory. Unfortunately, the existence of exponential bases on general domains remains largely unexplored. For example, we do not know if exponential bases on triangular and circular domains exist or not. The existence of exponential bases on finite unions of intervals in $\mathbb{R}$ and on multirectangular domains (i.e., unions of rectangles with sides parallel to the coordinate axes) of $\mathbb{R}^n$ is proved in [15, 16]. See also [23]. The bases produced in these papers and their frame bounds are not explicit, but see e.g. [8, 11, 14] and also [2] for explicit examples of bases. Further examples of domains with exponential bases can be found in [10, 12, 21] and [9, 18, 19] and their references. Additionally, sets with no exponential basis were found recently, see [17].

A classical problem in this area involves understanding the stability properties of exponential bases under various perturbations. Most existing research has primarily focused on the stability of these bases concerning small changes in the phase or frequency parameters. For examples, refer to [3, 4, 14, 27] and the references therein.

O. Asipchuk (✉)
Department of Mathematical Sciences, University of Cincinnati, Cincinnati, OH, USA
e-mail: asipchah@ucmail.uc.edu

D. Bilyk et al. (eds.), *Recent Advances in Approximation and Potential Theory*,
Applied and Numerical Harmonic Analysis,
https://doi.org/10.1007/978-3-031-95551-8_2

A less explored yet equally important aspect is the stability of exponential bases when the underlying domain itself is changed. Precisely, if $\mathcal{B} = \{e^{2\pi i\lambda_n x}\}$ is an exponential basis on a domain $D \subset \mathbb{R}^d$ and $\tilde{D} \subset \mathbb{R}^d$ is such that $|D \setminus \tilde{D}| < \epsilon$ is $\mathcal{B}$ still a basis on $\tilde{D}$? The problem does not have a simple answer even for very simple domains.

1.1 *The Problem*

In this article, we study the effect of domain modifications on exponential bases on unions of intervals of the real line. The domain choice is motivated by the fact that any set of finite measures in $\mathbb{R}$ can be approximated by finite unions of intervals with rational endpoints. Any such union of intervals can be mapped into a union of intervals of intervals with integer endpoints, and so it is natural to treat unions of intervals with integer endpoints as a representative model.

Let $s \in \mathbb{N}$ and $a_j \in \mathbb{N} \cup \{0\}$ for $j = 0, \ldots, s-1$. Also, we assume that $0 = a_0 < a_1 < \ldots < a_{s-1}$. We consider the set

$$\mathcal{I}_s = \bigcup_{j=0}^{s-1} [a_j, a_j + 1). \tag{1}$$

If $a_j \pmod{s}$ are distinct, then

$$\mathcal{B} = \left\{ e^{2\pi i \frac{nx}{s}} \right\}_{n\in\mathbb{Z}} = \bigcup_{j=0}^{s-1} \left\{ e^{2\pi i \left(n + \frac{j}{s}\right) x} \right\}_{n\in\mathbb{Z}}$$

is an orthogonal exponential basis for $L^2(\mathcal{I}_s)$ with $A = B = s$ as shown in Proposition 3. Next, we consider a set of perturbations $\mathcal{E} = \{\epsilon_j\}_{j=0}^{s-1} \subset \mathbb{Q}$ and new union of intervals

$$\mathcal{I}_{s,\mathcal{E}} = \bigcup_{j=0}^{s-1} [a_j + \epsilon_j, a_j + \epsilon_j + 1).$$

We keep the first interval in the same position, i.e. $\epsilon_0 = 0$. We show with Example 1 that for some choice of the ϵ_j, $\mathcal{B}$ is no longer a basis for $L^2(\mathcal{I}_{s,\mathcal{E}})$. However, the next theorem shows that we can obtain a basis on $I_{s,\mathcal{E}}$ from a small perturbation of the basis $\mathcal{B}$.

Theorem 1 *Let $s \in \mathbb{N}$ and $\{\epsilon_l\}_{l=0}^{s-1} \subset \mathbb{Q}$ with a property $\sup_l |\epsilon_l| < \frac{1}{2}$. Let N be the least common denominator of $\{\epsilon_l\}_{l=0}^{s-1}$ and β be a solution of the equation*

$$\frac{\sin(\pi s N \beta)}{\sin(\pi \beta)} = sN \sin \frac{1}{sN} \tag{2}$$

on the interval $\left(0, \frac{1}{sN}\right)$. *If*

$$\frac{1}{2s^2N^3m} \leq |\delta| \leq \frac{1}{sN^2m} - \frac{\beta}{Nm}, \tag{3}$$

where $m := N(a_{s-1} + \epsilon_{s-1})$ *and* a_j *(mod s) are distinct, then*

$$\mathcal{B}_{I_{\mathcal{E}},\delta} = \bigcup_{j=0}^{s-1} \left\{ e^{2\pi i \left(n + \frac{j}{s} + j\delta\right)x} \right\}_{n \in \mathbb{Z}}$$

is a basis for $L^2(\mathcal{I}_{s,\mathcal{E}})$ *with Riesz constants*

$$\frac{1}{N}\left(1 - sN \sin \frac{1}{sN}\right)\left(sN - \frac{\sin\left(\frac{\pi}{2Nm}\right)}{\sin\left(\frac{\pi}{2sN^2m}\right)}\right) \leq A$$

and

$$B \leq \frac{1}{N}\left(1 + sN \sin \frac{1}{sN}\right)\left(sN + \frac{\sin\left(\frac{\pi}{2Nm}\right)}{\sin\left(\frac{\pi}{2sN^2m}\right)}\right).$$

The existence of β that satisfies (2) and the existence of δ that satisfies (3) are justified by Lemma 6 and Proposition 1 in Sect. 2.

Next, we consider a basis on the union of the intervals $\mathcal{I}_M$ inside the interval $I_N = [0, N)$, the solution to which we provide in the next theorem. In Proposition 2 we prove that

$$\mathcal{B}_M = \bigcup_{j=1}^{M} \left\{ e^{2\pi i \left(n + \frac{j-1}{N}\right)x} \right\}_{n \in \mathbb{Z}}$$

is a basis for $L^2(\mathcal{I}_M)$. Estimating the Riesz constants is often the most challenging aspect of the problem, and we address this in the following theorem.

Theorem 2 *Let* $N, M \in \mathbb{N}$ *such that* $2 < M \leq \frac{N}{2}$. *Also, let* $u \in \mathbb{N}$ *and*

$$u > \begin{cases} \frac{N}{\pi} \cos^{-1}\left(M \sin \frac{1}{M}\right) & N \text{ is even} \\ \frac{N}{\pi} \cos^{-1}\left(M \sin \frac{1}{M} \cos \frac{\pi}{2N}\right) - \frac{1}{2} & N \text{ is odd} \end{cases}. \tag{4}$$

If all distinct a_k *and* $a_{k'}$ *satisfy the following inequality*

$$\min_{m\in\mathbb{Z}}\left|\frac{a_k-a_{k'}}{N}-m\right|-\min_{m\in\mathbb{Z}}\left|\frac{(a_k-a_{k'})M}{N}-m\right|>\frac{u}{N}, \tag{5}$$

then $\mathcal{B}_M$ *is a basis for* $L^2(\mathcal{I}_M)$ *with*

$$\begin{cases} M\left(1-\left|\cos\frac{\pi u}{N}\right|\right)\le A \text{ and } B\le M\left(1+\left|\cos\frac{\pi u}{N}\right|\right) & N \text{ is even}\\ M\left(1-\left|\frac{\cos\left(\frac{\pi}{2N}+\frac{\pi u}{N}\right)}{\cos\frac{\pi}{2N}}\right|\right)\le A \text{ and } B\le M\left(1+\left|\frac{\cos\left(\frac{\pi}{2N}+\frac{\pi u}{N}\right)}{\cos\frac{\pi}{2N}}\right|\right) & N \text{ is odd}\end{cases}.$$

The problem is considerably more difficult because the set of exponentials needs to be adjusted to satisfy the assumption of Beurling's density theorem [6]. In this paper, we consider only a model case but we plan to pursue the investigation of this problem in a subsequent work.

Theorem 3 *Let* $2<N\in\mathbb{N}$, $N-1>m\in\mathbb{N}$, $I_N=[0,N)$, *and* β *be a solution of the equation*

$$\frac{\sin(\pi(N-1)\beta)}{\sin(\pi\beta)}=(N-1)\sin\frac{1}{N-1}.$$

If $\frac{1}{2(N-1)^2}<\delta<\frac{1}{N-1}-\beta$ *then*

$$\mathcal{B}_\delta=\bigcup_{j=0}^{N-2}\left\{e^{2\pi i\left(n+\frac{j}{N-1}-j\delta\right)x}\right\}_{n\in\mathbb{Z}},$$

is a basis for $L^2(I_N\setminus(m,m+1))$ *with*

$$\left(1-(N-1)\sin\frac{1}{N-1}\right)\left(N-1-\frac{1}{\sin\frac{\pi}{2(N-1)}}\right)\le A$$

and

$$B\le\left(1+(N-1)\sin\frac{1}{N-1}\right)\left(N-1+\frac{1}{\sin\frac{\pi}{2(N-1)}}\right).$$

The structure of this paper is as follows: In Sect. 2, we provide a brief overview of the definitions and tools utilized in our research. Section 3 is dedicated to proving the main results. Finally, Sect. 4 presents a discussion of open problems and concluding remarks.

2 Preliminaries

We have used the excellent textbooks [13] and [29] for the definitions and some of the results presented in this section.

Let H be a separable Hilbert space with inner product $\langle\ ,\ \rangle$ and norm $||\ || = \sqrt{\langle\ ,\ \rangle}$. We will mostly work with $L^2(D)$, where $D \subset \mathbb{R}^d$. So, the norm will be $||f||^2_{L^2(D)} = \int_D |f(x)|^2 dx$. The characteristic function on D we denote as χ_D.

A sequence of vectors $\mathcal{V} = \{v_j\}_{j\in\mathbb{Z}}$ in H is a *Riesz basis* if there exist constants $A,\ B > 0$ such that, for any $w \in H$ and for all finite sequences $\{a_j\}_{j\in J} \subset \mathbb{C}$, the following inequalities hold.

$$A \sum_{j\in J} |a_j|^2 \leq \Big\| \sum_{j\in J} a_j v_j \Big\|^2 \leq B \sum_{j\in J} |a_j|^2. \tag{6}$$

$$A||w||^2 \leq \sum_{j=1}^{\infty} |\langle w, v_j\rangle|^2 \leq B||w||^2. \tag{7}$$

The constants A and B are called *frame constants* of the basis. The left inequality in (6) implies that $\mathcal{V}$ is linearly independent, and the left inequality in (7) implies that $\mathcal{V}$ is complete. If condition (6) holds we call $\mathcal{V}$ a Riesz sequence. We call $\mathcal{V}$ a frame if the condition (7) holds. If the condition (7) holds and $A = B$ then we call $\mathcal{V}$ a tight frame. If $A = B = 1$, then we have a Parseval frame. The following lemma is well-known, but for the reader's convenience, we will prove it.

Lemma 1 *If a sequence of vectors $\mathcal{V} = \{v_j\}_{j\in\mathbb{Z}}$ is a frame with upper constant B, then the right inequality in* (6) *holds, i.e. for all finite sequences $\{a_j\}_{j\in J} \subset \mathbb{C}$*

$$\Big\| \sum_{j\in J} a_j v_j \Big\|^2 \leq B \sum_{j\in J} |a_j|^2.$$

Proof $\mathcal{V}$ is a frame, so for all finite sequences $\{a_j\}_{j\in J} \subset \mathbb{C}$ there is $f \in H$ such that $f = \sum_{j\in J} a_j v_j$. So,

$$\begin{aligned} ||f||^2 &= \langle \sum_{j\in J} a_j v_j, f\rangle = \sum_{j\in J} a_j \langle v_j, f\rangle \\ &\leq \sqrt{\sum_{j\in J} |a_j|^2} \sqrt{\sum_{j\in J} |\langle v_j, f\rangle|^2} \leq \sqrt{\sum_{j\in J} |a_j|^2} \sqrt{B}||f||. \end{aligned}$$

Therefore,

$$\Big\| \sum_{j\in J} a_j v_j \Big\| \leq \sqrt{\sum_{j\in J} |a_j|^2} \sqrt{B}.$$

□

The next lemma shows that frames in $L^2(D)$ can be obtained from bases of $L^2(D)$, when $D' \supset D$.

Lemma 2 *If $\mathcal{B}$ is basis for $L^2(D')$ and $D \subset D'$, then $\mathcal{B}$ is a frame for $L^2(D)$ with at least the same frame constants. In particular, if $\mathcal{B}$ is an orthogonal basis for $L^2(D')$, then it is a tight frame for $L^2(D)$.*

Proof Let $\mathcal{B}$ is a basis for $L^2(D')$, then for all $w \in L^2(D')$

$$A||w||^2 \leq \sum_{j=1}^{\infty} |\langle w, v_j \rangle|^2 \leq B||w||^2.$$

Also, for any $f \in L^2(D)$, there is $w \in L^2(D')$ such that $f = w\chi_D$. So, the frame inequality holds for any $f \in L^2(D)$. Therefore, $\mathcal{B}$ is a frame for $L^2(D)$ with at least the same frame constants. □

Let $\mathbf{v} \in \mathbb{R}^d$ and $\rho > 0$; we denote with $d_\rho D = \{\rho x \ : \ x \in D\}$ and by $t_\mathbf{v} D = \{x + \mathbf{v} \ : \ x \in D\}$ the dilation and translation of D. Sometimes we will write $\mathbf{v} + D$ instead of $t_\mathbf{v} D$ when there is no risk of confusion.

The following lemma can easily be proved by applying the definitions (6) and (7) with $\mathcal{V} = \{e^{2\pi i \langle x, \lambda_n \rangle}\}_{n \in \mathbb{Z}}$ and a change of variables in the integrals.

Lemma 3 *Let $\mathbf{v} \in \mathbb{R}^d$ and $\rho > 0$. The set $\mathcal{V} = \{e^{2\pi i \langle x, \lambda_n \rangle}\}_{n \in \mathbb{Z}}$ is a Riesz basis for $L^2(D)$ with constants A and B if and only if the set $\{e^{2\pi i \langle x, \frac{1}{\rho} \lambda_n \rangle}\}_{n \in \mathbb{Z}}$ is a Riesz basis for $L^2(t_\mathbf{v}(d_\rho D))$ with constants $A\rho^d$ and $B\rho^d$.*

2.1 Exponential Bases and Vandermonde Matrices

One method for demonstrating the existence of an exponential basis and calculating their frame constants consists in associating a set of exponentials on the union of intervals with a matrix and showing how the properties of the matrix can be used to derive properties of the set of exponentials. Specifically, for a given $L \in \mathbb{N}$, we set $I_L = [0, L)$. For $N \in \mathbb{N}$, we choose $\{a_1, \ldots, a_N\} \subset I_L \cap \mathbb{Z}$. Next, we let

$$\mathcal{I}_N := \bigcup_{j=1}^{N} [a_j, a_j + 1).$$

The next theorem is proved in [8].

Theorem 4 *Let $\{a_1, \ldots, a_N\} \subset [0, L) \cap \mathbb{Z}$ and $\{\delta_1, \ldots, \delta_N\} \subset \mathbb{R}$. The set*

$$\mathcal{B} = \mathcal{B}(\delta_1, \ldots, \delta_N) = \bigcup_{j=1}^{N} \{e^{2\pi i(n+\delta_j)x}\}_{n\in\mathbb{Z}}$$

is a Riesz basis for $L^2(\mathcal{I}_N)$ if and only if the matrix $\boldsymbol{\Gamma} = \{e^{2\pi i \delta_j a_k}\}_{1\le j,k\le N}$ is nonsingular. The optimal frame constants of $\mathcal{B}$ are the maximum and minimum singular values of $\boldsymbol{\Gamma}$.

A more general version of this theorem is in [1]. Theorem 4 implies that estimating the Riesz constants for certain exponential bases on unions of intervals of the real line can be reduced to estimating the singular values of a corresponding matrix. This task can be challenging, especially for large matrices. Nonetheless, there are several methods to address this problem, one of which we discuss in the next subsection.

2.2 Clusterization of Matrices

Vandermonde matrices appear naturally in connection with problems on exponential bases.

Recall that a *Vandermonde matrix* $\boldsymbol{V}$ of size $L \times N$ with elements on the complex unit circle $\mathbb{T} = \mathbb{R}/\mathbb{Z}$ is of the form

$$\boldsymbol{V} := \boldsymbol{V}(L, X) := \left[e^{2\pi i j x_k} \right]_{\substack{0\le j<L-1, \\ 1\le k\le N}}$$

with nodes $X = \{x_k\}_{k=1}^{N} \subset \mathbb{R}$. It is well known that $\boldsymbol{V}$ has rank N if and only if X consists of N distinct elements of $\mathbb{T}$. See for example [22, 28]. As shown in [25] the condition number of Vandermonde matrices, i.e. ratio between the greatest and smallest singular values, greatly depends on the pairwise distance of the nodes according to the *wrap-around metric*

$$|t - s|_{\mathbb{T}} := \min_{n\in\mathbb{Z}} |t - s - n|.$$

We say that two nodes x_k and $x_{k'}$ are equivalent if $|x_k - x_{k'}|_{\mathbb{T}} = 0$.

One method for estimating the singular values of Vandermonde matrices is clustering its columns, i.e. grouping the columns of the matrix into groups (clusters) with enough distances (in the "wrap-around" sense) between them. This method is described for example in [5, 20] for different conditions. Rectangular matrices are more suitable for this method, although some results can also be applied to square matrices.

Let $\{L_k\}_k$ be a set of column vectors, we set

$$\angle_{\min}(L_k, L_{k'}) = \min_{k\ne k'} \cos^{-1}\left(\frac{|\langle L_k, L_{k'}\rangle|}{||L_k||\,||L_{k'}||} \right).$$

The following lemma, Lemma 5.1 from [5], shows how the singular values of a clustered matrix can be estimated in terms of the singular values of the clusters.

Lemma 4 *Let $A \in \mathbb{C}^{N\times N}$, such that A is given in the following block form*

$$A = [A_1, \ldots, A_M],$$

with $A_k \in \mathbb{C}^{N\times s_k}$ and $\sum_{k=1}^{M} s_k = N$. Let $L_k \subset \mathbb{C}^N$ be the subspace spanned by the columns of the sub-matrix A_k. Assume that for all $1 \le k, k' \le M$, $k \neq k'$, and $0 \le \alpha \le \frac{1}{N}$,

$$\angle_{\min}(L_k, L_{k'}) \ge \frac{\pi}{2} - \alpha. \tag{8}$$

Let

$$\sigma_1 \ge \ldots \ge \sigma_N$$

be the ordered collection of all the singular values of A, and let

$$\tilde{\sigma}_1 \ge \ldots \ge \tilde{\sigma}_N$$

be the ordered collection of all the singular values of the sub-matrices $\{A_k\}$. Then

$$\sqrt{1 - N\alpha}\tilde{\sigma}_j \le \sigma_j \le \sqrt{1 + N\alpha}\tilde{\sigma}_j.$$

Note that condition (8) can be modified as

$$\max_{k\neq k'}\langle L_k, L_{k'}\rangle \le N\sin\alpha \tag{9}$$

for $||L_k||_2^2 = N$.

Lemma 4 is quite general and applicable to any matrix. However, for matrices associated to exponential bases, the term $\langle L_k, L_{k'}\rangle$ involves a sum of exponents that is difficult to control. The structure of Vandermonde matrices allows us to express this sum in a more manageable form, as demonstrated in the following lemma.

Lemma 5 *Let $\{a_1, \ldots, a_m\} \subset \mathbb{R}$, $\delta \in \mathbb{R}$, and $L_k = [1, e^{2\pi i a_k \delta}, \ldots, e^{2\pi i a_k (m-1)\delta}]^T$, where T denotes the transpose. Then for all $k \neq k'$ and $\delta(a_k - a_{k'}) \notin \mathbb{Z}$*

$$|\langle L_k, L_{k'}\rangle| = \left|\frac{\sin(\pi m(a_k - a_{k'})\delta)}{\sin(\pi(a_k - a_{k'})\delta)}\right|. \tag{10}$$

Proof *For all $k \neq k'$ and $\delta(a_k - a_{k'}) \notin \mathbb{Z}$ we use the formula for the sum of a geometric series to write*

$$|\langle L_k, L_{k'}\rangle| = \left|\sum_{j=0}^{m-1} e^{2\pi i(a_k - a_{k'})\delta j}\right| = \left|\frac{1 - e^{2\pi i m(a_k - a_{k'})\delta}}{1 - e^{2\pi i(a_k - a_{k'})\delta}}\right|$$
$$= \left|\frac{\sin(\pi m(a_k - a_{k'})\delta)}{\sin(\pi(a_k - a_{k'})\delta)}\right|.$$

□

The ratio (10) has several properties that will simplify the process of placing nodes into different clusters.

For our problem, we consider matrices $\mathbf{\Gamma} = \{e^{2\pi i\delta(j-1)a_k}\}_{1\le j,k\le L}$. These are Vandermonde matrices with nodes $\{a_k\}_{1\le k\le L}$. We distribute these nodes in the following way: we put in one cluster all nodes for which the pairwise ratio (10) is large, and we make sure that the pairwise ratio of any two points from different clusters is small. In order to do this, we need to carefully estimate the ratio in (10), which we will do in the next subsection.

2.3 Three Useful Lemmas

In this subsection, we state and prove several technical lemmas that are instrumental for proving the main results. Each lemma explores the properties of the ratio between two sine functions with different arguments.

Lemma 6 *For every $M \in \mathbb{N} \setminus \{1\}$, the function $g_M(t) := \frac{\sin(\pi M t)}{\sin(\pi t)}$ is decreasing, non negative, and $g_M(t) < M$ on the interval $\left(0, \frac{1}{M}\right)$. Moreover, there is only one $\beta \in \left(0, \frac{1}{M}\right)$ that solve equation*

$$\frac{\sin(\pi M\beta)}{\sin(\pi\beta)} = M \sin\frac{1}{M}.$$

Proof We use induction. First, if $M = 2$, we have that

$$g_2(t) = \frac{\sin(2\pi t)}{\sin(\pi t)} = \frac{2\sin(\pi t)\cos(\pi t)}{\sin(\pi t)} = 2\cos(\pi t).$$

So, $g_2(t)$ is decreasing, non negative, and $g(t) < 2$ on the interval $\left(0, \frac{1}{2}\right)$. Next, we assume that the conclusions of the lemma hold for $M > 2$ and using the trigonometric identities $\sin(a+b) = \sin a\cos b + \sin b\cos a$, we can write

$$\begin{aligned} g_{M+1}(t) &= \frac{\sin(\pi(M+1)t)}{\sin(\pi t)} \\ &= \frac{\sin(\pi M t)\cos(\pi t)}{\sin(\pi t)} + \cos(\pi M t) \\ &= g_M(t)\cos(\pi t) + \cos(\pi M t). \end{aligned}$$

By assumption, $g_M(t) < M$, and so $g_{M+1}(t) < M\cos(\pi t) + \cos(\pi M t) \leq M + 1$. By assumption, $g_M(t)$ is decreasing, $g_M\left(\frac{1}{M}\right) = 0$, and $g_M(t) \to M$, when $t \to 0$, so the equation

$$\frac{\sin(\pi M t)}{\sin(\pi t)} = M \sin\frac{1}{M}$$

has only one solution on the interval $\left(0, \frac{1}{M}\right)$. □

Lemma 7 *For any $M \in \mathbb{N}\setminus\{1\}$ and $t \leq \frac{1}{2M^2}$ we have that*

$$\left|\frac{\sin(\pi M t)}{\sin(\pi(\frac{1}{M} - t))}\right| < M \sin\frac{1}{M}. \tag{11}$$

Proof By Lemma 6 for $0 < t < \frac{1}{M}$ we have that

$$\left|\frac{\sin(\pi M t)}{\sin(\pi(\frac{1}{M} - t))}\right| = \frac{\sin(\pi M t)}{\sin(\pi(\frac{1}{M} - t))} = \frac{\sin(\pi M(\frac{1}{M} - t))}{\sin(\pi(\frac{1}{M} - t))} > 0$$

and the function $t \to \frac{\sin(\pi M t)}{\sin(\pi(\frac{1}{M} - t))}$ is increasing. Using the inequality $\frac{2}{\pi}|x| \leq \sin|x| \leq |x|$, we obtain

$$\frac{\sin(\pi M t)}{\sin(\pi(\frac{1}{M} - t))} \leq \frac{\pi M t}{2(\frac{1}{M} - t)} = \frac{\pi M^2 t}{2(1 - Mt)}$$

for $0 < t < \frac{\pi}{2} - \frac{1}{M}$. Since the inequality

$$\frac{\pi M^2 t}{2(1 - Mt)} < M \sin\frac{1}{M}$$

is satisfied when $0 < t < \frac{2\sin\frac{1}{M}}{M(\pi + 2\sin\frac{1}{M})}$.

We show that

$$\frac{2\sin\frac{1}{M}}{M(\pi + 2\sin\frac{1}{M})} > \frac{1}{2M^2}, \tag{12}$$

whenever $M \geq 2$. For $M = 2$, direct calculation shows (12). For $M > 2$, we start with the simple estimation

$$\frac{2\sin\frac{1}{M}}{M(\pi+2\sin\frac{1}{M})}\geq\frac{2\sin\frac{1}{M}}{M(\pi+\frac{2}{M})}=\frac{M\sin\frac{1}{M}}{M(\frac{M\pi}{2}+1)}.$$

Next, we show that $v(M):=M\sin\frac{1}{M}$ is increasing. Using the Taylor series of $\sin t$ and $\cos t$, we can estimate $v'(M)$ as follows

$$v'(M)=\sin\frac{1}{M}-\frac{\cos\frac{1}{M}}{M}\geq\frac{1}{3M^3}-\frac{1}{30M^5}>0$$

for all $M>2$, so $v(M)$ is increasing. Thus,

$$\frac{M\sin\frac{1}{M}}{M(\frac{M\pi}{2}+1)}=\frac{v(M)}{M(\frac{M\pi}{2}+1)}\geq\frac{3\sin\frac{1}{3}}{M(\frac{M\pi}{2}+1)}\geq\frac{3\sin\frac{1}{3}}{M^2(\frac{\pi}{2}+\frac{1}{3})}.$$

Finally, because

$$\frac{3\sin\frac{1}{3}}{\frac{\pi}{2}+\frac{1}{3}}>\frac{1}{2},$$

we have

$$\frac{2\sin\frac{1}{M}}{M(\pi+2\sin\frac{1}{M})}>\frac{1}{2M^2},$$

for $M\geq 2$. Therefore, by (12) the inequality (11) holds for $0<t\leq\frac{1}{2M^2}$. □

Proposition 1 *Let $M\geq 2$ and $\beta\in\left(0,\frac{1}{M}\right)$ be a solution of the equation*

$$\frac{\sin(\pi M\beta)}{\sin(\pi\beta)}=M\sin\frac{1}{M}.$$

Then $\frac{1}{2M^2}<\frac{1}{M}-\beta$.

Proof Let $\beta\in\left(0,\frac{1}{M}\right)$ be a solution of

$$\frac{\sin(\pi M\beta)}{\sin(\pi\beta)}=M\sin\frac{1}{M}.$$

Then $\frac{1}{M}-\beta$ solves

$$\left|\frac{\sin(\pi Mt)}{\sin(\pi(\frac{1}{M}-t))}\right|=M\sin\frac{1}{M}.$$

Using Lemma 6 and simple change of variables, $t := \frac{1}{M} - t$, we can show that $\left|\frac{\sin(\pi M t)}{\sin(\pi(\frac{1}{M}-t))}\right|$ is decreasing on the interval $\left(0, \frac{1}{M}\right)$.

By Lemma 7 we have that for $t = \frac{1}{2M^2}$

$$\left|\frac{\sin(\pi M t)}{\sin(\pi(\frac{1}{M}-t))}\right| > M \sin \frac{1}{M}.$$

These three facts prove that $\frac{1}{2M^2} < \frac{1}{M} - \beta$. □

The final lemma will be used to prove Theorem 2.

Lemma 8 *Let $N \in N \setminus \{1\}$. The function*

$$\omega_N(t) = \frac{\sin \frac{\pi(t-u)}{N}}{\sin \frac{\pi t}{N}}$$

is increasing on the interval $0 < u < t < \frac{N}{2}$.

Proof We rewrite our function using simple trigonometric identity

$$\omega_N(t) = \cos \frac{\pi u}{N} - \cot \frac{\pi t}{N} \sin \frac{\pi u}{N}.$$

Since

$$\omega_N'(t) = \frac{\pi}{N} \csc^2 \frac{\pi t}{N} \sin \frac{\pi u}{N} > 0,$$

when $0 < u < \frac{N}{2}$, $\omega_N(t)$ is an increasing function on the interval $0 < u < t < \frac{N}{2}$. □

3 Proof of the Main Results

In this section, we prove Theorems 1, 2 and 3. The proofs are primarily built upon Theorem 4 and Lemma 4, and have the following key components:

1. Associating a set of exponentials on a given domain with a Vandermonde matrix using Theorem 4;
2. Distributing the columns of the matrix among different clusters using lemmas from Sect. 2.3;
3. Estimating the singular values of the matrix using Lemma 4.

3.1 Proof of Theorem 1

We start with a union of intervals

$$\mathcal{I}_{s,\mathcal{E}} = \bigcup_{k=0}^{s-1} [a_k + \epsilon_k, a_k + \epsilon_k + 1),$$

where $\mathcal{E} = \{\epsilon_k\}_{k=0}^{s-1} = \left\{\frac{p_k}{q_k}\right\}_{k=0}^{s-1} \subset \mathbb{Q}$ and the a_k (mod s) are distinct. Letting N the least common multiple of the denominators of $\{q_0, \ldots, q_{s-1}\}$, we extend $\mathcal{I}_{\mathcal{E}}$ to obtain a union of intervals with integer endpoints

$$\mathcal{I}_{sN,N\mathcal{E}} = \bigcup_{k=0}^{s-1} [(a_k + \epsilon_k)N, (a_k + \epsilon_k + 1)N).$$

By Lemma 3 $\mathcal{B} = \bigcup_{j=0}^{s-1} \left\{e^{2\pi i\left(n+j\left(\frac{1}{s}+N\delta\right)\right)x}\right\}$ is a basis for $L^2(\mathcal{I}_\epsilon)$ if and only if $\tilde{\mathcal{B}} = \bigcup_{j=0}^{sN-1} \left\{e^{2\pi i\left(n+j\left[\frac{1}{sN}+\delta\right]\right)x}\right\}$ is a basis for $L^2(\mathcal{I}_{sN,N\mathcal{E}})$. To show that $\tilde{\mathcal{B}}$ is a basis for $L^2(\mathcal{I}_{sN,N\mathcal{E}})$ we will use Theorem 4 and Lemma 4.

Indeed, $\mathcal{I}_{sN,N\mathcal{E}}$ has the nodes

$$p_k \in \{0, \ldots, N-1, (a_1+\epsilon_1)N, \ldots, (a_1+\epsilon_1)N-1, \ldots, (a_{s-1}+\epsilon_{s-1})N, \\ \ldots, (a_{s-1}+\epsilon_{s-1})N-1\}$$

for $k = 0, \ldots, sN-1$. We "wrap around" these nodes to the interval $[0, 1)$ by multiplying by $\frac{1}{sN} + \delta$, i.e $p'_k = p_k\left(\frac{1}{sN} + \delta\right)$. Next, we group the p_k into clusters consisting of either one or two points. To achieve this, we analyze the proximity of each p_k to other nodes using the wrap-around distance as the measure of closeness. Let's fix k. If for all $k \neq k'$ the ratio (10) is less than $sN \sin \frac{1}{sN}$, then p_k is far enough from other points and we can put it in the cluster of size 1. Similarly, if we fix distinct k and k' such that the ratio (10) greater or equal than $sN \sin \frac{1}{sN}$ and if for all k'' not equal k and k' the ratio (10) less than $sN \sin \frac{1}{sN}$, then p_k and $p_{k'}$ are far enough from other points and we can put them in the cluster of size 2. This approach can be used to form clusters of any size, however, the condition $|\epsilon_l| < \frac{1}{2}$ guarantees that we have clusters at most length 2. Let's explain why this is the case.

Let's examine the ratio in Eq. (10) more closely. In our context δ is $\frac{1}{sN} + \delta$, so we have

$$\left|\frac{\sin\left(\pi sN(p_k - p_{k'})(\frac{1}{sN} + \delta)\right)}{\sin\left(\pi(p_k - p_{k'})(\frac{1}{sN} + \delta)\right)}\right|.$$

Note that $p_k - p_{k'} \in \mathbb{Z}$. So, for small δ the closest nodes will be those for which there is $m_{k,k'} \in \mathbb{Z}$ such that $p_k - p_{k'} = m_{k,k'} sN$. In this case

$$\left| \frac{\sin(\pi s N (p_k - p_{k'})(\frac{1}{sN} + \delta)}{\sin(\pi (p_k - p_{k'})(\frac{1}{sN} + \delta))} \right| = \left| \frac{\sin(\pi m_{k,k'} s^2 N^2 \delta)}{\sin(\pi m_{k,k'} s N \delta)} \right| > sN \sin \frac{1}{sN}$$

for small enough δ. It means that p_k and $p_{k'}$ will be in one cluster if there is $m_{k,k'} \in \mathbb{Z}$ such that $p_k - p_{k'} = m_{k,k'} sN$. Therefore, we need to avoid situations when there is a k'' for which we can express $p_k - p_{k''} = m_{k,k''} sN$ and $p_{k'} - p_{k''} = m_{k,k''} sN$, with $m_{k,k'}, m_{k',k''} \in \mathbb{Z}$.

According to the condition of the Theorem 1 the $a_l (\text{mod } s)$ are distinct and wrap around in $[0, 1)$ perfectly (without intersections) when $\delta = 0$ and all $\epsilon_l = 0$. However, when we introduce perturbations to our intervals, the situation changes significantly. Those changes produce nodes that will be close to each other in a wrap-around sense, i.e. $p_k - p_{k'} = m_{k,k'} sN$. Therefore, we need to avoid scenarios where, after perturbation, intervals that were not "neighbors" in the wrap-around sense before the perturbation end up having nodes that are close to each other.

Without loss of generality, we consider $a_l = a_{l'} + 2(\text{mod } s)$. Since $I := [(a_l + \epsilon_l)N, (a_l + \epsilon_l + 1)N)$ and $I' := [(a_{l'} + \epsilon_{l'})N, (a_{l'} + \epsilon_{l'} + 1)N)$, the intervals I and I'contain the nodes

$$\{(a_l + \epsilon_l)N, (a_l + \epsilon_l)N + 1, \ldots, (a_l + \epsilon_l + 1)N - 1\}$$

and

$$\{(a_{l'} + \epsilon_{l'})N, (a_{l'} + \epsilon_{l'})N + 1, \ldots, (a_{l'} + \epsilon_{l'} + 1)N - 1\}$$

respectively. We need to show that there is no $m \in \mathbb{Z}$ such that

$$(a_l + \epsilon_l)N - (a_{l'} + \epsilon_{l'} + 1)N + 1 = msN.$$

It means that

$$\frac{a_l - a_{l'} - 1}{s} + \frac{\epsilon_l - \epsilon_{l'}}{s} + \frac{1}{sN} \notin \mathbb{Z}$$

and

$$\frac{(\epsilon_l - \epsilon_{l'} + 1)N + 1}{sN} \notin \mathbb{Z}.$$

So, if $\epsilon > 0$ and $|\epsilon_l| \leq \epsilon$ and $|\epsilon_{l'}| \leq \epsilon$, it is enough to find the smallest ϵ such that $\epsilon_{l'} = \epsilon$, $\epsilon_l = -\epsilon$, and

$$\frac{(-2\epsilon + 1)N + 1}{sN} > 0.$$

This is true for $\epsilon < \frac{N+1}{2N}$ or $\epsilon \le \frac{1}{2}$. Therefore, if $|\epsilon_l| \le \frac{1}{2}$, for $l = 1, \ldots, s-1$, then all nodes p_k can be sorted in the clusters with the maximal length 2.

Now, we continue with the main proof. Using Lemma 4 we represent our main matrix

$$\mathbf{\Gamma} = \left\{ e^{2\pi i p_k \left(\frac{1}{sN}+\delta\right)(j-1)} \right\}_{1\le k, j \le sN}$$

as

$$\mathbf{\Gamma} = [\mathbf{\Gamma}_1, \ldots, \mathbf{\Gamma}_{s'}, \mathbf{\Gamma}'_1, \ldots, \mathbf{\Gamma}'_{s'}],$$

where $\mathbf{\Gamma}_m$ are $1 \times sN$ matrices and $\mathbf{\Gamma}_{m'}$ are $2 \times sN$ matrices. All $\mathbf{\Gamma}_m$ are column vectors, so they have the singular value equal to the square norm of the vector, i.e.

$$\sigma_m^2 = ||\mathbf{\Gamma}_m||^2 = sN.$$

Note that according to our clusterization all $\mathbf{\Gamma}'_{m'}$ are associated to two nodes p_k and $p_{k'}$ with the property $\frac{p_k - p_{k'}}{sN} \in \mathbb{Z}$, so

$$\mathbf{\Gamma}'_{m'} = \begin{pmatrix} 1 & 1 \\ e^{2\pi i p_k(\frac{1}{sN}+\delta)} & e^{2\pi i p_{k'}(\frac{1}{sN}+\delta)} \\ \vdots & \vdots \\ e^{2\pi i p_k(\frac{1}{sN}+\delta)(sN-1)} & e^{2\pi i p_{k'}(\frac{1}{sN}+\delta)(sN-1)} \end{pmatrix}.$$

So, the singular values of $\mathbf{\Gamma}'_{m'}$ are the eigenvalues of the matrix

$$\mathbf{\Gamma}'^{*}_{m'}\mathbf{\Gamma}'_{m'} = \begin{pmatrix} sN & \sum_{j=0}^{sN-1} e^{2\pi i(p_k - p_{k'})(\frac{1}{sN}+\delta)j} \\ \sum_{j=0}^{sN-1} e^{-2\pi i(p_k - p_{k'})(\frac{1}{sN}+\delta)j} & sN \end{pmatrix}$$

which can be easily evaluated. Letting:

$$b = \sum_{j=0}^{sN-1} e^{2\pi i(p_k - p_{k'})(\frac{1}{sN}+\delta)j} = \frac{1 - e^{2\pi i(p_k - p_{k'})(\frac{1}{sN}+\delta)sN}}{1 - e^{2\pi i(p_k - p_{k'})(\frac{1}{sN}+\delta)}},$$

then the eigenvalues satisfy the following equation

$$\lambda^2 - 2sN\lambda + s^2N^2 - b^2 = 0.$$

Then

$$\sigma'^2_{1,2} = \lambda_{1,2} = sN \pm |b| = sN \pm \left| \frac{\sin(\pi sN(p_k - p_{k'})(\frac{1}{sN}+\delta)}{\sin(\pi(p_k - p_{k'})(\frac{1}{sN}+\delta))} \right|. \quad (13)$$

By Lemma 6 we know that the second term on the right-hand side of (13) is an increasing function of $(p_k - p_{k'})\delta$. Therefore, this term reaches its maximum value when $|p_k - p_{k'}|$ is at as small as possible value, i.e. $|p_k - p_{k'}| = 1$. Thus, the minimal and maximal singular values of the sub-matrices can be estimated as

$$sN - \frac{\sin(\pi s^2 N^2 |\delta|)}{\sin(\pi s N |\delta|)} \leq {\sigma'}^2_{sN} \text{ and } {\sigma'}^2_1 \leq sN + \frac{\sin(\pi s^2 N^2 |\delta|)}{\sin(\pi s N |\delta|)}.$$

Note that according to the condition (3)

$$\frac{1}{2s^2N^3m} \leq |\delta| \leq \frac{1}{sN^2m} - \frac{\beta}{Nm},$$

where we recall that β is the only one solution of the equation

$$\frac{\sin(\pi s N \beta)}{\sin(\pi\beta)} = sN \sin\frac{1}{sN}$$

and $m := a_{s-1} + \epsilon_{s-1}$. So, we can simplify our estimations to

$$sN - \frac{\sin\left(\frac{\pi}{2Nm}\right)}{\sin\left(\frac{\pi}{2sN^2m}\right)} \leq {\sigma'}^2_{sN} \text{ and } {\sigma'}^2_1 \leq sN + \frac{\sin\left(\frac{\pi}{2Nm}\right)}{\sin\left(\frac{\pi}{2sN^2m}\right)}.$$

Taking into account the fact that for all $m_{k,k'}$

$$m_{k,k'}sN + 1 \leq N(a_{s-1} + \epsilon_{s-1}) = Nm$$

and Lemma 7 we can show that for

$$\frac{1}{2s^2N^3m} \leq |\delta| \leq \frac{1}{sN^2m} - \frac{\beta}{Nm}$$

there is $\alpha < \frac{1}{sN}$ such that the maximal angle between sub-matrices is greater or equal than $\frac{\pi}{2} - \alpha$, i.e.

$$\angle_{\min}(L_k, L_{k'}) \geq \frac{\pi}{2} - \alpha,$$

where

$$\alpha = \frac{1}{sN}\frac{\sin(\pi s N^2 |\delta| m)}{\sin(\pi(\frac{1}{sN} - |\delta| N m))}. \tag{14}$$

Using Lemma 4 we obtain that

$$(1 - sN\alpha)\left(sN - \frac{\sin\left(\frac{\pi}{2Nm}\right)}{\sin\left(\frac{\pi}{2sN^2m}\right)}\right) \leq \sigma_{sN}^2 \text{ and } \sigma_1^2$$

$$\leq (1 + sN\alpha)\left(sN + \frac{\sin\left(\frac{\pi}{2Nm}\right)}{\sin\left(\frac{\pi}{2sN^2m}\right)}\right).$$

Using Lemma 7 and the substitution $t := \frac{1}{sN} - |\delta|Nm$ we can show that α is a decreasing function of $|\delta|$, using the bounds for delta (3) in (14) we obtain

$$\alpha \leq \frac{1}{sN}\frac{\sin\left(\pi sN^2\left(\frac{1}{sN^2m} - \frac{\beta}{Nm}\right)m\right)}{\sin\left(\pi\left(\frac{1}{sN} - \left(\frac{1}{sN^2m} - \frac{\beta}{Nm}\right)Nm\right)\right)} = \frac{1}{sN}\frac{\sin(\pi sN\beta)}{\sin(\pi\beta)} = \sin\frac{1}{sN}.$$

By Theorem 4 $\tilde{\mathcal{B}}$ is a basis for $L^2(\mathcal{I}_{sN,N\mathcal{E}})$ and the Riesz constants can be estimated as follows:

$$\left(1 - sN\sin\frac{1}{sN}\right)\left(sN - \frac{\sin\left(\frac{\pi}{2Nm}\right)}{\sin\left(\frac{\pi}{2sN^2m}\right)}\right) \leq A$$

and

$$B \leq \left(1 - sN\sin\frac{1}{sN}\right)\left(sN - \frac{\sin\left(\frac{\pi}{2Nm}\right)}{\sin\left(\frac{\pi}{2sN^2m}\right)}\right).$$

By Lemma 3 $\mathcal{B}$ is a basis for $L^2(\mathcal{I}_{s,\mathcal{E}})$ with Riesz constants A' and B', that satisfy

$$\frac{1}{N}\left(1 - sN\sin\frac{1}{sN}\right)\left(sN - \frac{\sin\left(\frac{\pi}{2Nm}\right)}{\sin\left(\frac{\pi}{2sN^2m}\right)}\right) \leq A'$$

and

$$B' \leq \frac{1}{N}\left(1 - sN\sin\frac{1}{sN}\right)\left(sN - \frac{\sin\left(\frac{\pi}{2Nm}\right)}{\sin\left(\frac{\pi}{2sN^2m}\right)}\right).$$

3.2 Proof of Theorem 2

We start with the interval $I_N = [0, N)$. The set

$$\mathcal{B}_N = \bigcup_{j=1}^{N} \left\{ e^{2\pi i \left(n + \frac{j-1}{N}\right)x} \right\}_{n\in\mathbb{Z}}$$

is an orthogonal basis for $L^2(I_N)$. From I_N we remove a union of intervals

$$\mathcal{I}_M = \bigcup_{k=1}^{M} [a_k, a_k + 1)$$

with a_k as in (5). We show that

$$\mathcal{B}_M = \bigcup_{j=1}^{M} \left\{ e^{2\pi i \left(n + \frac{j-1}{N}\right)x} \right\}_{n\in\mathbb{Z}}$$

is a basis for $L^2(I_M)$ and we estimate its Riesz constants using Lemma 4.

To do so, we associate our set of exponents $\mathcal{B}_M$ on the set of intervals with the matrix

$$\mathbf{\Gamma}_M = \left\{ e^{2\pi i a_k \frac{j-1}{N}} \right\}_{1\le j,k\le M}$$

as in Theorem 4. This is a Vandermonde matrix with distinct nodes $\frac{a_k}{N}$. Therefore, it is non-singular and by Theorem 4, $\mathcal{B}_M$ is a basis for $L^2(\mathcal{I}_M)$.

The conditions on a_k outlined in the theorem align with those in Lemma 4, and we employ the approach used in the proof of Theorem 1, with only minor adjustments to estimate the frame constants of $\mathcal{B}_M$.

We evaluate the inner product between two different column vectors v_k and $v_{k'}$ of the matrix $\mathbf{\Gamma}$, obtaining

$$|\langle v_k, v_{k'}\rangle| = \left| \sum_{j=1}^{M} e^{2\pi i (a_k - a_{k'}) \frac{j-1}{N}} \right| = \left| \frac{1 - e^{2\pi i (a_k - a_{k'}) \frac{M}{N}}}{1 - e^{2\pi i (a_k - a_{k'}) \frac{1}{N}}} \right|$$

$$= \left| \frac{\sin\left(\pi (a_k - a_{k'}) \frac{M}{N}\right)}{\sin\left(\pi (a_k - a_{k'}) \frac{1}{N}\right)} \right|.$$

The condition (5) guarantees that for all distinct a_k and $a_{k'}$ and for some positive integer l that satisfy the inequality $0 < u < l \le \frac{N}{2}$, the following inequality holds

$$\left|\frac{\sin\left(\pi(a_k - a_{k'})\frac{M}{N}\right)}{\sin\left(\pi(a_k - a_{k'})\frac{1}{N}\right)}\right| \leq \frac{\sin\left(\pi\frac{l-u}{N}\right)}{\sin\left(\pi\frac{l}{N}\right)}.$$

Next, we use Lemma 8 to show that this ratio is an increasing function of l, so

$$\frac{\sin\left(\pi\frac{l-u}{N}\right)}{\sin\left(\pi\frac{l}{N}\right)} \leq \frac{\sin\left(\pi\frac{\frac{N}{2}-\left\{\frac{N}{2}\right\}-u}{N}\right)}{\sin\left(\pi\frac{\frac{N}{2}-\left\{\frac{N}{2}\right\}}{N}\right)},$$

where $\left\{\frac{N}{2}\right\}$ is a decimal part of $\frac{N}{2}$. Thus, Eq. (4) guarantees that the following inequality holds

$$|\langle v_k, v_{k'}\rangle| < M\sin\frac{1}{M}.$$

This implies that all distinct columns of the matrix $\mathbf{\Gamma}_M$ are sufficiently well-separated, allowing us to apply Lemma 4 to estimate the singular values of the matrix and the frame constants of the basis. Therefore, $\mathcal{B}_M$ is a basis for $L^2(\mathcal{I}_M)$ with Riesz's constants

$$\begin{cases} M\left(1-\left|\cos\frac{\pi u}{N}\right|\right) \leq A \text{ and } B \leq M\left(1+\left|\cos\frac{\pi u}{N}\right|\right) & N \text{ is even} \\ M\left(1-\left|\frac{\cos\left(\frac{\pi}{2N}+\frac{\pi u}{N}\right)}{\cos\frac{\pi}{2N}}\right|\right) \leq A \text{ and } B \leq M\left(1+\left|\frac{\cos\left(\frac{\pi}{2N}+\frac{\pi u}{N}\right)}{\cos\frac{\pi}{2N}}\right|\right) & N \text{ is odd} \end{cases}.$$

Theorem 2 addresses the case where the nodes are distributed across M clusters, with each cluster containing exactly one node. Nevertheless, we can extend the approach outlined in the proof of Theorem 1 to handle the case of clusters containing at most two nodes each.

Corollary 1 *Let $N, M \in \mathbb{N}$ such that $2 < M \leq \frac{N}{2}$. Also, let $u \in \mathbb{N}$ satisfies (4). Let a_k can be distributed between clusters with at most 2 points in each of them. If all distinct a_k and $a_{k'}$ from different clusters satisfy (5), then $\mathcal{B}_M$ is a basis for $L^2(\mathcal{I}_M)$ with*

$$\begin{cases} (M-\alpha)\left(1-\left|\cos\frac{\pi u}{N}\right|\right) \leq A \text{ and } B \leq (M+\alpha)\left(1+\left|\cos\frac{\pi u}{N}\right|\right) & N \text{ even} \\ (M-\alpha)\left(1-\left|\frac{\cos\left(\frac{\pi}{2N}+\frac{\pi u}{N}\right)}{\cos\frac{\pi}{2N}}\right|\right) \leq A \text{ and } B \leq (M+\alpha)\left(1+\left|\frac{\cos\left(\frac{\pi}{2N}+\frac{\pi u}{N}\right)}{\cos\frac{\pi}{2N}}\right|\right) & N \text{ odd} \end{cases},$$

where

$$\alpha := \max_{a_k \neq a_{k'}} \left|\frac{\sin\left(\pi(a_k - a_{k'})\frac{M}{N}\right)}{\sin\left(\pi(a_k - a_{k'})\frac{1}{N}\right)}\right|.$$

Proof Using the proof of Theorem 2 we show that any two clusters are sufficiently well-separated to apply Lemma 4. Next, we calculate the smallest and largest singular values of the $2 \times M$ matrices corresponding to clusters. Since there are finitely many nodes, there are distinct a_k and $a_{k'}$ for which

$$\alpha := \max_{a_k \neq a_{k'}} \left| \frac{\sin\left(\pi(a_k - a_{k'})\frac{M}{N}\right)}{\sin\left(\pi(a_k - a_{k'})\frac{1}{N}\right)} \right| = \left| \frac{\sin\left(\pi(a_k - a_{k'})\frac{M}{N}\right)}{\sin\left(\pi(a_k - a_{k'})\frac{1}{N}\right)} \right|.$$

Such a_k and $a_{k'}$ produce a matrix that yields the largest and smallest singular values. A direct calculation yields

$$\sigma'_{min} = M - \alpha \text{ and } \sigma'_{max} = M + \alpha.$$

Therefore, by Lemma 4 and Theorem 4 $\mathcal{B}_M$ is a basis for $L^2(\mathcal{I}_M)$ with frame constants

$$\begin{cases} (M-\alpha)\left(1 - \left|\cos\frac{\pi u}{N}\right|\right) \le A \text{ and } B \le (M+\alpha)\left(1 + \left|\cos\frac{\pi u}{N}\right|\right) & N \text{ even} \\ (M-\alpha)\left(1 - \left|\frac{\cos\left(\frac{\pi}{2N} + \frac{\pi u}{N}\right)}{\cos\frac{\pi}{2N}}\right|\right) \le A \text{ and } B \le (M+\alpha)\left(1 + \left|\frac{\cos\left(\frac{\pi}{2N} + \frac{\pi u}{N}\right)}{\cos\frac{\pi}{2N}}\right|\right) & N \text{ odd} \end{cases}.$$

□

3.3 *Proof of Theorem 3*

We start with the interval $I_N = [0, N)$. Recall that the set

$$\mathcal{B} = \bigcup_{j=0}^{N-1} \left\{ e^{2\pi i \left(n + \frac{j}{N}\right)x} \right\}_{n \in \mathbb{Z}}$$

is an orthogonal basis for $L^2(I_N)$. From I_N we remove the interval $I_m = (m, m+1)$, where $m < N - 1$ and $m \in \mathbb{N}$. We find a basis for $L^2(I_N \setminus I_m)$ in the form

$$\mathcal{B}_\delta = \bigcup_{j=0}^{N-2} \left\{ e^{2\pi i \left(n + \frac{j}{N-1} - j\delta\right)x} \right\}_{n \in \mathbb{Z}},$$

where $|\delta|$ is sufficiently small. This is equivalent to showing that the matrix

$$\mathbf{\Gamma}_\delta = \left\{ e^{2\pi i k \left(\frac{j}{N-1} - j\delta\right)} \right\}_{0 \le j,k \le N-1}$$

is not singular.

First, we observe that for small δ we can form $N-1$ clusters. The first $N-2$ of them will contain only one point from 1 to $N-2$, and the last cluster will include the points 0 and $N-1$. Note that the distance between any two points from different clusters is $\frac{1}{N-1}-\delta$ in a wrap-around sens.

Let β be the solution of the equation

$$\frac{\sin(\pi(N-1)\beta)}{\sin(\pi\beta)} = (N-1)\sin\frac{1}{N-1}$$

on the interval $\left(0, \frac{1}{N-1}\right)$. Using Lemmas 6 and 7 we get that for $\frac{1}{2(N-1)^2} < \delta < \frac{1}{N-1} - \beta$ and for any $k \neq k'$ such that $|p_k - p_{k'}| < N-1$

$$\begin{aligned} |\langle L_k, L_{k'}\rangle| &= \left|\frac{\sin(\pi(p_k - p_{k'})(N-1)\delta)}{\sin\left(\pi(p_k - p_{k'})\left(\frac{1}{N-1}-\delta\right)\right)}\right| \leq \left|\frac{\sin(\pi(N-1)\delta)}{\sin\left(\pi\left(\frac{1}{N-1}-\delta\right)\right)}\right| \\ &< (N-1)\sin\frac{1}{N-1}. \end{aligned}$$

Next, we evaluate the singular values of the submatrices of $\mathbf{\Gamma}$ formed by the clusters. Matrices formed by one-node clusters, only have one singular value equal to $N-1$. The singular values of the $2 \times N-1$ matrices formed by two-nodes clusters can be evaluated as in the proof of Theorem 1. We obtain

$$\sigma'^2_{1,2} = \lambda_{1,2} = N - 1 \pm \left|\frac{\sin(\pi(N-1)^2\delta)}{\sin(\pi(N-1)\delta)}\right|.$$

Thus, by Lemma 4, $\mathcal{B}_\delta$ is a basis for $L^2(I_N \setminus I_m)$ with frame constants

$$\left(1 - \left|\frac{\sin(\pi(N-1)\delta)}{\sin\left(\pi\left(\frac{1}{N-1}-\delta\right)\right)}\right|\right)\left(N-1-\frac{\sin(\pi(N-1)^2\delta)}{\sin(\pi(N-1)\delta)}\right) \leq A$$

and

$$B \leq \left(1 + \left|\frac{\sin(\pi(N-1)\delta)}{\sin\left(\pi\left(\frac{1}{N-1}-\delta\right)\right)}\right|\right)\left(N-1+\frac{\sin(\pi(N-1)^2\delta)}{\sin(\pi(N-1)\delta)}\right),$$

whenever $\frac{1}{2(N-1)^2} < \delta < \frac{1}{N-1} - \beta$. Furthermore, using Lemmas 6 and 7, the constants can be simplified as follows:

$$\left(1-(N-1)\sin\frac{1}{N-1}\right)\left(N-1-\frac{1}{\sin\frac{\pi}{2(N-1)}}\right)\leq A$$

and

$$B\leq\left(1+(N-1)\sin\frac{1}{N-1}\right)\left(N-1+\frac{1}{\sin\frac{\pi}{2(N-1)}}\right).$$

4 Remarks and Open Problems

We begin this section by providing details on some of the examples that were mentioned in the introduction. The first example demonstrates how even minor perturbations in the domain can affect an exponential basis.

Example 1 Let $\mathcal{I}_2=[0,1)\cup[3,4)$ and $\mathcal{I}_{2,\frac{1}{N}}=[0,1)\cup\left[3-\frac{1}{N},4-\frac{1}{N}\right)$ for some $N\in\mathbb{N}$. Then

$$\mathcal{B}=\bigcup_{j=0}^{1}\left\{e^{2\pi i\left(n+\frac{j}{2}\right)x}\right\}_{n\in\mathbb{Z}}$$

is an orthogonal basis for $L^2(\mathcal{I}_2)$, but it is not a basis for $L^2(\mathcal{I}_{2,\epsilon})$.

By using Theorem 4 the matrix associated to $\mathcal{B}$ and I_2 is

$$\mathbf{\Gamma}=\begin{pmatrix}1&1\\1&-1\end{pmatrix}.$$

This matrix is orthogonal, so $\mathcal{B}$ is an orthogonal basis for $L^2(\mathcal{I}_2)$.

We apply the linear transformation $\mathcal{L}(x)=Nx$ to $\mathcal{I}_{2,1/N}$ and get $\mathcal{I}_{2N,1}=[0,N)\cup[3N-1,4N-1)$. Using Theorem 4 we form the matrix $\mathbf{\Gamma}$ for $\mathcal{I}_{2N,1}$ and to

$$\mathcal{B}=\bigcup_{j=0}^{2N-1}\left\{e^{2\pi i\left(n+\frac{j}{2N}\right)x}\right\}_{n\in\mathbb{Z}}.$$

$\mathbf{\Gamma}=\{e^{2\pi i\frac{p_k(j-1)}{2N}}\}_{1\leq j,k\leq L}$ is a Vandremonde matrix with endpoints $p_k=\{0,\ldots,N-1,3N-1,\ldots,4N-2\}$. The endpoints $N-1$ and $3N-1$ of $\mathcal{I}_{2N,1}$ correspond to the same node $\frac{N-1}{2N}$ because

$$\left|\frac{3N-1}{2N} - \frac{N-1}{2N}\right|_{\mathbb{T}} = 0.$$

Thus, $\mathbf{\Gamma}$ is singular and so $\mathcal{B}$ is not a basis for $L^2(\mathcal{I}_{2,\epsilon})$.

The following example shows that removing portions of the domain cannot be compensated solely by adjusting the density of exponents based on the measure of the new domain. So, additional perturbation of the phase may sometimes be required.

Example 2 The set $\mathcal{B}_3 = \bigcup_{j=1}^{3} \left\{e^{2\pi i\left(n+\frac{j-1}{3}\right)x}\right\}$ is the standard orthogonal basis for $L^2([0, 3))$. However, $\mathcal{B}_2 = \bigcup_{j=1}^{2} \left\{e^{2\pi i\left(n+\frac{j-1}{2}\right)x}\right\}$, is not a basis for $L^2([0, 1) \cup [2, 3))$.

By using Theorem 4 the matrix associated to $\mathcal{B}_2$ and $[0, 1) \cup [2, 3)$ is

$$\mathbf{\Gamma}_2 = \begin{pmatrix} 1 & 1 \\ 1 & 1 \end{pmatrix}.$$

It is easy to see that $\mathbf{\Gamma}_2$ is singular. Therefore, $\mathcal{B}_2$ is not a basis for $L^2([0, 1)\cup[2, 3))$.

The next proposition follows from Theorem 4. However, as was mentioned earlier the most difficult part of this research is to estimate the Riesz constants of the basis.

Proposition 2 *Let $M, N \in \mathbb{N}$ such that $M < N$. Then*

$$\mathcal{B}_M = \bigcup_{j=1}^{M} \left\{e^{2\pi i\left(n+\frac{j-1}{N}\right)x}\right\}_{n\in\mathbb{Z}},$$

is a basis for $L^2(\mathcal{I}_M)$, where $\mathcal{I}_M$ is as in (1).

Proof Using Theorem 4 we form a matrix

$$\mathbf{\Gamma} = \left\{e^{2\pi i\left(\frac{j-1}{N}\right)a_k}\right\}_{1\le j,k\le M}$$

that corresponds to the set $\mathcal{B}_M$ on $\mathcal{I}_M$. $\mathbf{\Gamma}$ is a Vandermonde matrix with distinct nodes, so it is non-singular, and $\mathcal{B}_M$ is a basis for $L^2(\mathcal{I}_M)$. □

The next proposition is well-known, but for the reader's convenience, we will prove it.

Proposition 3 *Let $\mathcal{I}_s$ be as in* (1). *If a_j (mod s) are distinct, then*

$$\mathcal{B} = \bigcup_{j=0}^{s-1} \left\{ e^{2\pi i \left(n + \frac{j}{s}\right)x} \right\}_{n\in\mathbb{Z}}$$

is an orthogonal exponential basis for $L^2(I_s)$ with $A = B = s$.

Proof Let a_j (mod s) are distinct. By using Theorem 4 the matrix associated to $\mathcal{B}$ and I_s is

$$\boldsymbol{\Gamma} = \left\{ e^{2\pi i \left(\frac{a_k(j-1)}{s}\right)} \right\}_{1\leq j,k\leq s} = \left\{ e^{2\pi i \left(\frac{(j-1)}{s}\right)} \right\}_{1\leq j,k\leq s}.$$

$\boldsymbol{\Gamma}$ is an orthogonal Vandermonde matrix with singular values equal to s, so $\mathcal{B}$ is an orthogonal exponential basis for $L^2(I_s)$ with the frame constants $A = B = s$. □

Theorem 1 demonstrates that if we slightly adjust the intervals in the domain, we can still obtain a basis by modifying a standard exponential basis on these intervals with a special δ, and the Riesz constants can be estimated accordingly. Moreover, when all $\epsilon_j = 0$, Theorem 1 provides a basis for the original domain too.

It is important to note that Theorem 1 applies only with standard exponential bases. However, a similar approach can be used for any domain with an exponential basis that can be associated with a Vandermonde matrix, as shown in Theorem 4. This method could potentially be applied to unions of intervals of length 1, but further research is needed, which we plan to pursue in the future.

In Theorem 2, we considered a standard exponential basis on union of intervals inside an interval of length N. The condition $2 < M \leq \frac{N}{2}$ illustrates that the union of intervals is just a "small" part of the interval $[0, N)$. The question, is how to find exponential bases on the "large" portion, i.e. is the complement of a "small" part of the interval $[0, N)$?

To answer this question we use Theorem 9 in [26] that we state below.

Theorem 5 *Let $\Delta > 0$ and $S \subset [0, \Delta)$. Suppose that for some $\Lambda \subset \frac{1}{\Delta}\mathbb{Z}$, $\mathcal{E}(\Lambda)$ is a Riesz basis for $L^2(S)$ with Riesz's constants A and B. Then $\mathcal{E}(\frac{1}{\Delta}\mathbb{Z} \setminus \Lambda)$ is a Riesz basis for $L^2([0, \Delta) \setminus S)$ with $A' = \Delta - B$ and $B' = \Delta - B$.*

The proof of such result can be found in [24, Proposition 2.1] and [7, Corollary 5.6]. Theorem 3 addresses the case of removing a single interval of length 1. However, the method presented in this article can be extended to handle the removal of multiple intervals, which we will explore in future work. Additionally, we will explore the application of the method in $\mathbb{R}^d$.

Acknowledgments First and foremost, I wish to express my deepest gratitude to my scientific advisor, Professor Laura De Carli, for her unwavering support and guidance throughout my research.

The author would also like to acknowledge the financial support provided by the Florida International University Graduate School Dissertation Fellowship.

References

1. O. Asipchuk, L. De Carli, and W. Li. Concerning the stability of exponential systems and Fourier matrices. *arXiv:2404.05469*, 2024.
2. O. Asipchuk and V. Drezels. Examples of exponential bases on union of intervals. *Canad. Math. Bull.*, 66(4):1296–1312, 2023.
3. S. A. Avdonin. On the question of Riesz bases of exponential functions in L^2. *Vestnik Leningrad. Univ.*, 154:5–12, 1974.
4. R. Balan. Stability theorems for Fourier frames and wavelet Riesz bases. Dedicated to the memory of Richard J. Duffin. *J. Fourier Anal. Appl. 3*, 3(5):499–504, 1997.
5. D. Batenkov, B. Diederichs, G. Goldman, and Y. Yomdin. The spectral properties of vandermonde matrices with clustered nodes. *Linear Algebra and its Applications*, 609:37–72, 2021.
6. A. Beurling. *The collected works of Arne Beurling. Vol. 2.* Contemporary Mathematicians. Birkhäuser Boston, Inc., Boston, MA, 1989. Harmonic analysis.
7. M. Bownik, P. Casazza, A. W. Marcus, and D. Speegle. Improved bounds in weaver and feichtinger conjectures. *Journal fur die Reine und Angewandte Mathematik*, 2019(749):267–293, 2019.
8. L. De Carli. *Concerning Exponential Bases on Multi-Rectangles of* $\mathbb{R}^d$, pages 65–85. Springer International Publishing, Cham, 2019.
9. L. De Carli and A. Kumar. Exponential bases on two dimensional trapezoids. *Proc. Amer. Math. Soc.*, 143(7):2893–2903, 2015.
10. A. Debernardi and N. Lev. Riesz bases of exponentials for convex polytopes with symmetric faces. *J. Eur. Math. Soc. (JEMS)*, 24(8):3017–3029, 2022.
11. B. Fuglede. Commuting self-adjoint partial differential operators and a group theoretic problem. *J. Functional Analysis*, 16:101–121, 1974.
12. S. Grepstad and N. Lev. Multi-tiling and Riesz bases. *Advances in Mathematics*, 252:1–6, 2014.
13. C. Heil. *A basis theory primer.* Applied and Numerical Harmonic Analysis. Birkhäuser/Springer, New York, expanded edition, 2011.
14. M. Ĭ. Kadec'. The exact value of the Paley-Wiener constant. *Dokl. Akad. Nauk SSSR*, 155:1253–1254, 1964.
15. G. Kozma and S. Nitzan. Combining Riesz bases. *Invent. Math.*, 199(1):267–285, 2015.
16. G. Kozma and S. Nitzan. Combining Riesz bases in $\mathbb{R}^d$. *Rev. Mat. Iberoam.*, 32(4):1393–1406, 2016.
17. G. Kozma, S. Nitzan, and A. Olevskii. A set with no Riesz basis of exponentials. *Rev. Mat. Iberoam*, 39(6):2007–2016, 2023.
18. D. G. Lee. Unions of exponential Riesz bases. *AIMS Math.*, 9(9):23890–23908, 2024.
19. D. G. Lee, G. E. Pfander, and D. Walnut. Bases of complex exponentials with restricted supports. *J. Math. Anal. Appl.*, 521(2):Paper No. 126917, 15, 2023.
20. W. Li and W. Liao. Stable super-resolution limit and smallest singular value of restricted Fourier matrices. *Applied and Computational Harmonic Analysis*, 51:118–156, 2021.
21. Y. Lyubarskii and A. Rashkovskii. Complete interpolating sequences for Fourier transforms supported by convex symmetric polygons. *Arkiv för matematik*, 38:139–170, 03 2000.
22. N. Macon and A. Spitzbart. Inverses of Vandermonde matrices. *Amer. Math. Monthly*, 65:95–100, 1958.
23. J. Marzo. Riesz basis of exponentials for a union of cubes in $\mathbb{R}^d$. *arXiv:math/0601288*, 2006.
24. B. Matei and Y. Meyer. A variant of compressed sensing. *Rev. Mat. Iberoam.*, 25(2):669–692, 2009.
25. A. Moitra. Super-resolution, extremal functions and the condition number of Vandermonde matrices. In *STOC'15—Proceedings of the 2015 ACM Symposium on Theory of Computing*, pages 821–830. ACM, New York, 2015.
26. G. Pfander, S. Revay, and D. Walnut. Exponential bases for partitions of intervals. *Applied and Computational Harmonic Analysis*, 68:101607, 2024.

27. W. Sun and X. Zhou. On the stability of multivariate trigonometric systems. *J. Math. Anal. Appl.*, 235(1):159–167, 1999.
28. L. R. Turner. Inverse of the vandermonde matrix with applications. In *NASA Technical Note D-3547, NASA Lewis Research Center, Cleveland, Ohio*, 1966.
29. R. M. Young. *An introduction to nonharmonic Fourier series*. Academic Press, Inc., San Diego, CA, first edition, 2001.

Positive Definite Singular Kernels on Two-Point Homogeneous Spaces

Dmitriy Bilyk and Peter J. Grabner

Dedicated to Edward B. Saff on the occasion of his 80th birthday.

1 Introduction

In the present paper we explore the connection between positive definiteness, energy minimization, and positivity of coefficients in the Jacobi expansions for singular kernels on two-point homogeneous spaces. While vast literature on this topic exists in the case of continuous or bounded kernels [8, 23], the general case of singular kernels is much less studied despite the ubiquity of such kernels—the most common example are the Riesz kernels with respect to various metrics, which arise naturally in harmonic analysis, potential theory, mathematical physics etc. Specific singular Riesz kernels of this type have been investigated in this context in numerous papers [1, 4, 9], but all of the prior results relied on ad hoc approximation tricks and a universal theory was still lacking (an attempt at a general study of energy minimization for singular kernels has been undertaken in [5]). The results presented in the current paper are stronger and broader. We shall demonstrate that most results about continuous positive definite kernels propagate to singular kernels under minimal integrability (and continuity away from the singularity) conditions and without any structural assumptions about the kernel.

In the following we will restrict ourselves to compact two-point homogeneous spaces. These are metric spaces (X, d) with an isometry group G acting on them such that for pairs of points (x_1, y_1) and (x_2, y_2) with $d(x_1, y_1) = d(x_2, y_2)$ there

D. Bilyk (✉)
School of Mathematics, University of Minnesota, Minneapolis, MN, USA
e-mail: dbilyk@umn.edu

P. J. Grabner
Institute of Analysis and Number Theory, Graz University of Technology, Graz, Austria
e-mail: peter.grabner@tugraz.at

D. Bilyk et al. (eds.), *Recent Advances in Approximation and Potential Theory*, Applied and Numerical Harmonic Analysis,
https://doi.org/10.1007/978-3-031-95551-8_3

exists a $g \in G$ such that $gx_1 = x_2$ and $gy_1 = y_2$. The two-point homogeneous compact manifolds have been classified (see [14, Chapter I.4]) to be the projective spaces $\mathbb{RP}^{d-1}$, $\mathbb{CP}^{d-1}$, $\mathbb{HP}^{d-1}$, $\mathbb{OP}^2$, and the spheres $\mathbb{S}^{d-1}$; in general, we write $\mathbb{FP}^{d-1}$ for the projective spaces over the ground "field" $\mathbb{F}$. We collect all the necessary facts about the harmonic analysis on these spaces in Sect. 2.

While most of our results are rather general, we shall pay special attention to Riesz kernels with respect to two special distances on these spaces: the chordal distance (which is an analogue of the Euclidean distance on the sphere) and the geodesic distance intrinsic to the space. Even though powers of these distances are equivalent at small scales, they behave quite differently with respect to positive definiteness and energy minimization, which has already been observed in [10]. This difference is particularly striking in the case of the projective spaces, and we further investigate and explain this curious phenomenon.

The outline of the paper is as follows. In Sect. 2 we present the necessary background on the harmonic analysis on two-point homogeneous spaces (zonal kernels, Jacobi polynomials, Poisson kernels). Section 3 explores the positive definiteness property in the case of singular kernels. In particular, in Theorems 1 and 2 we extend the classical results that go back to Schoenberg [23] (for projective spaces, see e.g. [10]) to the case of singular zonal kernels, i.e. we show that a singular zonal kernel is (strictly) positive definite if and only if all of its Jacobi coefficients are non-negative (positive) when only integrability with respect to the uniform (i.e. isometry invariant) measure and continuity away from the singularity are assumed. In Sect. 4 we demonstrate that the classical Schur's lemma generalizes to singular kernels: the product of positive definite kernels is again positive definite (as long as this product is integrable). In Sect. 5 we extend Schoenberg's [23] results on functions, which are positive definite on spheres of all dimensions to the case of projective spaces. For continuous functions such results have been obtained by Guella and Menegatto [11], we extend these to the case of singular integrable kernels. As a byproduct we obtain that no geodesic Riesz kernel can be positive definite on all projective spaces.

2 Harmonic Analysis on Two-Point Homogeneous Spaces

In this section we collect some basic facts about the harmonic analysis on the spheres $\mathbb{S}^{d-1}$ and the projective spaces $\mathbb{FP}^{d-1}$. It is convenient to set $\frac{\pi}{2\kappa}$ as the geodesic diameter of these spaces, where $\kappa = \frac{1}{2}$ for the spheres and $\kappa = 1$ for the projective spaces. We denote the geodesic distance by $\vartheta(x, y)$. Furthermore, we set

$$\alpha = \frac{d-1}{2}\dim_{\mathbb{R}}(\mathbb{F}) - 1 \quad \text{and } \beta = \begin{cases} \alpha & \text{for } X = \mathbb{S}^{d-1} \\ \frac{\dim_{\mathbb{R}}(\mathbb{F})}{2} - 1 & \text{for } X = \mathbb{FP}^{d-1}. \end{cases} \tag{1}$$

Throughout this paper we will denote the space associated to parameters α and β by $X_{\alpha,\beta}$ (but will sometimes drop the indices if it does not cause confusion). Furthermore, we denote by $D = 2\alpha + 2 = (d-1)\dim_{\mathbb{R}}$ the (real) dimension of the space.

It then follows that the surface area of the geodesic sphere of radius $r \in [0, \frac{\pi}{2\kappa}]$ is given by (see [12, Proposition 5.6])

$$A(r) = C_{\alpha,\beta}\kappa \sin(\kappa r)^{2\alpha+1} \cos(\kappa r)^{2\beta+1}, \tag{2}$$

where the constant is chosen as

$$C_{\alpha,\beta} = \frac{2\Gamma(\alpha+\beta+2)}{\Gamma(\alpha+1)\Gamma(\beta+1)}$$

to normalize the volume of the space

$$\int_0^{\frac{\pi}{2\kappa}} A(r)\, dr.$$

The Laplace operator for functions only depending on the distance to one fixed point ("zonal functions") is given by (see [13, Chapter X.7.4])

$$\Delta f = -\frac{1}{A(r)}\frac{\partial}{\partial r}\left(\frac{\partial}{\partial r} A(r) f\right). \tag{3}$$

This implies that the zonal eigenfunctions of the Laplace operator are given by

$$P_n^{(\alpha,\beta)}(\cos(2\kappa r)), \tag{4}$$

where $P_n^{(\alpha,\beta)}$ denotes the Jacobi polynomial. The corresponding eigenvalues are then

$$\lambda_n = 4\kappa^2 n(n+\alpha+\beta+1),$$

see [25, Theorem 4.2.2].

Let $\sigma_{\alpha,\beta}$ denote the normalized surface measure on $X_{\alpha,\beta}$, which is induced by the Haar measure on the group G. Then the equality

$$\begin{aligned}\int_{X_{\alpha,\beta}} F(\cos(2\kappa\vartheta(x,a)))\, d\sigma_{\alpha,\beta}(x) &= \int_0^{\frac{\pi}{2\kappa}} F(\cos(2\kappa\vartheta))\, d\nu_{\alpha,\beta}(\vartheta) \\ &= \int_{-1}^{1} F(t)\, d\mu_{\alpha,\beta}(t),\end{aligned} \tag{5}$$

where

$$d\nu_{\alpha,\beta}(\vartheta) = C_{\alpha,\beta}\kappa \sin(\kappa\vartheta)^{2\alpha+1}\cos(\kappa\vartheta)^{2\beta+1}\,d\vartheta,$$
$$d\mu_{\alpha,\beta}(\vartheta) = C_{\alpha,\beta}(1-t)^{\alpha}(1+t)^{\beta}dt,$$

holds for every function $F : [-1, 1] \to \mathbb{R}$, which is integrable with respect to $d\mu_{\alpha,\beta}$, or equivalently, $F(\cos 2\kappa\vartheta) \in L^1(\nu_{\alpha,\beta})$.

Since the Jacobi polynomials are orthogonal with respect to $d\mu_{\alpha,\beta}$, every function $F \in L^2(\mu_{\alpha,\beta})$ can be represented as a "Fourier series"

$$F(t) = \sum_{n=0}^{\infty} \widehat{F}(n) P_n^{(\alpha,\beta)}(t), \tag{6}$$

where

$$\widehat{F}(n) = \frac{m_n^{(\alpha,\beta)}}{\left(P_n^{(\alpha,\beta)}(1)\right)^2} \int_0^{\frac{\pi}{2\kappa}} F(\cos(2\kappa\vartheta)) P_n^{(\alpha,\beta)}(\cos(2\kappa\vartheta))\, d\nu_{\alpha,\beta}(\vartheta) \tag{7}$$

and

$$m_n^{(\alpha,\beta)} = \frac{2n+\alpha+\beta+1}{\alpha+\beta+1} \frac{(\alpha+\beta+1)_n(\alpha+1)_n}{n!(\beta+1)_n} \tag{8}$$

are the multiplicities of the eigenvalues λ_n.

Equality (6) is a priori only valid in the L^2-sense. For instance, if F is continuous and all coefficients $\widehat{F}(n)$ are non-negative, then the series converges absolutely and uniformly (which can be viewed as a consequence of Mercer's theorem, see [20]).

For the values α and β occurring in our context, the numbers $m_n^{(\alpha,\beta)}$ are integers and equal the dimension of the space of eigenfunctions of the Laplace operator (3) for the eigenvalue λ_n.

We define the following function on $[-1, 1]$ for $0 \le r \le 1$

$$\begin{aligned} \mathcal{P}_r(\cos 2\kappa\vartheta) &= \sum_{n=0}^{\infty} \frac{2n+\alpha+\beta+1}{\alpha+\beta+1} \frac{(\alpha+\beta+1)_n}{(\beta+1)_n} P_n^{(\alpha,\beta)}(\cos 2\kappa\vartheta) r^n \\ &= \frac{1-r}{(1+r)^{\alpha+\beta+2}}\, {}_2\mathrm{F}_1\!\left(\left. \begin{matrix} \frac{\alpha+\beta+2}{2}, \frac{\alpha+\beta+3}{2} \\ \beta+1 \end{matrix} \right| \frac{4r\cos^2\kappa\vartheta}{(1+r)^2} \right), \end{aligned} \tag{9}$$

which we will call the *Poisson kernel*. Here and throughout ${}_2F_1$ denotes the classical hypergeometric function (see, for instance [2]). For the special case $X = \mathbb{S}^{d-1}$ this is the classical Poisson kernel, which solves the Poisson equation for the Euclidean ball.

Equation (9) can be derived from the generating function (see [2, (6.4.7)])

$$\sum_{n=0}^{\infty} \frac{(\alpha+\beta+1)_n}{(\beta+1)_n} P_n^{(\alpha,\beta)}(\cos 2\kappa\vartheta) r^n$$

$$= \frac{1}{(1+r)^{\alpha+\beta+2}}\, {}_2\mathrm{F}_1\left(\left. \begin{matrix} \frac{\alpha+\beta+1}{2}, \frac{\alpha+\beta+2}{2} \\ \beta+1 \end{matrix} \right| \frac{4r\cos^2\kappa\vartheta}{(1+r)^2} \right)$$

by differentiating with respect to r and applying contiguous relations for the hypergeometric function.

We use the Poisson kernel to define the associated convolution operator acting on continuous functions, for which we use the same notation

$$\mathcal{P}_r(f)(x) = \int_{X_{\alpha,\beta}} \mathcal{P}_r(\cos 2\kappa\vartheta(x,y)) f(y)\, d\sigma_{\alpha,\beta}(y).$$

Furthermore, for $f \in \mathrm{C}(X_{\alpha,\beta})$ the limit relation

$$\lim_{r\to 1-} \mathcal{P}_r(f)(x) = f(x) \tag{10}$$

holds uniformly for $x \in X_{\alpha,\beta}$. The proof for this is completely analogous to the proof that Fourier series of continuous functions converge in the sense of Abel summation (see, for instance [26]).

If $F \in L^1(\mu_{\alpha,\beta}) \cap \mathrm{C}([-1,1))$, then

$$\mathcal{P}_r(F)(\cos 2\kappa\vartheta(x,z)) = \sum_{n=0}^{\infty} \widehat{F}(n) r^n P_n^{\alpha,\beta}(\cos 2\kappa\vartheta(x,z)). \tag{11}$$

Here the operator is applied to the zonal function $F(\cos 2\kappa\vartheta(x,\cdot))$. In this case (10) still holds uniformly on $\cos 2\kappa\vartheta(x,y) \in [-1, 1-\epsilon]$ for $\epsilon > 0$.

3 Positive Definite Functions

Positive definite kernels play an important role in potential theory, harmonic analysis, and operator theory on metric spaces. Unfortunately, for singular kernels the general theory is scarce. Even the standard definition does not apply in this case, although it is not hard to properly adjust it. Below we define positive definiteness for singular zonal kernels on the two-point homogeneous spaces $X_{\alpha,\beta}$, but the extension to the general case is straightforward.

Definition 1 A measurable function $F : [-1, 1] \to \mathbb{R}$ is called *positive definite* on $X_{\alpha,\beta}$, if for all signed Borel measures μ, for which $F(\cos 2\kappa\vartheta(\cdot, y)) \in L^1(|\mu|)$ for all $y \in X_{\alpha,\beta}$, the inequality

$$E_F(\mu) = \iint\limits_{X_{\alpha,\beta}\times X_{\alpha,\beta}} F(\cos 2\kappa\vartheta(x, y))\, d\mu(x)\, d\mu(y) \geq 0 \tag{12}$$

holds. If there is strict inequality in (12) for all $\mu \neq 0$, the kernel is called *strictly positive definite*. A kernel K is called *conditionally positive definite*, if (12) holds for all signed measures of zero total mass ($\mu(X) = 0$).

Remark 1 This definition is an immediate generalisation of the definition given by Bochner in [7]. This classical definition requires the matrix $(F(\cos 2\kappa\vartheta(x_i, x_j)))_{i,j}$ to be positive semidefinite for all choices of $x_i \in X_{\alpha,\beta}$ $(i = 1, \dots, N)$. It is obvious that this is equivalent to the fact that (12) holds for all finitely supported measures, and weak-$*$ density of such measures implies the equivalence of the two definitions for continuous kernels $F \in C[-1, 1]$.

The original definition forces the kernel F to be defined for all pairs x, y in the domain. A very prominent example, namely the Riesz kernel $\|x - y\|^{-s}$ defined for $X \subset \mathbb{R}^d$, does not satisfy this requirement when $x = y$. The definition above avoids this issue and applies to singular kernels. We mention that for the case of $\mathbb{R}^d$ singular positive kernels have been studied in [21].

We now demonstrate that, just like in the continuous case [3], positive definiteness on $X_{\alpha,\beta}$ amounts to non-negativity of the coefficients in the Jacobi expansion.

Theorem 1 *Let $F : [-1, 1) \to \mathbb{R}$ be a function in $L^1(\mu_{\alpha,\beta}) \cap C([-1, 1))$. Then F is positive definite if and only if all of its Fourier coefficients are non-negative.*

Proof We first prove that positive definiteness implies non-negativity of the Fourier coefficients. For this purpose we consider the signed measure $d\mu_z(x) = P_n^{(\alpha,\beta)}(\cos(2\kappa\vartheta(x, z)))\, d\sigma_{\alpha,\beta}(x)$ for fixed z and compute its energy

$$\begin{aligned} E_F(\mu_z) = \iint\limits_{X_{\alpha,\beta}\times X_{\alpha,\beta}} & F(\cos 2\kappa\vartheta(x, y)) P_n^{(\alpha,\beta)}(\cos(2\kappa\vartheta(x, z))) \\ & \times P_n^{(\alpha,\beta)}(\cos(2\kappa\vartheta(y, z)))\, d\sigma_{\alpha,\beta}(x)\, d\sigma_{\alpha,\beta}(y). \end{aligned} \tag{13}$$

Since F is positive definite and this energy is obviously finite, we have $E_F(\mu_z) \geq 0$.

We shall use the following relation (see e.g. [5, Lemma 9.1]), which can be seen as a projective analogue of the classical Funk–Hecke formula and is easily proved using the addition formula

$$\int_{X_{\alpha,\beta}} P_n^{(\alpha,\beta)}\big(\cos(2\kappa\vartheta(x, z))\big) P_n^{(\alpha,\beta)}\big(\cos(2\kappa\vartheta(y, z))\big) d\sigma_{\alpha,\beta}(z) \tag{14}$$

$$= \frac{P_n^{(\alpha,\beta)}(1)}{m_n^{(\alpha,\beta)}} P_n^{(\alpha,\beta)}\big(\cos(2\kappa\vartheta(x,y))\big).$$

According to (5) and (7), we see that

$$\widehat{F}(n) = \frac{m_n^{(\alpha,\beta)}}{\left(P_n^{(\alpha,\beta)}(1)\right)^2} \int_{X_{\alpha,\beta}} \int_{X_{\alpha,\beta}} F(\cos(2\kappa\vartheta(x,y)))$$

$$\times P_n^{(\alpha,\beta)}(\cos(2\kappa\vartheta(x,y)))\, d\sigma_{\alpha,\beta}(x)\, d\sigma_{\alpha,\beta}(y).$$

The introduction of the outer integral with respect to $d\sigma_{\alpha,\beta}(y)$ above is harmless since the inner integral is independent of $y \in X_{\alpha,\beta}$. Invoking (14) and interchanging the order of integration (which is justified since F is integrable with respect to $d\sigma_{\alpha,\beta}$ and $P_n^{(\alpha,\beta)}$ is bounded), we obtain

$$\widehat{F}(n) = \frac{\left(m_n^{(\alpha,\beta)}\right)^2}{\left(P_n^{(\alpha,\beta)}(1)\right)^3} \int_{X_{\alpha,\beta}} E_F(\mu_z)\, d\sigma_{\alpha,\beta}(z) \geq 0, \tag{15}$$

which proves the first implication.

Conversely, assume now that all Fourier coefficients of F are non-negative. If the series

$$\sum_{n=0}^{\infty} \widehat{F}(n) P_n^{(\alpha,\beta)}(1)$$

converges, then the Fourier series of F converges absolutely and uniformly and thus F is continuous, and the assertion follows from the classical result for continuous kernels. Thus we assume that the series diverges. Let μ be a signed measure such that $F(\cos 2\kappa\vartheta(\cdot, y)) \in L^1(|\mu|)$ for all $y \in X$. Such a measure is necessarily continuous (atomless) under our assumption on F.

Then for $0 < r < 1$ the function $\mathcal{P}_r(F)$ is continuous and has an absolutely and uniformly convergent Fourier series. Furthermore, $\mathcal{P}_r(F)$ is positive definite. The corresponding energy

$$E_{\mathcal{P}_r(F)}(\mu) = \iint\limits_{X_{\alpha,\beta}\times X_{\alpha,\beta}} \mathcal{P}_r(F)(\cos 2\kappa\vartheta(x,y))\, d\mu(x)\, d\mu(y)$$

$$= \iint\limits_{X_{\alpha,\beta}\times X_{\alpha,\beta}} \int\limits_{X_{\alpha,\beta}} F(\cos 2\kappa\vartheta(x,z))\mathcal{P}_r(\cos 2\kappa\vartheta(z,y))\, d\sigma_{\alpha,\beta}(z)\, d\mu(x)\, d\mu(y)$$

is then positive for $0 < r < 1$ and is an increasing function of r, which can be seen from (11), positivity of $\widehat{F}(n)$, and positive definiteness of $P_n^{(\alpha,\beta)}$. We interchange the order of integration and define

$$G(z) = \int_{X_{\alpha,\beta}} F(\cos 2\kappa\vartheta(x,z))\, d\mu(x).$$

The function G is then continuous under our assumptions: fix $z_0 \in X$ and let $\varepsilon > 0$. Then there exists a $\delta > 0$ such that

$$\forall z \in B(z_0,\delta) \quad \forall x \in X \setminus B(z_0, 2\delta) : |F(\cos 2\kappa\vartheta(x,z_0)) - F(\cos 2\kappa\vartheta(x,z))| < \varepsilon.$$

Since μ is continuous and $F(\cos 2\kappa\vartheta(\cdot, y)) \in L^1(|\mu|)$ for all $z \in X$, we can choose δ so small that

$$\int_{B(z_0,2\delta)} |F(\cos 2\kappa\vartheta(x,z))|\, d|\mu|(x) < \varepsilon$$

for all $z \in X$. Then we have for $z \in B(z_0,\delta)$

$$|G(z) - G(z_0)| \leq \int\limits_{B(z_0,2\delta)} (|F(\cos 2\kappa\vartheta(x,z))| + |F(\cos 2\kappa\vartheta(x,z_0))|)\, d|\mu|(x)$$

$$+ \int\limits_{X_{\alpha,\beta}\setminus B(z_0,2\delta)} |F(\cos 2\kappa\vartheta(x,z_0)) - F(\cos 2\kappa\vartheta(x,z))|\, d|\mu|(x) < (2+|\mu|(X_{\alpha,\beta}))\varepsilon,$$

which shows that G is continuous.

We then have

$$E_{\mathcal{P}_r(F)}(\mu) = \int\limits_{X_{\alpha,\beta}} \int\limits_{X_{\alpha,\beta}} G(z)\mathcal{P}_r(\cos 2\kappa\vartheta(y,z))\, d\sigma_{\alpha,\beta}(z)\, d\mu(y).$$

By the continuity of G the inner integral tends to $G(y)$ uniformly in y for $r \to 1$. From this we obtain

$$\lim_{r\to 1} E_{\mathcal{P}_r(F)}(\mu) = \int\limits_{X_{\alpha,\beta}} G(y)\, d\mu(y) = E_F(\mu).$$

Since $E_{\mathcal{P}_r(F)} \geq 0$ for $0 < r < 1$, we have obtained $E_F(\mu) \geq 0$. □

The following theorem is an immediate corollary of our proof using the observation that $E_{\mathcal{P}_r(F)}(\mu)$ is an increasing function of r.

Theorem 2 *Let $F : [-1, 1) \to \mathbb{R}$ be a function in $L^1(\mu_{\alpha,\beta}) \cap C([-1, 1))$. Then F is strictly positive definite if and only if all its Fourier coefficients are strictly positive.*

In addition, one can easily see that, since measures of mass zero annihilate constants, the proof easily adapts to yield a similar characterization holds for (strict) conditional positive definiteness.

Theorem 3 *Let $F : [-1, 1) \to \mathbb{R}$ be a function in $L^1(\mu_{\alpha,\beta}) \cap C([-1, 1))$. Then the kernel F is (strictly) conditionally positive definite if and only if its Fourier coefficients $\widehat{F}(n)$ are non-negative (strictly positive) for all $n \geq 1$.*

3.1 Positive Definiteness and Energy Minimization

In general, the connection between positive definiteness and minimizing energy integrals is classical [8]. However, general equivalence results for singular kernels are not found in the literature. We use characterizations proved above to yield the necessary and sufficient condition for the energy $E_F(\mu)$ with a singular kernel F to be minimized by the normalized Haar measure $\sigma_{\alpha,\beta}$ amongst all Borel probability measures.

Theorem 4 *Let $F : [-1, 1) \to \mathbb{R}$ be a function in $L^1(\mu_{\alpha,\beta}) \cap C([-1, 1))$. Then the energy integral*

$$E_F(\mu) = \iint\limits_{X_{\alpha,\beta} \times X_{\alpha,\beta}} F(\cos 2\kappa\vartheta(x, y))\, d\mu(x)\, d\mu(y) \tag{16}$$

is (uniquely, resp.) minimized by the normalized Haar measure $\sigma_{\alpha,\beta}$ amongst all Borel probability measures if and only if the kernel F is (strictly, resp.) conditionally positive definite.

This result, according to Theorem 3, implies that the fact that the uniform measure $\sigma_{\alpha,\beta}$ minimizes $E_F(\mu)$ (i.e. energy minimization leads to uniform distribution) amounts to checking the positivity of the Fourier (Jacobi) coefficients. In the case of continuous kernels, this result is well known and can be found in different variations in [4, 7, 9, 23], but it has not been stated in such generality for singular kernels. Certain ad hoc approximation tricks have been used for some particular (Riesz) kernels [1, 4, 9], and it was shown in [5, Theorem 4.2] that positivity of the coefficients is sufficient for a fairly general class of singular kernels, with certain restrictions on the behavior at the singularity. We demonstrate that positivity (non-negativity) of the coefficients is a necessary and sufficient condition, assuming merely the integrability of the kernel (i.e. finiteness of the energy $E_F(\sigma_{\alpha,\beta})$) and continuity away from the singularity.

Proof of Theorem 4 Let us first assume that F is conditionally positive definite. According to Theorem 3, this is equivalent to the fact that $\widehat{F}(n) \geq 0$ for $n \geq 1$. Since $\mathcal{P}_r(F)$ is continuous and conditionally positive definite, the aforementioned classical results imply that $E_{\mathcal{P}_r(F)}(\mu)$ is minimized by $\sigma_{\alpha,\beta}$. Arguing exactly as in the second part of Theorem 1, we find that for any Borel probability measure such that $E_F(\mu)$ is finite

$$E_F(\mu) = \lim_{r\to 1} E_{\mathcal{P}_r(F)}(\mu) \geq \lim_{r\to 1} E_{\mathcal{P}_r(F)}(\sigma_{\alpha,\beta}) = E_F(\sigma_{\alpha,\beta}), \tag{17}$$

i.e. E_F is minimized by σ. In the case of strict conditional positive definiteness, uniqueness of the minimizer follows from general potential theory, see [8, Theorem 4.2.7].

Conversely, if F is not conditionally positive definite, i.e. there exists $n \geq 1$ such that $\widehat{F}(n) < 0$, one can choose the measure

$$d\mu_z(x) = \left(1 + \varepsilon P_n^{(\alpha,\beta)}(\cos(2\kappa\vartheta(x,z)))\right) d\sigma_{\alpha,\beta}(x),$$

where $\varepsilon > 0$ is chosen small enough so that the measure is positive. It is standard to show, using the addition formula for Jacobi polynomials (see e.g. [1, Theorem 2.8]) that the value of the energy $E_F(\mu_z)$ is independent of $z \in X_{\alpha,\beta}$. Therefore, a computation almost identical to the one presented in the first part of the proof of Theorem 1 and based on (14) and (15) demonstrates that

$$E_F(\mu_z) = E_F(\sigma_{\alpha,\beta}) + \varepsilon^2 \frac{\left(P_n^{(\alpha,\beta)}(1)\right)^3}{\left(m_n^{(\alpha,\beta)}\right)^2} \widehat{F}(n) < E_F(\sigma_{\alpha,\beta}), \tag{18}$$

i.e. $\sigma_{\alpha,\beta}$ does not minimize the energy. If F is conditionally positive definite, but not strictly, then $\widehat{F}(n) = 0$ for some $n \geq 1$ and thus $E_F(\mu_z) = E_F(\sigma_{\alpha,\beta})$, hence $\sigma_{\alpha,\beta}$ cannot be a unique minimizer. □

4 Schur's Lemma for Singular Integrable Kernels

Schur's lemma is the most important structural result in the context of continuous positive definite functions. It asserts that the Hadamard product of two positive definite matrices is again positive definite. As a consequence the product of two continuous positive definite functions is again positive definite. This implication depends directly on the fact of continuity (see Remark 1). The following theorem gives an analogue of Schur's lemma in the context of singular positive definite kernels.

Theorem 5 *Let $F_1, F_2 : [-1, 1) \to \mathbb{R}$ be positive definite functions in $L^1(\mu_{\alpha,\beta}) \cap \mathrm{C}([-1, 1))$ such that their product is also in $L^1(\mu_{\alpha,\beta}) \cap \mathrm{C}([-1, 1))$. Then $F_1 F_2$ is also positive definite.*

Proof We start with proving the theorem for the restricted case with $F_1 \in L^1(\mu_{\alpha,\beta}) \cap \mathrm{C}([-1, 1)$ and $F_2 \in \mathrm{C}([-1, 1])$. We define

$$F_{1,r} = \mathcal{P}_r(F_1),$$

which is continuous and positive definite on $[-1, 1]$ for $0 < r < 1$.

Both functions are positive definite by assumption. Let μ be a signed measure such that $F_1(\cos(2\kappa\vartheta\cdot, y)) \in L^1(|\mu|)$. Then we have

$$\begin{aligned}0 &\le \int_{X_{\alpha,\beta}} \int_{X_{\alpha,\beta}} F_{1,r}(\cos 2\kappa\vartheta(x, y)) F_2(\cos 2\kappa\vartheta(x, y))\, d\mu(x)\, d\mu(y)\\ &= \int_{X_{\alpha,\beta}} \int_{X_{\alpha,\beta}} \int_{X_{\alpha,\beta}} F_1(\cos 2\kappa\vartheta(x, z)) \mathcal{P}_r(\cos 2\kappa\vartheta(z, y))\\ &\quad \times F_2(\cos 2\kappa\vartheta(x, y))\, d\sigma(z)\, d\mu(x)\, d\mu(y).\end{aligned}$$

We interchange the order of integration and set

$$G(y, z) = \int_{X_{\alpha,\beta}} F_1(x, z) F_2(x, y)\, d\mu(x).$$

This function is then continuous on $X_{\alpha,\beta} \times X_{\alpha,\beta}$ by the same argument as in the proof of Theorem 1. By the properties of the Poisson kernel

$$\int_X G(y, z)\mathcal{P}_r(\cos 2\kappa\vartheta(z, y))\, d\sigma(z) \to G(y, y)$$

for $r \to 1$ uniformly on $X_{\alpha,\beta}$. From this we obtain

$$\begin{aligned}0 &\le \int_{X_{\alpha,\beta}} \int_{X_{\alpha,\beta}} F_{1,r}(\cos 2\kappa\vartheta(x, y)) F_2(\cos 2\kappa\vartheta(x, y))\, d\mu(x)\, d\mu(y)\\ &\to \int_{X_{\alpha,\beta}} G(y, y)\, d\sigma_{\alpha,\beta}(y)\\ &= \int_X \int_X F_1(\cos 2\kappa\vartheta(x, y)) F_2(\cos 2\kappa\vartheta(x, y))\, d\mu(x)\, d\mu(y)\end{aligned}$$

for $r \to 1$.

For proving the general case we apply the assertion proved above to F_1 and $F_{2,t}$ (defined analogous to $F_{1,t}$) and again take the limit as $r \to 1$. □

5 Functions That Are Positive Definite for all $\mathbb{FP}^{d-1}$

In [23] Schoenberg characterizes positive definite functions on the spheres $\mathbb{S}^{d-1}$ by the positivity of the coefficients of their ultraspherical expansion. Furthermore, he gives a characterization of the functions that are positive definite on all spheres. He proves the following theorem:

Theorem 6 ([23, Theorem 2]) *A function $f : [-1, 1] \to \mathbb{R}$ is positive definite on all spheres $\mathbb{S}^{d-1}$, if and only if it can be expressed in the form*

$$f(\cos\vartheta) = \sum_{n=0}^{\infty} a_n \cos^n \vartheta$$

with $a_n \geq 0$ for all n and $\sum_{n=0}^{\infty} a_n < \infty$.

Remark 2 We note here that Guella and Menegatto [11] proved a similar theorem for series in terms of Jacobi polynomials $P_n^{(\alpha_m,\beta_m)}$ with $\alpha_m \to \infty$ and $\frac{\beta_m}{\alpha_m} \to c$ for $m \to \infty$.

In this section we will show the corresponding result for the projective spaces $\mathbb{FP}^{d-1}$ and also extend it to singular functions. It will be convenient to use normalized Jacobi polynomials in this section

$$p_n^{(\alpha,\beta)}(t) = \frac{P_n^{(\alpha,\beta)}(t)}{P_n^{(\alpha,\beta)}(1)}.$$

We first state and prove a lemma that will be essential for the proof of this theorem.

Lemma 1 *For every $n \in \mathbb{N}_0$ the limit relation*

$$\lim_{\alpha\to\infty} p_n^{(\alpha,\beta)}(\cos 2\vartheta) = \cos^{2n} \vartheta$$

holds uniformly for $\vartheta \in [0, \frac{\pi}{2}]$.

Proof of Lemma 1 We fix β and first notice that

$$p_n^{(\alpha,\beta)}(\cos 2\vartheta) = \sum_{m=0}^{n} (-1)^{n-m} \binom{n}{m} \frac{(\beta+m+1)(\beta+m+2)\cdots(\beta+n)}{(\alpha+1)(\alpha+2)\cdots(\alpha+n-m)} \times \left(\frac{1+\cos 2\vartheta}{2}\right)^m \left(\frac{1-\cos 2\vartheta}{2}\right)^{n-m}.$$

For fixed n every term tends to 0 uniformly in ϑ for $\alpha \to \infty$, except for the term with $m = n$, which equals $(\frac{1+\cos 2\vartheta}{2})^n = \cos^{2n}\vartheta$. This shows uniform convergence for $\vartheta \in [0, \frac{\pi}{2}]$ for all n.

□

Lemma 2 *For $\beta \geq 0$ and $\beta = -\frac{1}{2}$ the normalized Jacobi polynomials satisfy the following inequalities*

$$|p_n^{(\alpha,\beta)}(\cos 2\vartheta)| \leq \frac{\Gamma(\alpha+1)}{\Gamma(\beta+1)(n\sin^2\vartheta_0)^{\alpha-\beta}} \tag{19}$$

valid for $\vartheta_0 \leq \vartheta \leq \frac{\pi}{2}$ and

$$|p_n^{(\alpha,\beta)}(\cos 2\vartheta)| \leq \frac{\Gamma(\alpha+1)}{\Gamma(\alpha-\beta)} \frac{(\cos\vartheta_0)^{2n}}{(\alpha-\beta-1-n\tan^2\vartheta_0)^{\beta+1}} \tag{20}$$

valid for $\vartheta_0 \leq \vartheta \leq \frac{\pi}{2}$ and $n\tan^2\vartheta_0 < \alpha-\beta-1$.

Proof of Lemma 2 For $\beta \geq 0$ we use the formula (see [15])

$$\begin{aligned} p_n^{(\alpha,\beta)}(\cos 2\vartheta) = & \frac{2\Gamma(\alpha+1)}{\sqrt{\pi}\Gamma(\alpha-\beta)\Gamma(\beta+\frac{1}{2})} \\ & \times \int_0^1\int_0^\pi \Big(\cos^2\vartheta - r^2\sin^2\vartheta + ir\cos\phi\sin 2\vartheta\Big)^n \\ & \times (1-r^2)^{\alpha-\beta-1} r^{2\beta+1}\sin^{2\beta}\phi\, d\phi\, dr. \end{aligned} \tag{21}$$

Notice that

$$\begin{aligned} & \Big|\cos^2\vartheta - r^2\sin^2\vartheta + ir\cos\phi\sin 2\vartheta\Big|^2 \\ =& (\cos^2\vartheta + r^2\sin^2\vartheta)^2 - 4r^2\sin^2\vartheta\cos^2\vartheta\sin^2\phi \\ \leq& (\cos^2\vartheta + r^2\sin^2\vartheta)^2 = (1-(1-r^2)\sin^2\vartheta)^2. \end{aligned}$$

For proving (19) we estimate

$$\begin{aligned} |p_n^{(\alpha,\beta)}(\cos 2\vartheta)| \leq & \frac{2\Gamma(\alpha+1)}{\sqrt{\pi}\Gamma(\alpha-\beta)\Gamma(\beta+\frac{1}{2})} \\ \times \int_0^1\int_0^\pi & e^{-n(1-r^2)\sin^2\vartheta_0}(1-r^2)^{\alpha-\beta-1} r^{2\beta+1}\sin^{2\beta}\phi\, d\phi\, dr \\ \leq & \frac{2\Gamma(\alpha+1)}{\Gamma(\alpha-\beta)\Gamma(\beta+1)} \int_0^1 e^{-n(1-r^2)\sin^2\vartheta_0}(1-r^2)^{\alpha-\beta-1} r\, dr \end{aligned}$$

In the last integral we substitute $1 - r^2$ and then extend the resulting integral to ∞ to obtain (19).

For proving (20) we estimate

$$|p_n^{(\alpha,\beta)}(\cos 2\vartheta)| \le (\cos\vartheta_0)^{2n} \frac{2\Gamma(\alpha+1)}{\sqrt{\pi}\Gamma(\alpha-\beta)\Gamma(\beta+\frac{1}{2})}$$

$$\times \int_0^1 \int_0^\pi e^{-r^2(\alpha-\beta-1-n\tan^2\vartheta_0)} r^{2\beta+1} \sin^{2\beta}\phi\, d\phi\, dr$$

$$\le \frac{2\Gamma(\alpha+1)}{\Gamma(\alpha-\beta)\Gamma(\beta+1)} \int_0^1 e^{-r^2(\alpha-\beta-1-n\tan^2\vartheta_0)} r^{2\beta+1}\, dr.$$

We extend the range of integration to ∞ to obtain (20).

For $\beta = -\frac{1}{2}$ we use the representation

$$p_n^{(\alpha,-\frac{1}{2})}(\cos 2\vartheta) = \frac{\Gamma(\alpha+2)}{\sqrt{\pi}\Gamma(\alpha+\frac{3}{2})}$$

$$\times \int_0^1 \Re\left(\cos^2\vartheta - x^2\sin^2\vartheta + 2ix\cos\vartheta\sin\vartheta\right)^n (1-x^2)^{\alpha-\frac{1}{2}}\, dx.$$

This can be obtained from (21) by taking the limit $\beta \to -\frac{1}{2}$. □

Definition 2 The space of continuous functions with a (moderate) singularity at the right end point is denoted by

$$\mathrm{C}_{\text{sing}}([-1,1]) = \left\{f \in \mathrm{C}([-1,1)) \mid \exists s \in \mathbb{R}^+ : (1-t)^s f(t) \in \mathrm{C}([-1,1])\right\}.$$

Remark 3 Notice that every $f \in \mathrm{C}_{\text{sing}}([-1,1])$ is contained in $L^1(\mu_{\alpha,\beta})$ for α large enough.

Theorem 7 *A function $f \in \mathrm{C}_{\text{sing}}([-1,1])$ is positive definite for all spaces $\mathbb{FP}^{d-1}$ (for $\mathbb{F} \in \{\mathbb{R},\mathbb{C},\mathbb{H}\}$) with d large enough to ensure integrability, if and only if there are non-negative a_n ($n \in \mathbb{N}_0$) such that*

$$f(\cos 2\vartheta) = \sum_{n=0}^{\infty} a_n \cos^{2n}\vartheta. \tag{22}$$

Proof The sufficiency of the condition is obvious by the fact that

$$\cos^2\vartheta = \frac{1}{\alpha+\beta+2} P_1^{(\alpha,\beta)}(\cos 2\vartheta) + \frac{\beta+1}{\alpha+\beta+2} P_0^{(\alpha,\beta)}(\cos 2\vartheta).$$

Then Schur's lemma shows that all powers $\cos^{2n}\vartheta$ are positive definite.

The proof of the necessity will make use of the same smoothing procedure as the proofs of Theorems 1 and 5, this time taking care of the dependence on α. Throughout this proof we express the function f by its normalized Fourier expansion

$$f(t)=\sum_{n=0}^{\infty}a_n^{(\alpha,\beta)}p_n^{(\alpha,\beta)}(t).$$

Let $f\in C_{\mathrm{sing}}([-1,1])$. Then there exists $s\geq 0$ such that

$$(1-t)^s f(t)=g(t)\in C([-1,1]).$$

We can choose $s\in\mathbb{N}$ with this property. Assume that $|g(t)|\leq 1$ for all $t\in[-1,1]$, which we can do without loss of generality. Then we have

$$|\mathcal{P}_r(f)(1)|\leq C_{\alpha,\beta}\int_0^{\frac{\pi}{2}}\mathcal{P}_r(\cos 2\vartheta)(\sin\vartheta)^{2\alpha-2s+1}(\cos\vartheta)^{2\beta+1}\,d\vartheta$$

$$=\frac{1-r}{(1+r)^{\alpha+\beta+2}}C_{\alpha,\beta}\int_0^{\frac{\pi}{2}}{}_2F_1\left(\begin{matrix}\frac{\alpha+\beta+2}{2},\frac{\alpha+\beta+3}{2}\\ \beta+1\end{matrix}\,\middle|\,\frac{4r\cos^2\vartheta}{(1+r)^2}\right)(\sin\vartheta)^{2\alpha-2s+1}(\cos\vartheta)^{2\beta+1}d\vartheta$$

$$=\frac{\Gamma(\alpha+\beta+2)\Gamma(\alpha-s+1)}{\Gamma(\alpha+1)\Gamma(\alpha+\beta-s+2)}\frac{1-r}{(1+r)^{\alpha+\beta+2}}\,{}_2F_1\left(\begin{matrix}\frac{\alpha+\beta+2}{2},\frac{\alpha+\beta+3}{2}\\ \alpha+\beta-s+2\end{matrix}\,\middle|\,\frac{4r}{(1+r)^2}\right);$$

the last equation can be seen integrating term by term.

We now use two basic hypergeometric function identities (see [2, 3.1.9] and [19, p. 42]) to obtain

$$\frac{1-r}{(1+r)^{\alpha+\beta+2}}\,{}_2F_1\left(\begin{matrix}\frac{\alpha+\beta+2}{2},\frac{\alpha+\beta+3}{2}\\ \alpha+\beta-s+2\end{matrix}\,\middle|\,\frac{4r}{(1+r)^2}\right)$$

$$=(1-r)\,{}_2F_1\left(\begin{matrix}\alpha+\beta+2,s+1\\ \alpha+\beta+2-s\end{matrix}\,\middle|\,r\right)$$

$$=(1-r)^{-2s}\,{}_2F_1\left(\begin{matrix}-s,\alpha+\beta+1-2s\\ \alpha+\beta+2-s\end{matrix}\,\middle|\,r\right).$$

The last hypergeometric function is a polynomial, since s was assumed to be an integer. For $\alpha\geq s$ all coefficients of this polynomial are bounded as functions of α.

Summing up, we have obtained that

$$\mathcal{P}_r(f)(1)=\sum_{n=0}^{\infty}a_n^{(\alpha,\beta)}r^n\leq\frac{M}{(1-r)^{2s}}\tag{23}$$

for $\alpha \geq 2s$ and a constant M only depending on s and β.

We use now that the function f is integrable and positive definite for all $\alpha > \alpha_0$. This implies the non-negativity of all coefficients $a_n^{(\alpha,\beta)}$ for all $\alpha > \alpha_0$. Inserting $r = 1 - \frac{1}{n}$ into (23) gives the upper bound

$$a_n^{(\alpha,\beta)} \leq 2eMn^{2s} \tag{24}$$

for all $\alpha > \alpha_0$. This shows that there is an increasing sequence $(\alpha_k)_{k\in\mathbb{N}}$ tending to ∞, such that

$$\forall n \in \mathbb{N}_0 : \lim_{k\to\infty} a_n^{(\alpha_k,\beta)} = b_n$$

exists. Notice that then $b_n \leq 2eMn^{2s}$ for all $n \geq 1$.

It remains to show that

$$f(\cos 2\vartheta) = \sum_{n=0}^{\infty} b_n (\cos\vartheta)^{2n}.$$

We fix ϑ_0. For a given $\epsilon > 0$ we choose N so large that

$$\sum_{n=N+1}^{\infty} b_n (\cos\vartheta_0)^{2n} \leq 2eM \sum_{n=N+1}^{\infty} n^{2s} (\cos\vartheta_0)^{2n} < \epsilon.$$

Then we choose α_0 so large that for all $\alpha_k > \alpha_0$ we have

$$\left| \sum_{n=0}^{N} a_n^{(\alpha_k,\beta)} p_n^{(\alpha_k,\beta)}(\cos 2\vartheta) - \sum_{n=0}^{N} b_n (\cos\vartheta)^{2n} \right| < \epsilon$$

for $\vartheta \in [0, \frac{\pi}{2}]$.

Then we have

$$\left| f(\cos 2\vartheta) - \sum_{n=0}^{\infty} b_n (\cos\vartheta)^{2n} \right| < 2\epsilon + \sum_{n=N+1}^{\infty} a_n^{(\alpha_k,\beta)} |p_n^{(\alpha_k,\beta)}(\cos 2\vartheta)|$$

for $\vartheta \in [\vartheta_0, \frac{\pi}{2}]$ and $\alpha_k > \alpha_0$. It remains to bound this last sum. For this purpose we use the estimates (19) and (20) depending on the summation index

$$\sum_{n=N+1}^{\infty} a_n^{(\alpha_k,\beta)} |p_n^{(\alpha_k,\beta)}(\cos 2\vartheta)|$$

$$\leq \frac{2Me\Gamma(\alpha_k + 1)}{\Gamma(\alpha_k - \beta)} \sum_{N<n<C\alpha_k} n^{2s} \frac{(\cos\vartheta_0)^{2n}}{(\alpha_k - \beta - 1 - n\tan^2\vartheta_0)^{\beta+1}}$$

$$+\frac{2Me\Gamma(\alpha_k+1)}{\Gamma(\beta+1)}\sum_{n>C\alpha_k} n^{2s-\alpha_k+\beta},$$

where the constant C satisfies $C\tan^2\vartheta_0 < 1$ and will be chosen later. We estimate further by

$$\frac{2Me\Gamma(\alpha_k+1)}{\Gamma(\alpha_k-\beta)((1-C\tan^2\vartheta_0)\alpha_k-\beta-1)^{\beta+1}}\sum_{n=N+1}^{\infty} n^{2s}(\cos\vartheta_0)^{2n}$$
$$+\frac{2Me\Gamma(\alpha_k+1)}{(\alpha_k-\beta-2s)\Gamma(\beta+1)(\alpha_k-\beta-2s-1)}(C\alpha_k)^{2s-\alpha_k+\beta+1}.$$

The first sum is bounded by ϵ times a function bounded in α_k (notice that the quotient of Γ-functions behaves like $\alpha_k^{\beta+1}$, which cancels with the denominator). The second term can be analyses using the Stirling approximation for the Γ-function. This gives $(Ce\sin^2\vartheta_0)^{-\alpha_k}\alpha_k^{2s+\beta+\frac{1}{2}}$. We choose C so that $Ce\sin^2\vartheta_0 > 1$ still satisfying $C\tan^2\vartheta_0 < 1$. Then the second sum tends to 0 for $\alpha_k \to \infty$.

Summing up, we have shown that

$$f(\cos 2\vartheta) = \sum_{n=0}^{\infty} b_n(\cos\vartheta)^{2n}$$

uniformly on $[\vartheta_0, \frac{\pi}{2}]$. □

Remark 4 The proof shows that actually the limit

$$\lim_{\alpha\to\infty} a_n^{(\alpha,\beta)}$$

exists for all n, since the representation of f as a power series in $\cos^2\vartheta$ is unique.

Remark 5 Theorem 7 contains the generalization of Schoenberg's Theorem 6 to the projective spaces. Furthermore, the estimate (19) shows that for any $f \in C_{\mathrm{sing}}([-1,1])$ the Fourier series converges uniformly on $[\vartheta_0, \frac{\pi}{2}]$ for all α large enough.

6 Riesz Kernels for the Geodesic and Chordal Distance

The most prominent examples of singular kernels are the Riesz kernels given by negative powers of a distance. These kernels are the subject of investigation in the context of classical potential theory (see [16]).

The spaces X under consideration can be equipped with two "natural" metrics, which exhibit rather different behavior in this context: the geodesic metric given by the angle $\vartheta(x, y)$ and the chordal metric $\chi(x, y) = \sin(\kappa\vartheta(x, y))$. These metrics are, of course, equivalent; the geodesic metric is the metric coming from the structure of X as a Riemannian manifold, whereas the chordal metric comes from embedding the space X into a suitable Euclidean space (on the sphere, up to a constant factor, this is the Euclidean distance). It is this second property that turns out to be crucial for positive definiteness of the associated Riesz kernels. As we shall see below, (conditional) positive definiteness (and consequently energy minimization properties) behave very differently for these two metrics.

Before further discussing these two special cases, we begin by discussing general Riesz-type kernels on metric spaces. In [22] I. J. Schoenberg investigated the question, under which conditions a metric space (X, d) can be isometrically embedded into a Hilbert space. He found out that the correct condition is exactly the conditional positive definiteness of the function $-d(x, y)^2$. Furthermore, he found that a function $K(x, y)$ is conditionally positive definite, if and only if $e^{\lambda K(x,y)}$ is positive definite for all $\lambda > 0$. Thus the metric space (X, d) can be embedded to Hilbert space if and only if the Gaussians $e^{-\lambda d(x,y)^2}$ are positive definite for all $\lambda > 0$.

For $s \in \mathbb{R}$, define the Riesz kernel on the metric space (X, d) as $\operatorname{sgn}(s)d(x, y)^{-s}$ if $s \neq 0$ and $-\log d(x, y)$ if $s = 0$. We show that conditional positive definiteness for one value of the parameter implies conditional positive definiteness of the Riesz kernel for a whole interval of parameters.

Theorem 8 *Let (X, d) be a compact metric space of Hausdorff dimension $D > 0$. Assume that for some $s_0 < 0$ the Riesz kernel is conditionally positive definite, then the Riesz kernels are conditionally positive definite for all $s \in [s_0, D)$.*

Proof Let us assume that $s_0 < 0$, i.e. $-d(x, y)^{-s_0}$ is conditionally positive definite and hence $e^{-\lambda d(x,y)^{-s_0}}$ is positive definite for all $\lambda > 0$. From this we derive immediately that

$$-d(x, y)^{-\gamma s_0} = \frac{1}{\Gamma(-\gamma)} \int_0^\infty \left(e^{-\lambda d(x,y)^{-s_0}} - 1\right) \lambda^{-\gamma-1}\, d\lambda$$

is conditionally positive definite for $0 < \gamma < 1$. Indeed, for a Borel measure μ on X of total mass zero, one immediately sees that $E_F(\mu) \geq 0$ by interchanging the order of integration. This proves the statement of the theorem for $s_0 < s < 0$.

Furthermore, for $s = 0$, the logarithmic kernel $-\log d(x, y)$ is also conditionally positive definite by

$$-\log d(x, y) = \lim_{s \to 0+} \frac{d(x, y)^{-s} - 1}{s}.$$

We can treat the case $s > 0$ similarly. We find that

$$d(x, y)^{-s} = \frac{1}{\Gamma(-2s/s_0)} \int_0^\infty e^{-\lambda d(x,y)^{-s_0}} \lambda^{-\frac{s}{s_0}-1}\, d\lambda$$

is positive definite for $0 < s < D$, where D is the dimension of the space. This upper bound for s comes from the potential theoretic fact that X does not carry a non-zero measure μ with finite Riesz energy for $s \geq D$. □

Remark 6 If X is a compact two-point homogeneous manifold and, e.g., $d = \vartheta$ is the geodesic distance, then the statement of Theorem 8 continues to hold for $s_0 = 0$, i.e. positive definiteness for $s > 0$ can be derived directly from the logarithmic case. Indeed, assume that $-\log d(x, y)$ is conditionally positive definite and observe that

$$d(x, y)^{-s} = e^{s\left(-\log d(x,y)+C\right)} e^{-Cs}, \tag{25}$$

where the constant C is chosen so that the kernel $-\log d(x, y) + C$ is positive definite. While we cannot apply the aforementioned result of Schoenberg directly (since the kernels are singular), one can still see that the kernel on the right hand side is positive definite by expanding the exponential in power series and observing that $(-\log d(x, y) + C)^n$ is integrable with respect to the uniform measure for all $n \geq 1$ and therefore positive definite by the singular version of Schur's lemma (Theorem 5).

Remark 7 We conjecture that Theorem 8 should hold for any $s_0 < D$ on arbitrary compact metric spaces, but we haven't been able to treat the case $s_0 > 0$.

We now return to the setting of two-point homogeneous compact manifolds and first discuss the Riesz kernels with respect to the chordal distance $\chi(x, y)$. It is well known that these spaces isometrically embed into Euclidean spaces, and therefore $-\chi(x, y)^2$ is conditionally positive definite, i.e. the condition of the Theorem 8 hold with $s_0 = -2$. Therefore we obtain the following corollary:

Corollary 1 *Let X be a compact two-point homogeneous manifold with the chordal distance $\chi(x, y)$. Then the associated Riesz kernels are conditionally positive definite for all s with $-2 \leq s < D$, where D is the dimension of X as a real manifold.*

While this might be the most streamlined proof, this statement can be found in the literature. In the case of the sphere it is classical [6, 8], while the case of projective spaces has been investigated in [1, 5]. It is also known that the range is sharp: these kernels are not positive definite for $s < -2$.

The case of the geodesic distance is much more complicated due to the following geometric reason. Since the spaces X all have closed geodesics, there cannot exist an embedding of the metric space (X, ϑ) into a Hilbert space. Thus $-\vartheta(x, y)^2$ cannot be conditionally positive definite. Taking four equally spaced points on a closed geodesic even shows that $-\vartheta(x, y)^\gamma$ is not conditionally positive definite for $1 < \gamma \leq 2$. Beyond this range, the behavior becomes very different for the sphere and projective spaces.

On the sphere $\mathbb{S}^{d-1}$ the kernel $-\vartheta(x, y)$ is conditionally positive definite for any dimension $d \geq 2$ (see [10, 17, 24]). This can be immediately concluded from the easy direction of Theorem 6, see [23], since $\frac{\pi}{2} - \arccos(t)$ has only non-negative power series coefficients. Since t is positive definite, so are all of its powers are by Schur's lemma, and thus so is $\frac{\pi}{2} - \arccos(t)$.

Therefore, applying Theorem 8, we recover the following result [4]:

Corollary 2 *Let $X = \mathbb{S}^{d-1}$ be the sphere with the metric $\vartheta(x, y)$. Then the associated Riesz kernels are conditionally positive definite for all s with $-1 \leq s < d - 1$.*

In the case of projective spaces $\mathbb{FP}^{d-1}$ the situation is drastically different (unless, of course, $d = 2$, in which case these spaces are isometric to the sphere $\mathbb{S}^{\dim_{\mathbb{R}} \mathbb{F}}$). For $d \geq 2$, the geodesic Riesz kernel with $s = -1$ is not conditionally positive definite, see [10] or [5, Lemma 5.2]. Energy minimization of the geodesic Riesz kernel for $s = -1$ can be viewed as the continuous version of the Fejes Tóth conjecture on the sum of line angles in discrete geometry, which has attracted considerable interest in the recent years, thus also bringing attention to the geodesic distance Riesz energy on projective spaces for all values of s [5, 18].

We observe that a dimension independent statement akin to Corollary 2 cannot hold on projective spaces. Theorem 7 immediately implies that the Riesz kernel $\vartheta(x, y)^{-s}$ cannot be conditionally positive definite for all $\mathbb{FP}^{d-1}$ for any fixed $s > 0$, because it cannot be expressed in the form (22). Indeed, functions of this form extend differentiably to a function even around $\frac{\pi}{2}$, whereas ϑ^{-s} does not. (Theorem 7 treats positive definiteness, but the case of conditional positive definiteness is immediate: we just need to require that the coefficients a_n are non-negative for $n \geq 1$, i.e. a_0 could have any sign).

Let $s_d(\mathbb{F})$ be the maximal number such that for all $s < s_d(\mathbb{F})$ the geodesic Riesz kernel $\operatorname{sgn}(s)\vartheta^{-s}$ is not conditionally positive definite on $\mathbb{FP}^{d-1}$. Since (conditional) positive definiteness on $\mathbb{FP}^{d-1}$ obviously implies the same property in all projective spaces over $\mathbb{F}$ with smaller dimension, Theorem 7 and the discussion above shows that

$$\lim_{d \to \infty} s_d(\mathbb{F}) = \infty.$$

Moreover, if $s_d(\mathbb{F}) \leq 0$, then the Riesz kernel is conditionally positive definite for all $s \in \big(s_d(\mathbb{F}), d - 1\big)$, according to Theorem 8 and the discussion thereafter. In addition, if the conjecture in Remark 7 is true, Riesz kernels should be conditionally positive definite for all $s > s_d(\mathbb{F})$ independently of the sign of $s_d(\mathbb{F})$.

The exact value of the critical exponent $s_d(\mathbb{F})$ which indicates the phase transition of positive definiteness of the geodesic Riesz kernel on the projective spaces remains absolutely mysterious. It follows from [5] that $s_d(\mathbb{R}) \leq d - 3$, $s_d(\mathbb{C}) \leq 2d - 5$, $s_d(\mathbb{H}) \leq 4d - 9$, $s_3(\mathbb{O}) \leq 7$, since it has been shown that ϑ^{-s} is positive definite for $\mathbb{FP}^{d-1}$ for s above those values. However, numerical experiments indicate that positive definiteness holds for a larger interval for s. In particular, numerical computations suggest that $s_3(\mathbb{R}) \leq -0.59$, $s_4(\mathbb{R}) \leq -0.125$, while it appears that $s_d(\mathbb{R}) > 0$ for $d \geq 5$. Hence, whether the phase transitions happens in the negative or positive range of exponents seems to depend on the dimension.

Table 1 Definiteness of the kernel $-\log\vartheta(x, y)$

Space	Definiteness
$\mathbb{RP}^2$	Positive definite by [5]
$\mathbb{RP}^3$	Open, numerical computations suggest positive definiteness
$\mathbb{RP}^4$	Not definite, numerical computations show that $c_8 < 0$
$\mathbb{CP}^2$	Open, numerical computations suggest positive definiteness
$\mathbb{CP}^3$	Not definite, numerical computations show that $c_6 < 0$
$\mathbb{HP}^2$	Not definite, numerical computations show that $c_{10} < 0$
$\mathbb{OP}^2$	Not definite, numerical computations show that $c_8 < 0$

In an attempt to gain a better insight into this phenomenon, we explored the positive definiteness of the logarithmic geodesic Riesz kernel $-\log\vartheta(x, y)$ on $\mathbb{FP}^{d-1}$ corresponding to $s = 0$ since it provides indication as to whether $s_d(\mathbb{F})$ is positive or negative. We numerically computed the appropriate Jacobi coefficients c_n of this kernel and check their positivity. The results are presented in Table 1 and they seem to indicate that the critical value $s_d(\mathbb{F})$ is positive in all cases other than $\mathbb{RP}^2$ and $\mathbb{RP}^3$.

Acknowledgments Dmitriy Bilyk was supported by the National Science Foundation grant DMS 2054606 and is also grateful to Fulbright Austria and NAWI Graz. Peter Grabner is supported by the Austrian Science Fund FWF project I 6750.

References

1. A. Anderson, M. Dostert, P. J. Grabner, R. W. Matzke, and T. A. Stepaniuk, *Riesz and Green energy on projective spaces*, Trans. Amer. Math. Soc. Series B **10** (2023), 1039–1076.
2. G. E. Andrews, R. Askey, and R. Roy, *Special Functions*, Encyclopedia of Mathematics and its Applications, vol. 71, Cambridge University Press, Cambridge, 1999.
3. V. S. Barbosa and V. A. Menegatto, *Strictly positive definite kernels on compact two-point homogeneous spaces*, Math. Inequal. Appl. **19** (2016), no. 2, 743–756.
4. D. Bilyk and F. Dai, *Geodesic distance Riesz energy on the sphere*, Trans. Amer. Math. Soc. **372** (2019), no. 5, 3141–3166.
5. D. Bilyk, R. W. Matzke, and J. Nathe, *Geodesic Distance Riesz Energy on Projective Spaces*, 2024, https://arxiv.org/abs/2409.16508.
6. G Björck, *Distributions of positive mass, which maximize a certain generalized energy integral*, Ark. Mat. **3** (1956), 255–269.
7. S. Bochner, *Hilbert distances and positive definite functions*, Ann. of Math. (2) **42** (1941), 647–656.
8. S. V. Borodachov, D. P. Hardin, and E. B. Saff, *Discrete energy on rectifiable sets*, Springer Monographs in Mathematics, Springer, New York, 2019.
9. S. B. Damelin and P. J. Grabner, *Energy functionals, numerical integration and asymptotic equidistribution on the sphere*, J. Complexity **19** (2003), 231–246, Corrigendum, **20** (2004), 883–884.
10. R. Gangolli, *Positive definite kernels on homogeneous spaces and certain stochastic processes related to Lévy's Brownian motion of several parameters*, Ann. Inst. H. Poincaré Sect. B (N.S.) **3** (1967), 121–226.

11. J. C. Guella and V. A. Menegatto, *A limit formula for semigroups defined by Fourier-Jacobi series*, Proc. Amer. Math. Soc. **146** (2018), no. 5, 2027–2038.
12. S. Helgason, *The Radon transform for Euclidean spaces, compact two-point homogeneous spaces and Grassmann manifolds.*, Acta Mathematica (Stockholm) **113** (1965), 153–180.
13. S. Helgason, *Differential Geometry, Lie Groups, and Symmetric Spaces*, Academic Press, 1978.
14. S. Helgason, *Groups and Geometric Analysis, Integral Geometry, Invariant Differential Operators, and Spherical Functions*, Mathematical Surveys and Monographs, vol. 83, American Mathematical Society, 2000.
15. T. H. Koornwinder, *The addition formula for Jacobi polynomials. I. Summary of results*, Indag. Math. **34** (1972), 188–191, Nederl. Akad. Wetensch. Proc. Ser. A **75**.
16. N. S. Landkof, *Foundations of modern potential theory*, Die Grundlehren der mathematischen Wissenschaften, Band 180, Springer-Verlag, New York-Heidelberg, 1972, Translated from the Russian by A. P. Doohovskoy.
17. P. Lévy, *Processus stochastiques et mouvement brownien*, Gauthier-Villars & Cie, Paris, 1965, Suivi d'une note de M. Loève, Deuxième édition revue et augmentée.
18. T. Lim and R. J. McCann, *Maximizing expected powers of the angle between pairs of points in projective space*, Probab. Theory Related Fields **184** (2022), no. 3-4, 1197–1214, With an appendix by D. Bilyk, A. Glazyrin, R. Matzke, J. Park and O. Vlasiuk.
19. W. Magnus, F. Oberhettinger, and R. P. Soni, *Formulas and Theorems for the Special Functions of Mathematical Physics*, enlarged ed., Die Grundlehren der mathematischen Wissenschaften, Band 52, Springer-Verlag New York, Inc., New York, 1966.
20. J. Mercer, *Functions of positive and negative type and their connection with the theory of integral equations*, Philos. Trans. Royal Soc. London **209** (1909), 415–446.
21. T. R. L. Phillips, K. M. Schmidt, and A. Zhigljavsky, *Extension of the Schoenberg theorem to integrally conditionally positive definite functions*, J. Math. Anal. Appl. **470** (2019), no. 1, 659–678.
22. I. J. Schoenberg, *Metric spaces and positive definite functions*, Trans. Amer. Math. Soc. **44** (1938), no. 3, 522–536.
23. S. Helgason, *Positive definite functions on spheres*, Duke Math. J. **9** (1942), 96–108.
24. M. M. Skriganov, *Point distributions in two-point homogeneous spaces*, Mathematika **65** (2019), no. 3, 557–587.
25. G. Szegő, *Orthogonal Polynomials*, fourth ed., American Mathematical Society Colloquium Publications, Vol. XXIII, American Mathematical Society, Providence, R.I., 1975.
26. A. Zygmund, *Trigonometric series. Vol. I, II*, third ed., Cambridge Mathematical Library, Cambridge University Press, Cambridge, 2002, With a foreword by Robert A. Fefferman.

The Hörmander–Bernhardsson Extremal Function: A Preliminary Study

Andriy Bondarenko, Joaquim Ortega-Cerdà, Danylo Radchenko, and Kristian Seip

To Ed Saff on the occasion of his 80th birthday, with admiration and gratitude.

1 Introduction

In [3], Hörmander and Bernhardsson obtained the remarkable numerical approximation

$$0.5409288219 \leq \mathscr{C}_1 \leq 0.5409288220, \tag{1}$$

where $\mathscr{C}_p$ for $0 < p < \infty$ is the smallest positive constant C such that the inequality

$$|f(0)|^p \leq C\|f\|_{L^p(\mathbb{R})}^p \tag{2}$$

holds for every f in the Paley–Wiener space PW^p, i.e., the subspace of $L^p(\mathbb{R})$ consisting of entire functions of exponential type at most π. The problem of computing $\mathscr{C}_p$ for all p arose in [5, 6] and was studied in some depth in the recent

A. Bondarenko (✉) · K. Seip
Department of Mathematical Sciences, Norwegian University of Science and Technology (NTNU), Trondheim, Norway
e-mail: andrii.bondarenko@ntnu.no; kristian.seip@ntnu.no

J. Ortega-Cerdà
Department de Matemàtiques i Informàtica, Universitat de Barcelona and CRM, Centre de Recerca Matemàtica, Barcelona, Spain
e-mail: jortega@ub.edu

D. Radchenko
Laboratoire Paul Painlevé, Université de Lille, Lille, France
e-mail: danylo.radchenko@univ-lille.fr

D. Bilyk et al. (eds.), *Recent Advances in Approximation and Potential Theory*, Applied and Numerical Harmonic Analysis,
https://doi.org/10.1007/978-3-031-95551-8_4

paper [1]. We refer to [2] for a number theoretic motivation for the study of the case $p = 1$.

Surprisingly, in spite of the impressive precision of (1), Hörmander and Bernhardsson obtained very limited information about the associated extremal function, i.e., the unique function that solves

$$\frac{1}{\mathscr{C}_p} = \inf_{f \in PW^p} \left\{ \|f\|_{L^p(\mathbb{R})}^p \, : \, f(0) = 1 \right\} \tag{3}$$

in the case $p = 1$. A compactness argument shows that (3) has solutions for any $0 < p < \infty$, and a rescaling argument reveals that the type of these extremal functions is exactly π. Uniqueness holds in the convex range $1 \le p < \infty$ in which case the corresponding extremal function φ_p is of the form

$$\varphi_p(x) = \prod_{n=1}^{\infty} \left(1 - \frac{x^2}{t_n^2}\right),$$

with $(t_n)_{n \ge 1}$ a strictly increasing sequence of positive numbers (see [1, 3, 5]). Note that the case $p = 2$ is trivial; then $\varphi_2(x) = \operatorname{sinc}(\pi x)$ (the reproducing kernel of PW^2 at 0), where as usual $\operatorname{sinc} x := \frac{\sin x}{x}$. Hence $\mathscr{C}_2 = 1$ and $t_n = n$ for all n.

Hörmander and Bernhardsson's method of proof strongly suggests that t_n is asymptotically close to $n + \frac{1}{2}$ when $p = 1$. The following theorem confirms that this is indeed the case; here and in what follows, $\pm\, \tau_n$ denotes the zeros of Hörmander and Bernhardsson's extremal function φ_1.

Theorem 1 *The ℓ^2 norm of $(\tau_n - n - \frac{1}{2})$ is bounded by* 0.13.

Our proof gives an algorithm for computing τ_n in terms of a fixed point iteration. We have implemented this algorithm numerically and found, after 15 iterations, that for instance the first four zeros are close to 1.4417, 2.4657, 3.4756, 4.4811.

Little has been known about these zeros so far. Hörmander and Bernhardsson proved that all the zeros τ_n are simple; it was shown in [1] that $t_1 \ge \frac{1}{2}$ and that the zero set of φ_p is uniformly discrete for all $p \ge 1$. Some additional bounds on the separation of the zeros were also obtained in the range $p > 2$.

We expect that our method of proof will enable us to say considerably more about the zeros and hence the extremal function φ_1. We plan to pursue our study of φ_1 in a forthcoming publication, which will also explore the relation to Hörmander and Bernhardsson's work.

The rest of this note is devoted to the proof of Theorem 1. Briefly, our approach will be to start from an operator equation satisfied by the τ_n. The linear part of the operator in question is an "almost" unitary Hilbert-type matrix. Thanks to the favorable properties of this matrix, we are able to show, by Banach's fixed point theorem, that our operator equation has a unique solution whose ℓ^2 distance to the sequence $(n + \frac{1}{2})$ can be quantified.

2 The Zeros τ_n as the Solution of an Equation

We will use the notation $\|f\|_p := \|f\|_{L^p(\mathbb{R})}$ which we declare to be the PW^p norm of f should f belong to PW^p. We start from the well known fact that a nontrivial function φ in PW^p is the extremal function φ_p if and only if

$$f(0) = \int_{-\infty}^{\infty} f(x)\frac{|\varphi(x)|^{p-2}}{\|\varphi\|_p^p}\varphi(x)dx \tag{4}$$

holds for every f in PW^p (see [1, Thm. 36]). Note that (4) embodies the classical fact that $\varphi_2(x-y)$ is the reproducing kernel of PW^2 at y.

It is clear from (4) that the case $p = 1$ is also particularly favorable. Indeed, when φ is real with only real zeros, the function $|\varphi|^{p-2}\varphi$ will only take the values ± 1 and change sign at the zeros of φ. In other words, the latter function depends in a simple explicit way on the location of the zeros of φ.

Now let φ be a function in PW^1 of the form

$$\varphi(x) = \prod_{n=1}^{\infty}\left(1 - \frac{x^2}{t_n^2}\right), \quad 0 < t_1 < t_2 < \cdots, \tag{5}$$

and set

$$\Phi(x) := \frac{\varphi(x)}{|\varphi(x)|\|\varphi\|_1}.$$

Following an idea in [3, p. 188], we will relate Φ to the function

$$\Psi(x) := -\|\varphi\|_1^{-1}\operatorname{sgn}\cos\pi x.$$

We define the Fourier transform of an L^1 function f as

$$\widehat{f}(\xi) := \int_{-\infty}^{\infty} f(x)e^{-i\xi x}dx$$

and find that the distributional Fourier transform of Ψ is

$$\widehat{\Psi}(\xi) = \pi\widehat{g}(\xi)\sum_{n=-\infty}^{\infty}\delta_{\pi n}(\xi),$$

where $\widehat{g}$ is the Fourier transform of $\chi_{(-\pi,\pi)}\Psi$. Since $\widehat{g}(0) = 0$, we see that $\widehat{\Psi}$ vanishes on $(-\pi,\pi)$, as observed in [3, p. 188].

Lemma 1 *Let φ be of the form* (5) *and assume that $(t_n - n - \frac{1}{2})$ is in ℓ^2. Then $(\Phi - \Psi) * \varphi_2$ is in PW^2 and*

$$\frac{1}{\|\varphi\|_1} \le \mathscr{C}_1 \le \frac{1}{\|\varphi\|_1} + \|\varphi_2 - (\Phi - \Psi) * \varphi_2\|_\infty. \tag{6}$$

Proof We begin by noting that

$$\begin{aligned}\frac{((\Phi - \Psi) * \varphi_2)(y)}{2\|\varphi\|_1} &= \int_{-\frac{1}{2}}^{\frac{1}{2}} \varphi_2(x-y)dx \\ &+ \sum_{n=1}^{\infty} (-1)^n \int_{t_n}^{n+\frac{1}{2}} (\varphi_2(x-y) + \varphi_2(x+y))\, dx.\end{aligned}$$

By duality, we therefore find that

$$\frac{\|(\Phi - \Psi) * \varphi_2\|_2}{2\|\varphi\|_1} = \sup_{\substack{f \in PW^2 \\ \|f\|_2 \le 1}} \left| \int_{-\frac{1}{2}}^{\frac{1}{2}} f(x)dx + \sum_{n=1}^{\infty} (-1)^n \int_{t_n}^{n+\frac{1}{2}} (f(x) + f(-x))\, dx \right|.$$

Here we may assume that f is real-valued, and so the expression on the right-hand side can be written as

$$f(x_0) + \sum_{n=1}^{\infty} (-1)^n \big(n + \frac{1}{2} - t_n\big) \left(f(x_n) + f(x_{-n})\right)$$

by the mean value theorem, where $|x_0| < \frac{1}{2}$ and $x_{\pm n}$ lie between $\pm\, (n + \frac{1}{2})$ and $\pm\, t_n$. Now applying the Cauchy–Schwarz inequality to the sum and invoking the Plancherel–Pólya inequality (see e.g. [4, Lect. 7]), we conclude that $(\Phi - \Psi) * \varphi_2$ is in PW^2.

The left inequality in (6) is trivial. To prove the right inequality, we set

$$F := \varphi_2 + \Phi - (\Phi - \Psi) * \varphi_2$$

and let f be a function in $PW^1 \cap \mathscr{S}$, where $\mathscr{S}$ is the Schwartz space. Since $PW^1 \cap \mathscr{S}$ is dense in PW^1, we have

$$\mathscr{C}_1 = \sup_{\substack{f \in PW^1 \cap \mathscr{S} \\ \|f\|_1 = 1}} |f(0)|.$$

But since $\widehat{\Psi}$ vanishes on $(-\pi, \pi)$, we also have $f(0) = (F * f)(0)$, and so

$$\mathscr{C}_1 \le \|\varphi\|_1^{-1} + \|\varphi_2 - (\Phi - \Psi) * \varphi_2\|_\infty$$

by the triangular inequality for the $L^\infty(\mathbb{R})$ norm. □

It follows from the above lemma that $\varphi = \varphi_1$ if $(\Phi - \Psi) * \varphi_2 = \varphi_2$. In fact, we may observe that this would still hold if we merely had $(\Phi - \Psi) * \varphi_2 = c\varphi_2$ for some constant c. Indeed, the Fourier transform of $\Phi - \Psi$ would then be constant on $(-\pi, \pi)$, and so

$$\int_{-\infty}^{\infty} ((\Phi - \Psi) * \varphi_2)(x)\varphi(x)dx = \int_{-\infty}^{\infty} (\Phi(x) - \Psi(x))\varphi(x)dx$$
$$= \int_{-\infty}^{\infty} \Phi(x)\varphi(x)dx = \varphi(0),$$

whence $\varphi_2 - (\Phi - \Psi) * \varphi_2 = 0$, and consequently $c = 1$. Since a function in PW^2 is determined by its values at the integers, we would therefore obtain τ_n as $n + \frac{1}{2} - \delta_n$ if

$$\int_{-\frac{1}{2}}^{\frac{1}{2}} \operatorname{sinc} \pi(x - k)dx + \sum_{n=1}^{\infty}(-1)^n \int_{n+\frac{1}{2}-\delta_n}^{n+\frac{1}{2}} (\operatorname{sinc} \pi(x - k) + \operatorname{sinc} \pi(x + k))\, dx = 0 \tag{7}$$

for all positive integers k with $\delta := (\delta_1, \delta_2, \ldots)$ in ℓ^2. We prefer to rewrite the left-hand side of this equation by making the following integration by parts:

$$\int_{-\frac{1}{2}}^{\frac{1}{2}} \operatorname{sinc} \pi(x - k)dx = \frac{(-1)^{k+1}}{\pi^2} \int_{-\frac{1}{2}}^{\frac{1}{2}} \frac{\cos \pi x}{(x - k)^2}dx.$$

Then (7) takes the form

$$\sum_{n=1}^{\infty} \int_0^{\delta_n} \cos \pi y\Big(\frac{1}{n + \frac{1}{2} - y - k} + \frac{1}{n + \frac{1}{2} - y + k}\Big)dy = \frac{1}{\pi} \int_{-\frac{1}{2}}^{\frac{1}{2}} \frac{\cos \pi x}{(x - k)^2}dx \tag{8}$$

for $k = 1, 2, \ldots$. The above discussion shows that Theorem 1 is equivalent to the following assertion. (Here and in what follows, $\|\cdot\|$ denotes the ℓ^2 norm.)

Theorem 2 *Equation* (8) *has a solution* δ *satisfying* $\|\delta\| \le 0.13$.

The next two sections will give the proof of this theorem.

3 The Linear Part of (8)

Performing a linear approximation of the operator on the left-hand side of (8), we may write this equation as

$$A\delta + Q\delta = w, \tag{9}$$

where

$$w_k := \frac{1}{\pi}\int_{-\frac{1}{2}}^{\frac{1}{2}} \frac{\cos \pi x}{(x-k)^2}dx;$$

$$(A\delta)_k := \sum_{n=1}^{\infty}\left(\frac{1}{n+k+\frac{1}{2}} + \frac{1}{n-k+\frac{1}{2}}\right)\delta_n;$$

$$(Q\delta)_k := \sum_{n=1}^{\infty}\int_0^{\delta_n}\Big(\frac{\cos \pi y}{n+\frac{1}{2}-y+k} - \frac{1}{n+\frac{1}{2}+k}$$

$$+ \frac{\cos \pi y}{n+\frac{1}{2}-y-k} - \frac{1}{n+\frac{1}{2}-k}\Big)dy.$$

We will use the notation $a_{k,n} := \frac{1}{n+k+1/2} + \frac{1}{n-k+1/2}$ for $n \geq 1$ so that

$$(Ax)_k = \sum_{n=1}^{\infty} a_{k,n}x_n.$$

The proof of Theorem 2 will rely on the fact that the infinite matrix $\frac{1}{\pi}A$ is "almost" unitary.

Lemma 2 *The operator A maps ℓ^2 to ℓ^2 and satisfies*

$$2\sqrt{2}\|x\| \leq \|Ax\| \leq \pi\,\|x\|$$

for all x in ℓ^2.

The proof of Lemma 2 will employ the following fact.

Lemma 3 *For $r \geq 2$, $a_1, \ldots, a_r \in \mathbb{Z}$, set*

$$C(a_1, \ldots, a_r) := \sum_{n\in\mathbb{Z}} \frac{1}{(n+a_1+\frac{1}{2})\ldots(n+a_r+\frac{1}{2})}.$$

Then $C(a_1, \ldots, a_r) = 0$ whenever $a_1, \ldots, a_r$ are distinct integers. Moreover, we have

$$C(a,a) = \pi^2, \qquad C(a,b,b) = \frac{\pi^2}{a-b}, \qquad C(a,a,b,b) = \frac{2\pi^2}{(a-b)^2}.$$

Proof The result is immediate by a partial fraction decomposition of the summands. □

Proof of Lemma 2 Using Lemma 3, we get

$$\frac{2}{(n+\frac{1}{2})(m+\frac{1}{2})} + \sum_{k=1}^{\infty} a_{k,n}a_{k,m} = C(n,m) - C(n,-m) = \pi^2 \delta_{n,m} .$$

Therefore

$$\|Ax\|^2 + 2\Big(\sum_{n\geq 1} \frac{x_n}{n+\frac{1}{2}}\Big)^2 = \pi^2 \|x\|^2 ,$$

which immediately implies that $\|Ax\| \leq \pi\|x\|$. To get the lower bound $\|Ax\| \geq 2\sqrt{2}\|x\|$, we apply the Cauchy–Schwarz inequality to the second term on the right-hand side and use that

$$\sum_{n=1}^{\infty} \frac{1}{(n+\frac{1}{2})^2} = \frac{\pi^2}{2} - 4.$$

□

Note that the above proof reveals the interesting fact that $\frac{1}{\pi}A$ restricted to the orthogonal complement of the sequence $((n+\frac{1}{2})^{-1})$ is an isometry.

We next show that A is invertible by identifying its inverse B. Lemma 2 shows that $\|Bx\| \leq (2\sqrt{2})^{-1}\|x\|$, and this is the only information needed about B in the proof of Theorem 2. However, it is useful for our numerical work to have an explicit expression for B.

Proposition 1 *The operator B with matrix*

$$b_{n,k} = \frac{1-(2n+1)^{-2}}{\pi^2(1-(2k)^{-2})} a_{k,n}$$

satisfies $AB = BA = I$.

Proof Note that $a_{k,n} = a_{-k,n} = -a_{k,-1-n}$. Using this, we find that

$$\sum_{n=1}^{\infty} a_{k_1,n}a_{k_2,n}(1-(2n+1)^{-2}) = \frac{1}{2}\sum_{n\in\mathbb{Z}} a_{k_1,n}a_{k_2,n}(1-(2n+1)^{-2})$$

$$= \frac{1}{2}\Big(C(k_1,k_2) + C(k_1,-k_2) + C(-k_1,k_2)$$

$$+ C(-k_1,-k_2) - C(k_1,k_2,-k_1,-k_2)\Big)$$

which clearly vanishes for $k_1 \neq k_2$, and for $k_1 = k_2 = k$ we get $\pi^2(1-(2k)^{-2})$, from which we see that $BA = I$. Similarly, we calculate

$$\begin{aligned}\sum_{k=1}^{\infty} \frac{a_{k,m}a_{k,n}}{1-(2k)^{-2}} &= \frac{1}{2}\sum_{k\in\mathbb{Z}} a_{k,m}a_{k,n}\Big(1+\frac{\frac{1}{4}}{k-\frac{1}{2}}-\frac{\frac{1}{4}}{k+\frac{1}{2}}\Big)\\ &= \frac{1}{2}\Big(C(m,n)-C(m,-1-n)-C(-1-m,n)+C(-1-m,-1-n)\Big)\\ &+\frac{1}{8}\Big(C(-1,m,n)-C(-1,m,-1-n)-C(-1,-1-m,n)\\ &+C(-1,-1-m,-1-n)\Big)\\ &-\frac{1}{8}\Big(C(0,m,n)-C(0,m,-1-n)-C(0,-1-m,n)\\ &+C(0,-1-m,-1-n)\Big),\end{aligned}$$

which after simplifications is easily seen to imply $AB = I$. □

4 Proof of Theorem 2

Retaining the notation B for the inverse of A and setting $x := \delta - Bw$, we see from (9) that our task is to find a fixed point of the map

$$Gx := -BQ(x + Bw).$$

We will prove that a unique solution to this problem can be found in the ball

$$K := \Big\{x = (x_k) : \; \|x\| \le 0.042\Big\}.$$

Theorem 3 *G maps K into itself and $\|G(x) - G(y)\| \le 0.73\,\|x - y\|$ for all x, y in K.*

In view of Banach's fixed point theorem, Theorem 2 follows from Theorem 3 and the next lemma.

Lemma 4 *We have $\|Bw\| \le 0.088$.*

Proof We see that

$$\frac{1}{\pi}\int_{-\frac{1}{2}}^{\frac{1}{2}} \frac{\cos \pi x}{(x-n)^2}dx \le \frac{1}{\pi n^2}\int_{-\frac{1}{2}}^{\frac{1}{2}} \frac{\cos \pi x}{(1-x)^2}dx,$$

and so

$$\|w\|_2 \le \frac{\sqrt{\zeta(4)}}{\pi} \int_{-\frac{1}{2}}^{\frac{1}{2}} \frac{\cos \pi x}{(x-1)^2} dx.$$

By Lemma 2, we then get the desired bound

$$\|Bw\|_2 \le \frac{\pi}{12\sqrt{5}} \int_{-\frac{1}{2}}^{\frac{1}{2}} \frac{\cos \pi x}{(x-1)^2} dx \le 0.088.$$

□

The proof of Theorem 3 splits naturally into the following two steps.

Step 1: Proof that G maps K into itself. We set $\xi := x + Bw$ and $\varepsilon := \|\xi\|$. We begin by noting that

$$\frac{\cos \pi y}{t-y} = \frac{1}{t}\left(1 + \frac{y}{t} + \frac{\frac{y^2}{t^2}}{1-\frac{y}{t}}\right)\cos \pi y,$$

whence

$$\frac{\cos \pi y}{t-y} - \frac{1}{t} = \frac{\cos \pi y - 1}{t} + \frac{1}{t}\left(\frac{y}{t} + \frac{\frac{y^2}{t^2}}{1-\frac{y}{t}}\right)\cos \pi y.$$

If we set

$$\alpha_n := \int_0^{\xi_n} \big(\cos \pi y - 1\big) dy = \pi^{-1}\sin(\pi \xi_n) - \xi_n \tag{10}$$

and

$$\beta_{n,k}^{\pm} := \int_0^{\xi_n} \left(1 + \frac{\frac{y}{n+\frac{1}{2}\pm k}}{1 - \frac{y}{n+\frac{1}{2}\pm k}}\right) y \cos \pi y dy,$$

then we may write

$$(Q\xi)_k = (A\alpha)_k + \sum_{n=1}^{\infty}\left(\frac{\beta_{n,k}^{+}}{(n+\frac{1}{2}+k)^2} + \frac{\beta_{n,k}^{-}}{(n+\frac{1}{2}-k)^2}\right).$$

We see that

$$|\beta_{n,k}^{\pm}| \le \frac{\xi_n^2}{2} + \frac{2|\xi_n|^3}{3(1-2|\xi_n|)} \le \left(\frac{1}{2} + \frac{2\varepsilon}{3(1-2\varepsilon)}\right)\xi_n^2 =: c(\varepsilon)\xi_n^2.$$

Setting $D := (d_{n,k})$ with $d_{n,k} := \frac{1}{(n+\frac{1}{2}+k)^2} + \frac{1}{(n+\frac{1}{2}-k)^2}$, we therefore find that

$$\|BQ\xi\| \leq \|\alpha\| + (2\sqrt{2})^{-1}c(\varepsilon)\|D(\xi_n^2)\|. \tag{11}$$

Since

$$|\alpha_n| \leq \frac{\pi^2}{6}|\xi_n|^3,$$

we have

$$\|\alpha\| \leq \zeta(2)\varepsilon^3. \tag{12}$$

By duality and an application of the Cauchy–Schwarz inequality,

$$\begin{aligned} \|D(\xi_n^2)\| &\leq \sup_{q:\|q\|=1} \sum_{k=1}^{\infty}\sum_{n=1}^{\infty}\left(\frac{1}{(n+\frac{1}{2}+k)^2} + \frac{1}{(n+\frac{1}{2}-k)^2}\right) q_k\xi_n^2 \\ &\leq C\|(\xi_n^2)\| \leq C\varepsilon^2, \end{aligned} \tag{13}$$

where

$$C := \sum_{n=2}^{\infty}\frac{1}{(n+\frac{1}{2})^2} + 2\sum_{k=0}^{\infty}\frac{1}{(k+\frac{1}{2})^2} \leq \frac{1}{2} + 6\zeta(2).$$

Inserting (12) and (13) into (11), we conclude that

$$\|BQ\xi\| \leq \left(\zeta(2)\varepsilon + \frac{3}{\sqrt{2}}c(\varepsilon)\left(\zeta(2) + \frac{1}{12}\right)\right)\varepsilon^2 \leq 0.042$$

if $\varepsilon \leq 0.13$. This gives the desired result since $\|Bw\| \leq 0.088$. □

Step 2: Proof of the contractivity of G. Let x and z be two arbitrary points in K, and set $\xi := x + Bw$ and $\eta := z + Bw$. We set $\varepsilon := \max(\|\xi\|, \|\eta\|)$ and argue as in the preceding proof to get

$$(Q\xi)_k - (Q\eta)_k = (A\alpha)_k + \sum_{n=1}^{\infty}\left(\frac{\beta_{n,k}^+}{(n+\frac{1}{2}+k)^2} + \frac{\beta_{n,k}^-}{(n+\frac{1}{2}-k)^2}\right),$$

where now

$$\alpha_n := \int_{\eta_n}^{\xi_n}\big(\cos\pi y - 1\big)dy = \pi^{-1}\big(\sin(\pi\xi_n) - \sin(\pi\eta_n) - (\xi_n - \eta_n)\big)$$

and

$$\beta_{n,k}^{\pm} = \int_{\eta_n}^{\xi_n} \left(1 + \frac{\frac{y}{n+\frac{1}{2}\pm k}}{1 - \frac{y}{n+\frac{1}{2}\pm k}} \right) y \cos \pi y dy.$$

In the present case, we have

$$|\alpha_n| \le \frac{\pi^2}{6} |\xi_n^3 - \eta_n^3| \le \frac{\pi^2 \varepsilon^2}{2} |x_n - z_n|$$

and

$$|\beta_{n,k}^{\pm}| \le \frac{|\xi_n^2 - \eta_n^2|}{2} + \frac{2|\xi_n^3 - \eta_n^3|}{3(1-2\varepsilon)} \le \left(\varepsilon + \frac{2\varepsilon^2}{1-2\varepsilon} \right) |x_n - z_n|.$$

Estimating as before, we get that

$$\|BQ\xi - BQ\eta\| \le \left(\frac{\pi^2\varepsilon^2}{2} + \frac{\left(\frac{1}{2} + \pi^2\right)}{2\sqrt{2}} \left(\varepsilon + \frac{2\varepsilon^2}{1-2\varepsilon}\right) \right) \|x - z\|,$$

where

$$\frac{\pi^2\varepsilon^2}{2} + \frac{\left(\frac{1}{2} + \pi^2\right)}{2\sqrt{2}} \left(\varepsilon + \frac{2\varepsilon^2}{1-2\varepsilon}\right) \le 0.73$$

when $\varepsilon \le 0.13$. □

Acknowledgments Bondarenko and Seip were supported in part by Grant 334466 of the Research Council of Norway. Ortega-Cerdà has been supported by grants PID2021-123405NB-I00 and CEX2020-001084-M by the Agencia Estatal de Investigación and by the Departament de Recerca i Universitats, grant 2021 SGR 00087 and ICREA Academia. Radchenko acknowledges funding by the European Union (ERC, FourIntExP, 101078782).

References

1. Brevig, O.-F., Chirre, A., Ortega-Cerdà, J., Seip, K.: Point evaluation in Paley–Wiener spaces. **153**, 595–670 (2024)
2. Carneiro, E., Milinovich, M. B., Soundararajan, K.: Fourier optimization and prime gaps. Comment. Math. Helv. **94**, 533–568 (2019)
3. Hörmander, L., Bernhardsson, B.: An extension of Bohr's inequality. In: Lions J.-L., Baiochhi C. (eds.) Boundary Value Problems for Partial Differential Equations and Applications, RMA Res. Notes Appl. Math., vol. 29, pp. 179–194. Masson, Paris (1993)

4. Levin, B. Ya.: Lectures on Entire Functions. Kluwer, Translations of Mathematical Monographs, vol. 150, American Mathematical Society, Providence, RI, In collaboration with and with a preface by Yu. Lyubarskii, M. Sodin and V. Tkachenko, Translated from the Russian manuscript by Tkachenko (1996)
5. Levin, E., Lubinsky, D.: L_p Christoffel functions, L_p universality, and Paley-Wiener spaces. J. Anal. Math. **125**, 243–283 (2015)
6. Lubinsky, D. S.: Scaling limits of polynomials and entire functions of exponential type. In: Fasshauer, G., Schumaker, L. (eds) Approximation Theory XV: San Antonio 2016. Springer Proceedings in Mathematics & Statistics, vol. 201, pp. 219–238. Springer, Cham (2016)

Absolute Minima of Potentials of a Certain Class of Spherical Designs

Sergiy Borodachov

1 Introduction and Setting of the Problem

Let $S^d := \{(x_1, \ldots, x_{d+1}) \in \mathbb{R}^{d+1} : x_1^2 + \ldots + x_{d+1}^2 = 1\}$ denote the unit sphere in $\mathbb{R}^{d+1}$. The problem of finding absolute minimum points for the total potential over S^d of a point configuration (regular in some sense) was earlier studied in a number of works. In particular, papers by Stolarsky [29, 30] and Nikolov and Rafailov [27, 28] focused on a regular N-gon on S^1, a regular simplex, regular cross-polytope, and cube on S^d, and the Riesz potential. In talk [7] and in paper [5] (see Theorems 1 and 2 below), we proved extensions of the results of these papers to the case of a completely monotone potential of the distance squared. Furthermore, we extended the case of the absolute minimum of these results to a general class of spherical designs (see Definition 2) which contains, for example, the 24-cell, and as we will further see, many other known configurations. We used a general approach to characterizing the absolute minima of the potential by first finding the Hermite interpolating polynomial for the potential function and then applying properties of Gegenbauer polynomials. This is the classical Delsarte-Yudin (or linear programming) method, see the work by Cohn and Kumar [15], book [10], and references therein. We also remark that configurations satisfying Definition 2 are exactly the ones satisfying the Fazekas-Levenshtein bound for covering [19] for designs of an odd strength.

S. Borodachov (✉)
Department of Mathematics, Towson University, Towson, MD, USA
e-mail: sborodachov@towson.edu

D. Bilyk et al. (eds.), *Recent Advances in Approximation and Potential Theory*, Applied and Numerical Harmonic Analysis,
https://doi.org/10.1007/978-3-031-95551-8_5

The results in [5, 7][1] mentioned above stimulated further research by another group [11, 12].[2] A light modification of the proofs from [5, 7] yields a universal lower bound for the potential energy of a given spherical design which is attained exactly by designs satisfying Definition 2 or by strongly sharp configurations, see [11, 12] for details. A significant contribution of works [11, 12] is showing that many known sharp configurations satify Definition 2 and providing a certain information about locations of absolute minima of their potentials. Universal minima of the icosahedron, dodecahedron, of the set of roots of E_8 lattice (see [3]), and of the 2_{41}-polytope (see [4]) were also characterized by the author. Paper [12][3] showed that these four configurations attain the universal lower bounds and that so does the set of minimal vectors of the Leech lattice as well. All these configurations do not satisfy Definition 2 and different methods were required.

The purpose of this work is to publish the proof from talk [7] (it is Theorem 1 here) and to characterize absolute minima of the potential of six new spherical designs (see Theorems 4, 5, and 6).

One important application of this problem is the maximal polarization problem on the sphere, which requires maximizing the absolute minimum over S^d of the total potential of N points lying on S^d. When the solution to the maximal polarization problem is expected to be a regular configuration, knowing the minimum value of its potential over the sphere would provide a sharp upper bound for the polarization constant. Reviews of known results on polarization can be found, for example, in book [10, Chapter 14] with the most recent work reviewed in, e.g., [5].

The paper is structured as follows. In Sect. 2, we state the problem rigorously and give a more detailed review of early results. Section 3 contains known facts and definitions used in the paper while Sect. 4 reviews more extensively further results on universal minima. In Sect. 5, we state the general results, Theorems 1 and 2, and in Sect. 6, we give their immediate consequences. We present proofs of Theorems 1 and 2 in Sect. 7. Section 8 contains a known auxiliary statement on the distribution of dot products that a universal minimum point forms with points of a given configuration satisfying Definition 2. In Sects. 9, 10, and 11, we characterize universal minima of the six new spherical designs.

2 Rigorous Setting and Early Work

Let $f : [0, 4] \to (-\infty, \infty]$ be a function finite and continuous on the interval $(0, 4]$ with $\lim_{t \to 0^+} f(t) = f(0)$. We call f a *potential function*. Later, we will

[1] Theorem 1 was announced in talk [6] in May 2021.

[2] Talk [7] was given in January 2022 at ESI (recording is available on YouTube), paper [5] was posted to ArXiv in March 2022, and papers [11, 12] are on ArXiv since July and October 2022, respectively.

[3] Paper [3] was announced in talk [8] in August 2022 and posted to Arxiv on October 9, 2022, paper [12] is on ArXiv since October 31, 2022, and paper [4] is on ArXiv since January 2023.

specify additional assumptions on its derivative. Points $\mathbf{x}$ and $\mathbf{y}$ on S^d interact via the kernel of the form $K(\mathbf{x}, \mathbf{y}) = f(|\mathbf{x} - \mathbf{y}|^2)$. Given a point configuration $\omega_N = \{\mathbf{x}_1, \ldots, \mathbf{x}_N\} \subset S^d$, we define *the f-potential of ω_N* as

$$p_f(\mathbf{x}, \omega_N) := \sum_{i=1}^{N} f(|\mathbf{x} - \mathbf{x}_i|^2), \quad \mathbf{x} \in S^d, \tag{1}$$

and study the following problem.

Problem 1 Find the quantity

$$P_f(\omega_N; S^d) := \min_{\mathbf{x} \in S^d} p_f(\mathbf{x}, \omega_N), \tag{2}$$

and determine all points $\mathbf{x}^* \in S^d$, for which the minimum is attained in (2).

Let ω_N^* be the configuration of the vertices of a regular N-gon inscribed in S^1. We will call dual for ω_N^* the regular N-gon formed by the midpoints of the arcs that join any two consecutive vertices of ω_N^*. Let ω_{d+1}^* be the set of vertices of a regular simplex inscribed in S^{d-1}. The configuration $-\omega_{d+1}^*$ will be called its dual. Let $\omega_{2d}^* := \{\pm\mathbf{e}_1, \ldots, \pm\mathbf{e}_d\}$, $d \geq 3$, where $\{\mathbf{e}_1, \ldots, \mathbf{e}_d\}$ is the standard basis in $\mathbb{R}^d$, be the set of vertices of the regular cross-polytope inscribed in S^{d-1}, and U_d be the set $\left\{\left(\pm\frac{1}{\sqrt{d}}, \ldots, \pm\frac{1}{\sqrt{d}}\right)\right\} \subset \mathbb{R}^d$ of 2^d vertices of the cube inscribed in S^{d-1}. The configurations ω_{2d}^* and U_d will be called dual for each other. Recall that ω_N^* is a spherical $(N-1)$-design, ω_{d+1}^* is a 2-design, while ω_{2d}^* and U_d are both 3-designs.

Stolarsky [29, 30] found the absolute minima and maxima for the Riesz potential function $f(t) = -t^{-s/2}$ for a certain range of $s < 0$ and the vertices of a regular polygon, regular simplex, regular cross-polytope, and a cube as well as absolute minimum for $f(t) = t^{-s/2}$, $s > 0$, and the vertices of a regular cross-polytope and a cube. Later, Nikolov and Rafailov [27, 28] extended these results to all values of $s \neq 0$ for all four configurations ω_N^*, ω_{d+1}^*, ω_{2d}^*, and U_d and potential function

$$f(t) = \begin{cases} (t+C)^{-s/2}, & s > 0, \\ -(t+C)^{-s/2}, & s < 0, \end{cases} \quad C \geq 0. \tag{3}$$

One of the following three cases was proved to be always true: the potential of the configuration achieves its absolute minimum over S^d at the points of the dual configuration and its absolute maximum at the points of the configuration itself, or vice versa, or the potential of the configuration is constant over S^d. Which case holds depends on whether the derivative of order greater by one than the strengh of the spherical design is positive, negative, or zero on $(0, 4]$.

3 Preliminaries

In this section, we recall relevant definitions and known facts. An infinitely differentiable function $f : (a, b) \to \mathbb{R}$ is called *completely monotone* on (a, b) if $(-1)^k f^{(k)}$ is non-negative on (a, b) for all $k \geq 0$ and *strictly completely monotone* if $(-1)^k f^{(k)}$ is strictly positive on (a, b) for all $k \geq 0$. Such are potential functions (3) for $s > 0$ and $\widetilde{f}(t) = e^{-at}$ for $a > 0$. Function (3) for $C = 0$ and $s > 0$ defines the Riesz s-kernel $K_s(\mathbf{x}, \mathbf{y}) := |\mathbf{x} - \mathbf{y}|^{-s}$ and function $\widetilde{f}$ the Gaussian kernel $K(\mathbf{x}, \mathbf{y}) = e^{-a|\mathbf{x}-\mathbf{y}|^2}$. Furthermore, potential functions (3) for $-2 < s < 0$ and $f_{\log}(t) = \frac{1}{2} \ln \frac{1}{t}$ become strictly completely monotone on $(0, 4)$ after adding an appropriate positive constant. Function (3) for $C = 0$ and $s < 0$ gives a rise to the Riesz kernel $K_s(\mathbf{x}, \mathbf{y}) := -|\mathbf{x} - \mathbf{y}|^{-s}$ and function $f_{\log}$ to the logarithmic kernel $K_{\log}(\mathbf{x}, \mathbf{y}) := \ln \frac{1}{|\mathbf{x}-\mathbf{y}|}$.

Completely monotone f on $(0, 4)$ (and $f + c$, where c is any constant) have *convex* derivatives of even orders and *concave* derivatives of odd orders. Extended to $[0, 4]$ by limits at the endpoints, they satisfy the assumptions of Theorems 1 and 2. Denote

$$w_d(t) := \gamma_d(1 - t^2)^{d/2-1},$$

where the constant γ_d is chosen so that w_d is a probability density on $[-1, 1]$. We recall that *the Gegenbauer (or ultraspherical) orthogonal polynomials* corresponding to the sphere S^d in $\mathbb{R}^{d+1}$ are terms of the sequence $\{P_n^{(d)}\}_{n=0}^{\infty}$ of univariate polynomials such that $\deg P_n^{(d)} = n$, $n \geq 0$, and

$$\int_{-1}^{1} P_i^{(d)}(t) P_j^{(d)}(t) w_d(t)\, dt = 0, \quad i \neq j,$$

normalized so that $P_n^{(d)}(1) = 1$, $n \geq 0$ (see [31, Chapter 4] or [10, Chapter 5]). When $d = 1$ or $d = 2$, we obtain the classical Chebyshev polynomials of the first kind or the Legendre polynomials, respectively. When $d = 3$, we have the Chebyshev polynomials of the second kind. It is a well-known fact that $P_n^{(d)}$ has exactly n distinct zeros in $(-1, 1)$ (see [31, Theorem 3.3.1]), which we will denote by

$$-1 < \kappa_1^n < \ldots < \kappa_n^n < 1. \tag{4}$$

Since $P_n^{(d)}$ is even for even n and odd for odd n, its set of roots is symmetric about 0.

Let σ_d denote the d-dimensional area measure on S^d normalized to be a probability measure; that is, $\sigma_d(\cdot) = \frac{1}{\mathcal{H}_d(S^d)} \mathcal{H}_d|_{S^d}(\cdot)$, where $\mathcal{H}_d$ is the d-dimensional Hausdorff measure in $\mathbb{R}^{d+1}$. Following the work by Delsarte et al. [17], we call a point configuration ω_N a *spherical n-design* if, for every polynomial q on $\mathbb{R}^{d+1}$ of degree at most n,

$$\frac{1}{N}\sum_{i=1}^{N} q(\mathbf{x}_i) = \int_{S^d} q(\mathbf{x})\, d\sigma_d(\mathbf{x}). \tag{5}$$

The number n is called the *strength* of the spherical design ω_N.

Let $\mathbb{P}_n$ denote the space of all univariate polynomials of degree $\leq n$. An equivalent definition of a spherical n-design, $n \geq 1$, on S^d is the fact [17] that the equality

$$\sum_{i=1}^{N} p(\mathbf{x}\cdot\mathbf{x}_i) = a_0(p)N, \quad \mathbf{x} \in S^d, \tag{6}$$

holds for every polynomial $p \in \mathbb{P}_n$. Here, $a_0(p)$ is the 0-th Gegenbauer coefficient of p.

Let $m \in \mathbb{N}$, $\mathbf{z} \in S^d$, and $-1 \leq t_1 < \ldots < t_m \leq 1$ be the distinct values of the dot product that $\mathbf{z}$ forms with points of the n-design ω_N. Denote by M_j the number of points in ω_N that form dot product t_j with $\mathbf{z}$, $j = 1, \ldots, m$. Substituting $\mathbf{z}$ for $\mathbf{x}$ in (6) we obtain a quadrature with nodes t_j and weights $\beta_j := M_j/N$ satisfying

$$a_0(p) = \int_{-1}^{1} p(t) w_d(t)\, dt = \sum_{j=1}^{m} \beta_j p(t_j), \quad p \in \mathbb{P}_n. \tag{7}$$

Let $\{\varphi_1, \ldots, \varphi_m\}$ be the fundamental set of polynomials for nodes $t_1, \ldots, t_m$; i.e., the set such that $\varphi_k \in \mathbb{P}_{m-1}$, $\varphi_k(t_k) = 1$, and $\varphi_k(t_i) = 0$, $i \neq k$, $k = 1, \ldots, m$.

When $n \geq m - 1$, the weights β_j in (7) are uniquely defined (see, e.g., [22]): $\beta_j = a_0(\varphi_j)$, $j = 1, \ldots, m$. To see this, simply let $p = \varphi_j$ in (7). Such a quadrature (7) is called interpolatory. Thus, $M_j = a_0(\varphi_j)N$, $j = 1, \ldots, m$; i.e., frequencies M_j are determined by dot products $t_1, \ldots, t_m$ and the cardinality N and do not depend on $\mathbf{z}$. Frequencies M_j can also be found as solutions to a certain Vandermonde linear system, see [13, Theorem 2.1].

We next recall a known classification of quadratures with the highest algebraic degree of precision and dot product distributions for the corresponding classes of spherical designs (cf. [17]). These quadratures are special cases of Gauss-Jacobi mechanical quadrature, see [31, Theorem 3.4.1]. The strict positivity of weights β_j (called Christoffel numbers) in quadratures described below follows from [31, Theorem 3.4.2].

Case I: $n = 2m - 1$ We will call ω_N m-stiff in this case, see Definition 2. Quadrature (7) is now exact on $\mathbb{P}_{2m-1}$. The set $\{t_1, \ldots, t_m\}$ is the set of zeros of $P_m^{(d)}$. To see this, simply let $p(t) = \varphi_k(t) P_m^{(d)}(t)$ and use the equality in (7). Quadrature (7) is called the Gauss-Gegenbauer quadrature in this case.

Remark 1 From the above mentioned facts, one can deduce the following [17, 19]. For every point $\mathbf{z} \in S^d$ that forms at most m distinct dot products with points of a given m-stiff configuration, the number of dot products must be exactly m, their values must be zeros of the m-th Gegenbauer polynomial $P_m^{(d)}$, and their frequencies must be independent of $\mathbf{z}$ and can be computed as $M_j = a_0(\varphi_j)N$.

Case II: $n = 2m - 2$, $m \geq 2$, and $\mathbf{z} \in \omega_N$ Then quadrature (7) is exact on $\mathbb{P}_{2m-2}$ with $t_m = 1$ and $\beta_m = 1/N$. The nodes $t_1, \ldots, t_{m-1}$ are zeros of the Jacobi orthogonal polynomial $J_{m-1}^{(\alpha,\beta)}$ of degree $m-1$, where $\alpha = d/2$ and $\beta = d/2 - 1$. To see this, let $p(t) = \varphi_k(t) J_{m-1}^{(\alpha,\beta)}(t)$, $k = 1, \ldots, m-1$, in (7) (which becomes 0) and move the factor $1 - t$ from φ_k to the weight w_d in the integral.

Case III: $n = 2m - 3$, $m \geq 2$, $\mathbf{z} \in \omega_N$, and ω_N is antipodal Then quadrature (7) is exact on $\mathbb{P}_{2m-3}$ with $t_1 = -1$, $t_m = 1$, and $\beta_1 = \beta_m = 1/N$. For $m \geq 3$, the nodes $t_2, \ldots, t_{m-1}$ are zeros of the Gegenbauer orthogonal polynomial $P_{m-2}^{(d+2)}$. To see this, let $p(t) = \varphi_k(t) P_{m-2}^{(d+2)}(t)$, $k = 2, \ldots, m-1$, in (7) and move factors $1 - t$ and $1 + t$ from φ_k to the weight in the integral.

Remark 2 (See [17, 19]) If ω_N is a strongly m-sharp or m-sharp antipodal configuration on S^d (see Definition 3), then each point in ω_N forms the same set of m dot products with other points of ω_N with the same frequencies (which can be computed as described above). This set of dot products is the set of zeros of the Jacobi polynomial $J_m^{(\alpha,\beta)}$, where $\alpha = d/2$ and $\beta = d/2 - 1$, when ω_N is strongly m-sharp and is the set of zeros of the Gegenbauer polynomial $P_{m-1}^{(d+2)}$ together with -1 when ω_N is m-sharp antipodal.

4 Review of Further Results on Universal Extrema

A number of recent works were devoted to finding extrema of general potentials of regular configurations. Among kernels of the form $K(\mathbf{x}, \mathbf{y}) = f(|\mathbf{x} - \mathbf{y}|^2)$, of greatest interest to us are the ones with a completely monotone potential function f on $(0, 4)$ or with f that becomes completely monotone after adding a constant.

Definition 1 Given a configuration $\omega_N \subset S^d$, a point $\mathbf{x}^* \in S^d$ is called a universal minimum point (of the potential) of ω_N if $\mathbf{x}^*$ attains the absolute minimum in (2) for every potential function f completely monotone on $(0, 4]$.

For universal minimality results obtained here, we will require only that the derivative of f of a certain even order be convex or the derivative of a certain odd order be concave. Hardin et al. [21] proved that absolute minima of the potential of ω_N^* on S^1 with respect to a decreasing and convex function of the geodesic distance are attained at points of the dual regular N-gon. This partially solved the problem about universal minima of ω_N^*. The set of universal minima of regular simplex ω_{d+1}^* was shown in [9] to be $-\omega_{d+1}^*$.

Fazekas and Levenshtein [19] proved a sharp upper bound for the covering radius of a spherical design of a given strength. If a design has strength $(2m-1)$, then it attains this bound if and only if some point on the sphere forms as few as m distinct dot products with points of the design; i.e., it is m-stiff (see Definition 2 and Remark 1). If a design has strength $2m$, then this bound is attained if and only if it is a strongly m-sharp configuration. Basic examples of stiff configurations are given by a regular $2m$-gon on S^1, cube and cross-polytope on S^{d-1}, and the 24-cell on S^3.

During talk [7], the author of this paper proved the following theorem: given an m-stiff configuration ω_N, its universal minima are those and only those points on S^d that form m distinct dot products with points of ω_N. We state this result in Theorem 1. This theorem immediately implies the following. The set of universal minima of a regular $2m$-gon on S^1 is the set of midpoints of the arcs joining its neighboring points. The set of universal minima of a regular cross-polytope is the dual cube and the one for a cube is the dual cross-polytope. The set of universal minima of the 24-cell is the dual 24-cell. We state these corollaries in Sect. 6.

The above mentioned theorem on stiff configurations is a straightforward consequence of Lemma 3.5 in [5] stated here as Lemma 7. Lemma 3.5 in [5] has three other similar immediate consequences. All four follow if one just uses the well-known error formula for Hermite interpolation (see, e.g., [22, Section 2.2] or [18, Theorem 4.5]) or its "slight" modifications to include interpolation at the endpoint(s) (see, e.g., [16, Theorem 3.5.1]). One of those consequences is that, for any strongly sharp configuration ω_N (see Definition 3), the set of its universal minima is the set of antipods of ω_N. We state it in Theorem 2. Therefore, the set of universal minima of a regular $(2m+1)$-gon is the dual regular $(2m+1)$-gon and the sets of universal minima of the Schläfli configuration on S^5 and of the McLaughlin configuration on S^{21} are at the antipods of points of these configurations.

The remaining two consequences of [5, Lemma 3.5] are that universal maxima of any strongly sharp or sharp antipodal configuration (see Definition 3) are (only) at the points of the configuration itself (cf. [5]). In particular, universal maxima of a regular N-gon, regular simplex, regular cross-polytope, regular icosahedron, and E_8 and Leech lattices are at points of these configurations.

The universal maxima of the 600-cell on S^3 (which is not sharp) were shown by Boyvalenkov et al. [11] to be also at points of the 600-cell itself. A number of new stiff configurations (which are also sharp) were found by Boyvalenkov et al. [12]. The result of [7] implies that their universal minima are at points of the corresponding dual configurations (see Definition 2). Paper [12] further provides some information on these dual configurations.

Results of [12, Theorem 1.4] (see also [11]) provide lower and upper bounds for the potential of any configuration ω_N on S^d when it is a design of a given strength. When ω_N is stiff, strongly sharp, or sharp antipodal, these bounds become sharp. Papers [11, 12] assume positiveness (negativeness) of the derivative of the potential function f of a certain order while we assume here the convexity (concavity) of the derivative of f of the order smaller by two.

Some remarkable regular configurations are not stiff or strongly sharp. The author [3][4] found and characterized all universal minima for the icosahedron (minima are only at the vertices of the dual dodecahedron), dodecahedron (at the vertices of the dual icosahedron), for the E_8 lattice on S^7 (at the vertices of the dual 2_{41} polytope), and [4] for the vertices of the 2_{41} polytope (at the vertices of the dual E_8 lattice). This was done by proving a general theorem for $(2m-3)$-designs with vanishing $(2m-1)$-th and $2m$-th moments contained in m parallel hyperplanes (the "skip one add two" technique).

One universal minimum and the universal minimum value of the potential of the Leech lattice on S^{23} was found by Boyvalenkov, Dragnev, Hardin, Saff, and Stoyanova who also proved a general statement for the "skip one add two" case. Their result provides the value of the Gaussian-type quadrature for the corresponding polynomial space as a lower bound. It gives a somewhat different proof for the result of [3] on universal minima of the potential of icosahedron, dodecahedron, and E_8 lattice, since this lower bound is sharp for these codes.

Critical points of the total potential of finite configurations of charges were also analysed (see [2, 20] and references therein). Those works are related to the known Maxwell's conjecture.

We remark that the problem about universal minima of the 600-cell remains open. The problem about universal maxima of stiff configurations that are not sharp antipodal is being studied in our upcoming work.

Using Theorems 1 and 2, we find in this paper all universal minima of six stiff configurations, for which universal minima have not been characterized before, see Sects. 9, 10, and 11. It is interesting that these configurations form three dual pairs.

Table 1 shows the current state of the art in finding universal extrema for the classes of configurations considered in this paper and for the 600-cell. Table 2 summarizes results of our work in [3–5, 7] and in this paper for particular configurations. Last three lines of Table 3 contain the six stiff configurations, for which we find here universal minima (the first three lines follow immediately from Theorem 1).

Table 1 Current progress on the solution of the problem about universal extrema of classes of stiff and sharp configurations and of the 600-cell

Class of configurations	Universal maxima	Universal minima
Strongly sharp	Solved [5]	Solved [5], see Theorem 2
Sharp antipodal, non-stiff	Solved [5]	Solved [3, 12] for a subclass
Sharp antipodal and stiff	Solved [5]	Solved [7], see Theorem 1
Stiff, not sharp antipodal	Solved for a subclass (upcoming)	Solved [7], see Theorem 1
600-cell	Solved [11]	Open

[4] The proof of the special case of icosahedron (dodecahedron) was earlier described in talk [7] in January 2022.

Table 2 Summary of our results in [3–5, 7] and in this paper giving all universal extrema of certain configurations (see the last column). Entry with a dash indicates that the property does not hold. We remark that all universal maxima of the 600-cell were found in [11] and some universal minima (without characterization) were found in [12] for the 5-demicube, the kissing 56-point configuration on S^6, and for Leech lattice on S^{23}. Paper [12] studies universal minima for a number of sharp configurations not listed in this table and shows that they are stiff

Name	N	Sphere	Design strength	Sharpness	Stiffness	We find
Regular $(2m+1)$-gon	$2m+1$	S^1	$2m$	Strongly m-sharp	–	Both
Regular $2m$-gon	$2m$	S^1	$2m-1$	m-sharp	m-stiff	Both
Regular d-simplex, $d \geq 3$	$d+1$	S^{d-1}	2	Strongly 1-sharp	–	Both
Regular cross-polytope, $d \geq 3$	$2d$	S^{d-1}	3	2-sharp	2-stiff	Both
Cube, $d \geq 3$	2^d	S^{d-1}	3	–	2-stiff	Min
d-demicube, $d \geq 5$	2^{d-1}	S^{d-1}	3	–	2-stiff	Min
Symmetrized $(2\ell+1)$-simplex	$4\ell+4$	$S^{2\ell}$	3	–	2-stiff	Min
Dual symm.$(2\ell+1)$-simpl., $\ell \geq 2$	$\binom{2\ell+2}{\ell+1}$	$S^{2\ell}$	3	–	2-stiff	Min
Icosahedron	12	S^2	5	3-sharp	–	Both
Dodecahedron	20	S^2	5	–	–	Min
24-cell	24	S^3	5	–	3-stiff	Min
Schläfli	27	S^5	4	Strongly 2-sharp	–	Both
Symmetrized Schläfli	54	S^5	5	–	3-stiff	Min
Roots of E_6 lattice	72	S^5	5	–	3-stiff	Min
Kissing	56	S^6	5	3-sharp	3-stiff	Both
E_7 lattice	126	S^6	5	–	3-stiff	Min
E_8 lattice	240	S^7	7	4-sharp	–	Both
2_{41} polytope	2160	S^7	7	–	–	Min
McLaughlin	275	S^{21}	4	Strongly 2-sharp	–	Both
Leech lattice	196,560	S^{23}	11	6-sharp	–	Max

When proving our main results, we use the Delsarte method (the Delsarte-Yudin bound, linear programming method, or polynomial method), see the work by Delsarte et al. [17]. This approach has been extensively used for problems on best-packing, kissing number, and minimal energy, as well as on estimation of the cardinality of a spherical design, see [10, 14, 15, 23–26] and references therein.

Table 3 Pairs of mutually dual stiff configurations. Universal minima for the first three lines are immediate consequences of Theorem 1, see Table 4. Universal minima of configurations in the last three lines are obtained in Sects. 9, 10, and 11

Sphere	Pairs of mutually dual stiff configurations	
S^1	Regular $2m$-gon	Regular $2m$-gon rotated by $\frac{\pi}{2m}$ rad
$S^{d-1}, d \geq 3$	Regular cross-polytope ω_{2d}^*	Cube U_d
S^3	24-cell $\overline{\omega}_{24}$	24-cell $\widetilde{\omega}_{24}$
$S^{d-1}, d \geq 3$ odd	Symmetrized regular d-simplex $\overline{\omega}_{2d+2}$	$\overline{\omega}_{\kappa(d)}$ defined by (12)
S^5	Symmetrized Schläfli configuration $\overline{\omega}_{54}$	Root system of E_6-lattice $\overline{\omega}_{72}$
S^6	Kissing configuration $\overline{\omega}_{56}$	Minimal vectors of E_7-lattice $\overline{\omega}_{126}$

5 General Results

We start by stating the solution to Problem 1 for stiff configurations. Throughout the rest of the paper, $\mathbf{x}_1, \ldots, \mathbf{x}_N$ will denote the points in a configuration ω_N on S^d whose total potential we are minimizing. For a given $m \in \mathbb{N}$ and a given configuration $\omega_N \subset S^d$, denote by $\mathcal{D}_m(\omega_N)$ the set of all points $\mathbf{z} \in S^d$ for which the set of dot products

$$D(\mathbf{z}, \omega_N) := \{\mathbf{z} \cdot \mathbf{x}_i : i = 1, \ldots, N\}$$

has at most m distinct elements. The set $D(\mathbf{z}, \omega_N)$ is allowed to contain 1, which occurs when $\mathbf{z} \in \omega_N$. If $\mathcal{D}_m(\omega_N) \neq \emptyset$, then ω_N is contained in a union of at most m parallel hyperplanes.

Definition 2 We call a point configuration $\omega_N \subset S^d$ *m-stiff*, $d, m \geq 1$, if ω_N is a spherical $(2m-1)$-design and the set $\mathcal{D}_m(\omega_N)$ is non-empty. The set $\mathcal{D}_m(\omega_N)$ of a given m-stiff configuration ω_N will be called *the dual configuration* for ω_N.

Remark 3 If ω_N is an m-stiff configuration on S^d, $d, m \geq 1$, then every point in $\mathcal{D}_m(\omega_N)$ forms exactly m distinct dot products with points of ω_N. If it did not, then ω_N would be contained in $k < m$ parallel hyperplanes $\mathbf{x} \cdot \mathbf{v} = \alpha_i$, $i = 1, \ldots, k$, and the polynomial $u(\mathbf{x}) := (\mathbf{x} \cdot \mathbf{v} - \alpha_1)^2 \cdots (\mathbf{x} \cdot \mathbf{v} - \alpha_k)^2$ of degree $2k < 2m - 1$ would vanish on ω_N while having a positive integral over S^d.

In view of this remark, the dual configuration for an m-stiff ω_N can be equivalently defined as the set of all points on S^d that form exactly m distinct dot products with points of ω_N.

There are many examples of stiff configurations other than the basic ones mentioned in Sect. 4. First, there are two interesting examples on S^{d-1} for $d \geq 3$ odd (both are 2-stiff), two more on S^5, and two more on S^6 (all are 3-stiff). We describe these configurations in Sects. 9, 10, and 11. Furthermore, paper [12] analyses one example on S^4 (5-demicube) and a number of examples on higher-dimensional spheres arising from the Golay code (these configurations are also sharp).

The following result holds, see [7].[5]

Theorem 1 *Let* $m \geq 1$, $d \geq 1$, $f : [0, 4] \to (-\infty, \infty]$ *be a function continuous on* $(0, 4]$ *with* $f(0) = \lim_{t \to 0^+} f(t)$, *differentiable on* $(0, 4)$ *with a convex derivative* $f^{(2m-2)}$ *on* $(0, 4)$. *If* $\omega_N = \{\mathbf{x}_1, \ldots, \mathbf{x}_N\}$ *is an* m*-stiff configuration on the sphere* S^d, *then the potential*

$$p_f(\mathbf{x}, \omega_N) = \sum_{i=1}^{N} f(|\mathbf{x} - \mathbf{x}_i|^2), \quad \mathbf{x} \in S^d,$$

attains its absolute minimum over S^d *at every point of the set* $\mathcal{D}_m(\omega_N)$.

If, in addition, $f^{(2m-2)}$ *is strictly convex on* $(0, 4)$, *then* $\mathcal{D}_m(\omega_N)$ *contains all points of absolute minimum of the potential* $p_f(\cdot, \omega_N)$ *on* S^d.

No point from an m-stiff configuration ω_N (or $-\omega_N$) can be in $\mathcal{D}_m(\omega_N)$, see Remark 1 or Lemma 6. Therefore, a point from ω_N in Theorem 1 cannot be a minimizer of the potential of ω_N when $f^{(2m-2)}$ is strictly convex.

The "slightly" less general case of Theorem 1 when f has a positive derivative $f^{(2m)}$ on $(0, 4]$ is an important special case of [5, Lemma 3.5] restated here as Lemma 7. To see this, use the well-known error formula for Hermite interpolation [22, Section 2.2] or [18, Theorem 4.5] at the set of nodes given by dot products that a point from $\mathcal{D}_m(\omega_N)$ forms with points from ω_N. The "full" case of Theorem 1 follows on applying Lemma 5 instead of the error formula for Hermite interpolation, see details in Sect. 7.

We next consider another class of regular configurations on the sphere, for which universal minima can be found. If the set $\mathcal{D}_m(\omega_N)$ contains a point from $-\omega_N$, then it is enough to require that ω_N be a $(2m-2)$-design. These assumptions are satisfied by any sharp configuration, which is an even order design. Sharp configuration were defined in [15].

Definition 3 We call a point configuration $\omega_N \subset S^d$ *m-sharp*, $m \geq 1$, if ω_N is a spherical $(2m-1)$-design and there are m distinct values of the dot product that can occur between any two distinct vectors from ω_N. We call an m-sharp configuration *strongly m-sharp* if it is a spherical $2m$-design.

Examples of strongly sharp configurations are vertices of a regular $(2m+1)$-gon inscribed in S^1, $m \in \mathbb{N}$, the set of vertices of a regular d-simplex inscribed in S^{d-1} $(m = 1)$, the Schläfli configuration of $N = 27$ points on S^5 $(m = 2)$, and the McLaughlin configuration of $N = 275$ points on S^{21} $(m = 2)$. There are no strongly m-sharp configurations on S^d for $d \geq 2$ and $m \geq 3$. This follows from the results of [17] and [1].

[5] See the 30-th minute of talk [7]. We will present a complete proof of Theorem 1, since its proof was only given in a talk before.

The classes of stiff and sharp configurations overlap. For example, sets of vertices of a regular cross-polytope on S^{d-1} and of a regular $2m$-gon inscribed in S^1 are both stiff and sharp as well as the kissing configuration $\overline{\omega}_{56}$ on S^6 and some examples mentioned in [12, Table 2]. However, there are stiff configurations, which are not sharp (e.g., the cube in $\mathbb{R}^d$ for $d \geq 3$ or the 24-cell on S^3). There are also sharp configurations, which are not stiff, e.g., the regular $(2m+1)$-gon inscribed in S^1, the regular icosahedron on S^2, and the minimal vectors of E_8 lattice on S^7. In addition, classes of strongly sharp and stiff configurations are disjoint (see Proposition 2).

The following result was proved in [5] for universal maxima. For completeness, we state here its version for universal minima and give the proof.

Theorem 2 *Let* $m \geq 1$, $d \geq 1$, $f : [0, 4] \to (-\infty, \infty]$ *be a function continuous on* $(0, 4]$ *with* $f(0) = \lim_{t\to 0^+} f(t)$, *differentiable on* $(0, 4)$ *with a concave derivative* $f^{(2m-1)}$ *on* $(0, 4)$. *If* $\omega_N = \{\mathbf{x}_1, \ldots, \mathbf{x}_N\} \subset S^d$ *is a strongly* m*-sharp configuration, then the potential*

$$p_f(\mathbf{x}, \omega_N) = \sum_{i=1}^{N} f(|\mathbf{x} - \mathbf{x}_i|^2), \quad \mathbf{x} \in S^d,$$

attains its absolute minimum over S^d *at every point of* $-\omega_N$.

If the concavity of $f^{(2m-1)}$ *on* $(0, 4)$ *is strict, then* $-\omega_N$ *contains all points of absolute minimum of the potential* $p_f(\cdot, \omega_N)$ *over* S^d.

Theorem 2 when $g(t) = f(2 - 2t)$ has a positive derivative $g^{(2m+1)}$ on $[-1, 1)$ can be derived from [5, Lemma 3.5] by recalling the error formula [16, Theorem 3.5.1] for interpolation to the second order at the nodes opposite to dot products between distinct points in ω_N and to the first order at $t = -1$. The "full" case of Theorem 2 follows on applying Lemma 5 instead of this error formula, see details in Sect. 7. In the proofs of Theorems 1 and 2, we do not need to show the positive definiteness of the interpolating polynomial when utilizing the linear programming approach of [15, 17], since the strength of the design we consider equals the degree of the interpolating polynomial for f.

6 Immediate Consequences of Theorems 1 and 2

This section discusses universal minima given by Theorem 1 for regular $2m$-gon, regular cross-polytope, cube, and 24-cell (finding the dual for each configuration is elementary) and universal minima given by Theorem 2 for a regular $(2m + 1)$-gon, regular simplex,[6] and Schläfli and McLaughlin configurations. For power potentials (3), the first five lines of Table 4 were proved in [27–30].

[6] For regular simplex, universal minima were found in [9] (here we provide an alternative proof).

The following auxiliary statements yield dual configurations. Let $\widetilde{\omega}_N$ denote the set of midpoints of arcs on S^1 that join the neighboring points in the regular N-gon ω_N^* inscribed in S^1.

Lemma 1 *For $N = 2m$, $m \geq 1$, we have $\mathcal{D}_m(\omega_N^*) = \widetilde{\omega}_N$.*

Proof Every point from $\widetilde{\omega}_N$, $N = 2m$, forms m distinct dot products with points of ω_N^*, while every point from ω_N^* itself forms $m + 1$ dot products and every other point on S^1 forms $2m$ dot products with points of ω_N^*. □

Lemma 2 *For every $d \geq 2$, we have $\mathcal{D}_2(\omega_{2d}^*) = U_d$.*

Proof The inclusion $U_d \subset \mathcal{D}_2(\omega_{2d}^*)$ is immediate. Let x_i be a non-zero coordinate of $\mathbf{x} := (x_1, \dots, x_d) \in \mathcal{D}_2(\omega_{2d}^*)$. Then $\mathbf{x} \cdot (\pm \mathbf{e}_i) = \pm x_i$. Every other coordinate x_j must have the same absolute value as x_i (otherwise $\pm x_j$ would give extra dot product(s)). Then $|x_i| = \frac{1}{\sqrt{d}}$, $i = \overline{1, d}$; that is, $\mathbf{x} \in U_d$. □

Lemma 3 *For every $d \geq 2$, we have $\mathcal{D}_2(U_d) = \omega_{2d}^*$.*

Proof The inclusion $\omega_{2d}^* \subset \mathcal{D}_2(U_d)$ is immediate. If $\mathbf{x} = (x_1, \dots, x_d) \in \mathcal{D}_2(U_d)$ had at least two non-zero coordinates, say x_1 and x_2, then dot product $\mathbf{x} \cdot \left(\pm\frac{1}{\sqrt{d}}, \pm\frac{1}{\sqrt{d}}, \frac{1}{\sqrt{d}}, \dots, \frac{1}{\sqrt{d}}\right) = \frac{\pm x_1 \pm x_2}{\sqrt{d}} + \frac{x_3}{\sqrt{d}} + \dots + \frac{x_d}{\sqrt{d}}$ would have at least three distinct values. Therefore, only one coordinate of $\mathbf{x}$ is non-zero; i.e., $\mathbf{x} \in \omega_{2d}^*$. □

The 24-cell $\overline{\omega}_{24} \subset S^3$ is the set of $N = 24$ points with two coordinates equal 0 and each of the remaining two coordinates being $1/\sqrt{2}$ or $-1/\sqrt{2}$. The 24-cell is known to be a 5-design. At the same time, every vector $\pm \mathbf{e}_i$, $i = \overline{1, 4}$, where $\{\mathbf{e}_1, \mathbf{e}_2, \mathbf{e}_3, \mathbf{e}_4\}$ is the standard basis in $\mathbb{R}^4$, forms only 3 distinct dot products with vectors of the 24-cell, which are $\pm 1/\sqrt{2}$ and 0. Furthermore, each of the 16 vectors of the form $\left(\pm\frac{1}{2}, \pm\frac{1}{2}, \pm\frac{1}{2}, \pm\frac{1}{2}\right)$ forms these three values of the dot product with vectors from $\overline{\omega}_{24}$ as well. Therefore, $\overline{\omega}_{24}$ is 3-stiff. Vectors $\pm \mathbf{e}_i$, $i = \overline{1, 4}$, and $\left(\pm\frac{1}{2}, \pm\frac{1}{2}, \pm\frac{1}{2}, \pm\frac{1}{2}\right)$ form another 24-cell denoted by $\widetilde{\omega}_{24}$.

Lemma 4 $\mathcal{D}_3(\overline{\omega}_{24}) = \widetilde{\omega}_{24}$.

Proof The inclusion $\widetilde{\omega}_{24} \subset \mathcal{D}_3(\overline{\omega}_{24})$ is immediate. Let $\mathbf{x} = (x_1, x_2, x_3, x_4) \in \mathcal{D}_3(\overline{\omega}_{24})$. There are at most three distinct numbers among all dot products $\frac{\pm x_i \pm x_j}{\sqrt{2}}$, $1 \leq i \neq j \leq 4$. Then all non-zero coordinates of $\mathbf{x}$ must have the same absolute value (denoted by a). If $\mathbf{x}$ had two or three non-zero coordinates, it would form dot products $0, \pm\frac{a}{\sqrt{2}}, \pm\sqrt{2}a$ with vectors from $\overline{\omega}_{24}$. Therefore, $\mathbf{x}$ has one non-zero coordinate ($a = 1$) or all its coordinates equal $\pm a$ ($a = 1/2$); i.e., $\mathbf{x} \in \widetilde{\omega}_{24}$. □

Immediate consequences of Theorems 1 and 2 are listed in Table 4.

Table 4 Immediate consequences of Theorems 1 and 2. We assume that $f : [0,4] \to (-\infty, \infty]$ is continuous on $(0,4]$ with $f(0) = \lim_{t\to 0^+} f(t)$ and differentiable on $(0,4)$ with an additional assumption on f specified on $(0,4)$ in the third column. Then the potential $p_f(\cdot, \omega_N)$ attains its absolute minimum over the sphere of the specified dimension at points of the set in the fourth column. If the assumption in the third column holds strictly, then no other absolute minima exist

Configuration ω_N	Sphere	Assumption	Absolute minima
Regular $2m$-gon	S^1	$f^{(2m-2)}$ is convex	Regular $2m$-gon rotated by $\frac{\pi}{2m}$ rad
Regular $(2m+1)$-gon	S^1	$f^{(2m-1)}$ is concave	Antipod of regular $(2m+1)$-gon
Regular d-simplex ω^*_{d+1}	S^{d-1}	f' is concave	Regular d-simplex $-\omega^*_{d+1}$
Cross-polytope ω^*_{2d}, $d \geq 3$	S^{d-1}	f'' is convex	Cube U_d
Cube U_d, $d \geq 3$	S^{d-1}	f'' is convex	Cross-polytope ω^*_{2d}
24-cell $\overline{\omega}_{24}$	S^3	$f^{(4)}$ is convex	24-cell $\widetilde{\omega}_{24}$
Schläfli $\overline{\omega}_{27}$	S^5	f''' is concave	$-\overline{\omega}_{27}$
McLaughlin $\overline{\omega}_{275}$	S^{21}	f''' is concave	$-\overline{\omega}_{275}$

7 Proofs of Theorems 1 and 2

We first establish the following interpolation lemma for the potential function. It is a "slight" modification of [5, Lemma 3.1]. The case $\nu = 0$ corresponds to stiff configurations and Theorem 1. The case $\nu = 1$ corresponds to strongly sharp configurations and Theorem 2.

Lemma 5 *Suppose $\nu = 0$ or 1. Let $n \geq 1+\nu$, $L := 2n-2-\nu$, and $-1 \leq t_1 < \ldots < t_n < 1$ be arbitrary nodes such that $t_1 > -1$ if $\nu = 0$ and $t_1 = -1$ if $\nu = 1$. Suppose also that $g : [-1,1] \to (-\infty, \infty]$ is a function continuous on $[-1,1)$ with $g(1) = \lim_{t\to 1^-} g(t)$ and differentiable on $(-1,1)$ with the derivative $g^{(L)}$ being convex on $(-1,1)$. Let $q \in \mathbb{P}_{L+1}$ be the polynomial such that*

$$q(t_i) = g(t_i), \quad i = 1, \ldots, n, \quad \text{and} \quad q'(t_i) = g'(t_i), \quad i = 1+\nu, \ldots, n. \tag{8}$$

Then

$$q(t) \leq g(t), \quad t \in [-1,1]. \tag{9}$$

If, in addition, the derivative $g^{(L)}$ is strictly convex on $(-1,1)$, then the inequality in (9) is strict for $t \in [-1,1] \setminus \{t_1, \ldots, t_n\}$.

Proof The assertion of the lemma follows immediately when $L = 0$; i.e., when $n = 1$ and $\nu = 0$. Therefore, we assume that $L \geq 1$ (then $n \geq 2$). Inequality (9) holds trivially for $t = t_1, \ldots, t_n$ (as equality) and, if $g(1) = \infty$, at $t = 1$ (as a strict inequality). Therefore, we choose arbitrary $x \in [-1,1] \setminus \{t_1, \ldots, t_n\}$ assuming additionally that $x < 1$ if $g(1) = \infty$. Denote

$$h(t) := q(t) + b(t - t_1)^{2-\nu} \prod_{i=2}^{n} (t - t_i)^2,$$

where the constant b is chosen so that $h(x) = g(x)$. Observe that the polynomial h also satisfies interpolation conditions (8). The function $\tau(t) := g(t) - h(t)$ has at least $n + 1$ distinct zeros in $[-1, 1]$ at least $n - \nu$ of which have multiplicity two and are located in $(-1, 1)$. By the Rolle's theorem, the derivative τ' has at least $2n - \nu = L + 2$ distinct zeros in $(-1, 1)$, which implies that the derivative $\tau^{(L)}$ has at least three distinct zeros in $(-1, 1)$. The polynomial $h^{(L)}$ has degree at most two with the coefficient of t^2 being $\frac{(L+2)!}{2}b$. It interpolates the convex function $g^{(L)}$ at least three distinct points in $(-1, 1)$. Then $b \geq 0$. Since $(x - t_1)^{2-\nu} > 0$ both for $\nu = 0$ and $\nu = 1$, we have

$$g(x) = h(x) = q(x) + b(x - t_1)^{2-\nu} \prod_{i=2}^{n} (x - t_i)^2 \geq q(x), \tag{10}$$

which proves (9). If $g^{(L)}$ is strictly convex on $(-1, 1)$, then $b \neq 0$ (otherwise $g^{(L)}$ would be interpolated by the polynomial $h^{(L)}$ of degree at most one at three distinct points). Therefore, $b > 0$ and the inequality in (9) is strict on $[-1, 1] \setminus \{t_1, \ldots, t_n\}$. □

Lemma 6 *Let ω_N be an m-stiff configuration on S^d, $d, m \geq 1$. Then there is no point in $\mathcal{D}_m(\omega_N)$ that forms dot products 1 or -1 with points from ω_N. Let ω_N be strongly m-sharp. Then ω_N contains no antipodal pair and $\mathcal{D}_m(\omega_N) = \emptyset$.*

Proof Assume to the contrary that there is a point $\mathbf{y} \in \mathcal{D}_m(\omega_N)$ such that $\mathbf{y} \cdot \mathbf{x}_\ell = 1$ or -1 for some $\mathbf{x}_\ell \in \omega_N$. Let $-1 \leq t_1 < \ldots < t_k \leq 1$, $k \leq m$, be the distinct dot products $\mathbf{y}$ forms with vectors from ω_N. If $\mathbf{y} \cdot \mathbf{x}_\ell = 1$, then $t_k = 1$ and we let $p(t) := (t - t_1)^2 \cdots (t - t_{k-1})^2(1 - t)$. If $\mathbf{y} \cdot \mathbf{x}_\ell = -1$, then $t_1 = -1$ and we let $p(t) := (t + 1)(t - t_2)^2 \cdots (t - t_k)^2$. Define $u(\mathbf{x}) := p(\mathbf{x} \cdot \mathbf{y})$, $\mathbf{x} \in \mathbb{R}^{d+1}$. Then u has degree at most $2m - 1$, $u(\mathbf{x}_i) = 0$, $i = 1, \ldots, N$, and $u(\mathbf{x}) > 0$ for σ_d-almost all $\mathbf{x} \in S^d$. Then, since ω_N is a $(2m - 1)$-design,

$$0 = \frac{1}{N} \sum_{i=1}^{N} u(\mathbf{x}_i) = \int_{S^d} u(\mathbf{x}) d\sigma_d(\mathbf{x}) > 0.$$

This contradiction proves the first part of the lemma.

Let now ω_N be strongly m-sharp. Assume to the contrary that ω_N contains an antipodal pair $\mathbf{x}_\ell = -\mathbf{x}_i$. Let $-1 = t_1 < t_2 < \ldots < t_m < 1$ be dot products between distinct points in ω_N. Letting $p(t) := (t + 1)(t - t_2)^2 \cdots (t - t_m)^2(1 - t)$ and $u(\mathbf{x}) := p(\mathbf{x} \cdot \mathbf{x}_\ell)$ and using a similar argument, we arrive at a contradiction proving that ω_N has no antipodal pair.

Finally, assume to the contrary that there is a point $\mathbf{z} \in \mathcal{D}_m(\omega_N)$. Let $-1 \leq t_1 < \ldots < t_k \leq 1$, $k \leq m$, be distinct elements in $D(\mathbf{z}, \omega_N)$. Letting $p(t) :=$

$(t-t_1)^2\cdots(t-t_k)^2$ and $u(\mathbf{x}) := p(\mathbf{x}\cdot\mathbf{z})$ and using a similar argument again, we arrive at a contradiction proving that $\mathcal{D}_m(\omega_N)$ is empty. □

Replacing g and q with $-g$ and $-q$ in [5, Lemma 3.5] we obtain the following.

Lemma 7 *Suppose that $d, n \geq 1$ and that a function $g : [-1, 1] \to (-\infty, \infty]$ and a set A consisting of numbers $-1 \leq t_1 < \ldots < t_n \leq 1$ are arbitrary. Suppose also that q is a polynomial such that $q(t_i) = g(t_i)$, $i = 1, \ldots, n$, and*

$$q(t) \leq g(t), \quad t \in [-1, 1]. \tag{11}$$

If $\omega_N = \{\mathbf{x}_1, \ldots, \mathbf{x}_N\} \subset S^d$ is a spherical M-design, where $M \geq \deg q$, and $\mathbf{y} \in S^d$ is any point such that $D(\mathbf{y}, \omega_N) \subset A$, then the potential

$$p^g(\mathbf{x}, \omega_N) := \sum_{i=1}^{N} g(\mathbf{x}\cdot\mathbf{x}_i), \quad \mathbf{x} \in S^d,$$

attains its absolute minimum over S^d at point $\mathbf{y}$.

If, in addition, the inequality in (11) is strict for every $t \in [-1, 1] \setminus A$, then any point $\mathbf{z} \in S^d$ such that $D(\mathbf{z}, \omega_N) \not\subset A$ is not a point of absolute minimum of $p^g(\cdot, \omega_N)$ on S^d.

We are now ready to prove Theorems 1 and 2.

Proof of Theorem 1 Let $\mathbf{y}$ be any point from $\mathcal{D}_m(\omega_N)$. Let $-1 < \tau_1 < \ldots < \tau_m < 1$ be distinct elements of the set $D(\mathbf{y}, \omega_N) = \{\mathbf{y}\cdot\mathbf{x}_i : 1 \leq i \leq N\}$. There are exactly m dot products there in view of Remark 3 with $\tau_1 > -1$ and $\tau_m < 1$ in view of Lemma 6. We apply Lemma 5 with $\nu = 0$, $n = m$, and $g(t) = f(2-2t)$ so that $g^{(L)}$ is convex on $(-1, 1)$, where $L = 2m-2$. Then the Hermite interpolating polynomial $q \in \mathbb{P}_{2m-1}$ for g at the nodes τ_i satisfies $q(t) \leq g(t)$, $t \in [-1, 1]$. Since ω_N is a spherical $(2m-1)$-design and $D(\mathbf{y}, \omega_N) = \{\tau_1, \ldots, \tau_m\}$, by Lemma 7 (with $n = m$ and $M = 2m-1$), the potential $p_f(\cdot, \omega_N) = p^g(\cdot, \omega_N)$ attains its absolute minimum over S^d at $\mathbf{y}$.

If $f^{(2m-2)}$ is strictly convex on $(0, 4)$, then so is $g^{(2m-2)}$ on $(-1, 1)$ and, by Lemma 5, $q(t) < g(t)$, $t \in [-1, 1] \setminus \{\tau_1, \ldots, \tau_m\}$. If $\mathbf{z} \in S^d$ is any point not in $\mathcal{D}_m(\omega_N)$, then the set $D(\mathbf{z}, \omega_N)$ contains at least $m+1$ distinct dot products. Hence, $D(\mathbf{z}, \omega_N) \not\subset \{\tau_1, \ldots, \tau_m\}$. In view of Lemma 7, the point $\mathbf{z}$ is not a point of absolute minimum of $p^g(\cdot, \omega_N)$ on S^d, and, hence, of $p_f(\cdot, \omega_N)$. Thus, $\mathcal{D}_m(\omega_N)$ is the set of all points of absolute minimum of $p_f(\cdot, \omega_N)$ on S^d. □

Proof of Theorem 2 Let $-1 < \tau_1 < \ldots < \tau_m < \tau_{m+1} = 1$ be the distinct values of the dot product between points in ω_N. Here, we have $\tau_1 > -1$ in view of Lemma 6. We apply Lemma 5 with $\nu = 1$, $n = m+1$, $t_1 = -\tau_{m+1} = -1$, $t_2 = -\tau_m, \ldots, t_{m+1} = -\tau_1$, and $g(t) = f(2-2t)$ so that $g^{(L)}$ is convex on $(-1, 1)$, where $L = 2m-1$. Then the polynomial $q \in \mathbb{P}_{2m}$, which interpolates both g and g' at $t_2, \ldots, t_{m+1}$ and only g at t_1 satisfies $q(t) \leq g(t)$, $t \in [-1, 1]$. Choose any point $\mathbf{y}$ in $-\omega_N$. Then $D(\mathbf{y}, \omega_N) \subset \{t_1, \dots, t_{m+1}\}$. Since ω_N is a $2m$-design, by Lemma 7 (with $n = m + 1$ and $M = 2m$) the potential $p_f(\cdot, \omega_N) = p^g(\cdot, \omega_N)$ attains its absolute minimum over S^d at $\mathbf{y}$.

Let now $f^{(2m-1)}$ be strictly concave on $(0, 4)$. Then $g^{(2m-1)}$ is strictly convex on $(-1, 1)$. Let $\mathbf{z}$ be an arbitrary point in $S^d \setminus (-\omega_N)$. Assume to the contrary that $D(\mathbf{z}, \omega_N) \subset \{t_1, \dots, t_{m+1}\}$. If $D(\mathbf{z}, \omega_N)$ did not contain $t_1 = -1$, then it would contain at most m distinct elements making the set $\mathcal{D}_m(\omega_N)$ non-empty. By Lemma 6, this is not possible. Consequently, $D(\mathbf{z}, \omega_N)$ contains -1. Then $\mathbf{z} \in -\omega_N$ contradicting its choice. Thus, $D(\mathbf{z}, \omega_N) \not\subset \{t_1, \dots, t_{m+1}\}$. By Lemma 5, we have $q(t) < g(t)$, $t \in [-1, 1] \setminus \{t_1, \dots, t_{m+1}\}$. Then by Lemma 7, the point $\mathbf{z}$ is not a point of absolute minimum of the potential $p_f(\cdot, \omega_N) = p^g(\cdot, \omega_N)$ on S^d. □

8 Dot Product Distribution Related to Stiff Configurations

In this section, we study the distribution of dot products between vectors from $\mathcal{D}_m(\omega_N)$ and ω_N. Spherical designs can be characterized using Gegenbauer polynomials as follows.

Theorem 3 (See [17] or [10, Theorems 5.2.2 and 5.4.2]) *Let $d, n \geq 1$ and $\omega_N = \{\mathbf{x}_1, \dots, \mathbf{x}_N\}$ be a point configuration on S^d. The following are equivalent:*

(i) ω_N *is a spherical n-design;*

(ii) $\sum\limits_{i=1}^{N} \sum\limits_{j=1}^{N} P_k^{(d)}(\mathbf{x}_i \cdot \mathbf{x}_j) = 0, \quad k = 1, \dots, n;$

(iii) For every polynomial $q \in \mathbb{P}_n$, *we have* $p^q(\mathbf{y}, \omega_N) := \sum_{i=1}^N q(\mathbf{y} \cdot \mathbf{x}_i) = a_0(q)N$, $\mathbf{y} \in S^d$, *where*

$$a_0(q) := \int_{-1}^{1} q(t) w_d(t)\, dt$$

is the 0*-th Gegenbauer coefficient of polynomial* q.

One can use Theorem 3 part (ii) to find the strength of every spherical design mentioned in this paper. We next restate Remark 1 formally and, for completeness, provide its proof. Here, $\varphi_1, \dots, \varphi_m$ are the fundamental polynomials for set of nodes (4); that is, $\varphi_i \in \mathbb{P}_{m-1}$, $\varphi_i(\kappa_i^m) = 1$, and $\varphi_i(\kappa_j^m) = 0$, $j \neq i$, $i = 1, \dots, m$.

Proposition 1 ([17, 19]) *If ω_N is an m-stiff configuration on S^d, then for every $\mathbf{z} \in \mathcal{D}_m(\omega_N)$, the set $D(\mathbf{z}, \omega_N)$ contains exactly m distinct elements located in $(-1, 1)$, which are the zeros $\kappa_1^m, \dots, \kappa_m^m$ of Gegenbauer polynomial $P_m^{(d)}$. Furthermore, the number of indices i such that $\mathbf{z} \cdot \mathbf{x}_i = \kappa_j^m$ does not depend on $\mathbf{z}$ and equals $a_0(\varphi_j)N$, $j = 1, \dots, m$.*

In particular, if $m = 2$*, then for every* $\mathbf{z} \in \mathcal{D}_2(\omega_N)$*, we have* $D(\mathbf{z}, \omega_N) = \left\{-\frac{1}{\sqrt{d+1}}, \frac{1}{\sqrt{d+1}}\right\}$*, and, if* $m = 3$*, then for every* $\mathbf{z} \in \mathcal{D}_3(\omega_N)$*, we have* $D(\mathbf{z}, \omega_N) = \left\{\pm\sqrt{\frac{3}{d+3}}, 0\right\}$*.*

Proof Denote by $-1 \le t_1 < \ldots < t_m \le 1$ the distinct elements in $D(\mathbf{z}, \omega_N)$ (in view of Remark 3, there are exactly m of them). Let $M_j := \#\{i : \mathbf{z} \cdot \mathbf{x}_i = t_j\}$, $j = 1, \ldots, m$. Denote $P(t) := (t - t_1) \cdots (t - t_m)$. Since ω_N is a $(2m - 1)$-design, using item (iii) of Theorem 3, for every polynomial $p \in \mathbb{P}_{m-1}$, we have

$$a_0(pP) = \frac{1}{N} \sum_{i=1}^{N} p(\mathbf{z} \cdot \mathbf{x}_i) P(\mathbf{z} \cdot \mathbf{x}_i) = \frac{1}{N} \sum_{j=1}^{m} M_j p(t_j) P(t_j) = 0.$$

Then $P \perp \mathbb{P}_{m-1}$ on $[-1, 1]$ with weight w_d; that is, P is a constant multiple of the Gegenbauer polynomial $P_m^{(d)}$. Then $t_j = \kappa_j^m$, $j = 1, \ldots, m$, and we have

$$a_0(\varphi_k) = \frac{1}{N} \sum_{i=1}^{N} \varphi_k(\mathbf{z} \cdot \mathbf{x}_i) = \frac{1}{N} \sum_{j=1}^{m} M_j \varphi_k(\kappa_j^m) = \frac{M_k}{N};$$

that is, $M_k = a_0(\varphi_k) N$, $k = 1, \ldots, m$.

Since $P_2^{(d)}(t) = \frac{d+1}{d} t^2 - \frac{1}{d}$ and $P_3^{(d)}(t) = \frac{d+3}{d} t^3 - \frac{3}{d} t$, the sets of zeros of these polynomials give the set $D(\mathbf{z}, \omega_N)$ for $m = 2$ and 3. □

We conclude this section with the following consequence of Lemma 6.

Proposition 2 *The classes of stiff and strongly sharp configurations on* S^d*,* $d \ge 1$*, are disjoint.*

Proof of Proposition 2 Assume to the contrary that there exists a configuration $\omega_N \subset S^d$ which is k-stiff and strongly m-sharp for some $k, m \ge 1$. By Lemma 6, the set $\mathcal{D}_m(\omega_N)$ is empty. Then $k \ge m + 1$. Since ω_N is k-stiff, it must be at least a $(2m + 1)$-design. If ω_N is not $(m + 1)$-stiff, then, by definition, $\mathcal{D}_{m+1}(\omega_N) = \emptyset$. If ω_N is $(m + 1)$-stiff, then by Lemma 6, the set $\mathcal{D}_{m+1}(\omega_N)$ does not contain any point from ω_N. However, since ω_N is strongly m-sharp, every point from ω_N is in $\mathcal{D}_{m+1}(\omega_N)$. This contradiction proves the proposition. □

9 Symmetrized Regular Simplex and Its Dual

In this section, we describe universal minima for the first pair of stiff configurations. Let $d \ge 3$ be odd. Denote $\overline{\omega}_{2d+2} = \omega_{d+1}^* \cup (-\omega_{d+1}^*)$, where $\omega_{d+1}^* = \{\mathbf{x}_0, \mathbf{x}_1 \ldots, \mathbf{x}_d\}$ is the set of vertices of a regular d-simplex inscribed in S^{d-1}. Since ω_{d+1}^* is a spherical 2-design, the symmetrized regular simplex $\overline{\omega}_{2d+2}$ is a spherical 3-design. Denote $\kappa(d) := \binom{d+1}{\frac{d+1}{2}}$ and let

$$\overline{\omega}_{\kappa(d)} := \left\{ \frac{2\sqrt{d}}{d+1} \sum_{i \in I} \mathbf{x}_i : I \subset \{0, 1, \dots, d\},\ \#I = \frac{d+1}{2} \right\}. \tag{12}$$

It is not difficult to see that $\overline{\omega}_{\kappa(d)} \subset S^{d-1}$. By Lemma 8 below, $\mathcal{D}_2(\overline{\omega}_{2d+2}) = \overline{\omega}_{\kappa(d)}$. Thus, $\overline{\omega}_{2d+2}$ is 2-stiff. The configuration $\overline{\omega}_{\kappa(d)}$, in turn, is a spherical 3-design (see Lemma 9 below) with $\mathcal{D}_2(\overline{\omega}_{\kappa(d)}) = \overline{\omega}_{2d+2}$ in view of Lemma 10. Then $\overline{\omega}_{\kappa(d)}$ is also 2-stiff. We remark that for $d = 2$, the configuration $\overline{\omega}_{2d+2}$ is a regular hexagon on S^1, which is 3-stiff, and its dual is also a regular hexagon.

Proposition 3 *For $d \geq 4$ even, $\mathcal{D}_2(\overline{\omega}_{2d+2}) = \emptyset$ and $\overline{\omega}_{2d+2}$ is not stiff.*

Proof The configuration $\overline{\omega}_{2d+2}$ is a 3-design. Assume to the contrary that

$$\mathcal{D}_2(\overline{\omega}_{2d+2}) \neq \emptyset.$$

Then $\overline{\omega}_{2d+2}$ is 2-stiff. By Proposition 1, there is a point $\mathbf{x} \in S^{d-1}$ such that $\mathbf{x} \cdot \mathbf{x}_i = \pm\frac{1}{\sqrt{d}}$, $i = 0, 1, \dots, d$. Let k be the number of indices i such that $\mathbf{x} \cdot \mathbf{x}_i = \frac{1}{\sqrt{d}}$. Then, since $d+1$ is odd,

$$\mathbf{x} \cdot \sum_{i=0}^{d} \mathbf{x}_i = \sum_{i=0}^{d} \mathbf{x} \cdot \mathbf{x}_i = \frac{k}{\sqrt{d}} - \frac{d+1-k}{\sqrt{d}} = \frac{2k-(d+1)}{\sqrt{d}} \neq 0$$

contradicting the fact that the center of mass of ω^*_{d+1} is at the origin. Therefore, $\mathcal{D}_2(\overline{\omega}_{2d+2}) = \emptyset$, and $\overline{\omega}_{2d+2}$ cannot be 1- or 2-stiff. Since $d \geq 4$, $\overline{\omega}_{2d+2}$ is not a 4-design. Then $\overline{\omega}_{2d+2}$ cannot be m-stiff for any $m \geq 3$. □

Lemma 8 *For every $d \geq 3$ odd, $\mathcal{D}_2(\overline{\omega}_{2d+2}) = \overline{\omega}_{\kappa(d)}$.*

Proof Let $\mathbf{x}_0, \mathbf{x}_1, \dots, \mathbf{x}_d$ be vertices of a regular d-simplex on S^{d-1} such that $\overline{\omega}_{2d+2} = \{\pm\mathbf{x}_i : i = 0, 1, \dots, d\}$. The inclusion $\overline{\omega}_{\kappa(d)} \subset \mathcal{D}_2(\overline{\omega}_{2d+2})$ holds. Indeed, choose any $\frac{d+1}{2}$-element subset $I \subset \{0, 1, \dots, d\}$. If k is an index in I, then

$$\left(\frac{2\sqrt{d}}{d+1} \sum_{i \in I} \mathbf{x}_i\right) \cdot \mathbf{x}_k = \frac{2\sqrt{d}}{d+1}\left(1 - \frac{1}{d} \cdot \frac{d-1}{2}\right) = \frac{1}{\sqrt{d}}. \tag{13}$$

If k is an index not in I, then

$$\left(\frac{2\sqrt{d}}{d+1} \sum_{i \in I} \mathbf{x}_i\right) \cdot \mathbf{x}_k = \frac{2\sqrt{d}}{d+1} \cdot \left(-\frac{1}{d} \cdot \frac{d+1}{2}\right) = -\frac{1}{\sqrt{d}}. \tag{14}$$

Then every point from $\overline{\omega}_{\kappa(d)}$ forms only two distinct values of the dot product with points from $\overline{\omega}_{2d+2}$ (which are $\pm\frac{1}{\sqrt{d}}$) and, hence, is contained in $\mathcal{D}_2(\overline{\omega}_{2d+2})$.

To show the opposite inclusion, choose any $\mathbf{y} \in \mathcal{D}_2(\overline{\omega}_{2d+2})$. Since points $\mathbf{x}_0, \mathbf{x}_1, \dots, \mathbf{x}_d$ form a 2-design, the set $\overline{\omega}_{2d+2}$ is a 3-design. Since $\mathcal{D}_2(\overline{\omega}_{2d+2}) \neq \emptyset$

(it contains $\overline{\omega}_{\kappa(d)}$), the set $\overline{\omega}_{2d+2}$ is 2-stiff. By Proposition 1, we then have $\mathbf{y} \cdot \mathbf{x}_i = \pm\frac{1}{\sqrt{d}}$, $i = 0, 1, \ldots, d$. Since $d + 1$ is even and

$$\sum_{i=0}^{d} \mathbf{y} \cdot \mathbf{x}_i = \mathbf{y} \cdot \sum_{i=0}^{d} \mathbf{x}_i = \mathbf{y} \cdot \mathbf{0} = 0,$$

exactly half of dot products $\mathbf{y} \cdot \mathbf{x}_i$ equal $\frac{1}{\sqrt{d}}$ and the other half equals $-\frac{1}{\sqrt{d}}$. Let $I \subset \{0, 1, \ldots, d\}$ be the $\frac{d+1}{2}$-element subset such that $\mathbf{y} \cdot \mathbf{x}_i = \frac{1}{\sqrt{d}}$, $i \in I$. Pick some index $m \in I$ and some index $k \in J := \{0, 1, \ldots, d\} \setminus I$. Then for every $i \in I \setminus \{m\}$, we have $\mathbf{y} \cdot (\mathbf{x}_i - \mathbf{x}_m) = 0$ and, for every $i \in J \setminus \{k\}$, we have $\mathbf{y} \cdot (\mathbf{x}_i - \mathbf{x}_k) = 0$. Let us show that the set

$$T := \{\mathbf{x}_i - \mathbf{x}_m\}_{i \in I \setminus \{m\}} \cup \{\mathbf{x}_i - \mathbf{x}_k\}_{i \in J \setminus \{k\}}$$

of $d - 1$ vectors in $\mathbb{R}^d$ (orthogonal to $\mathbf{y}$) is linearly independent. Let c_i, $i \in \{0, 1, \ldots, d\} \setminus \{m, k\}$ be any numbers such that

$$\sum_{i \in I \setminus \{m\}} c_i(\mathbf{x}_i - \mathbf{x}_m) + \sum_{i \in J \setminus \{k\}} c_i(\mathbf{x}_i - \mathbf{x}_k) = \mathbf{0}.$$

Then

$$\sum_{\substack{i=0 \\ i \neq m,k}}^{d} c_i(\mathbf{x}_i - \mathbf{x}_m) - \left(\sum_{i \in J \setminus \{k\}} c_i\right)(\mathbf{x}_k - \mathbf{x}_m) = \mathbf{0}. \tag{15}$$

Since $\mathbf{x}_0, \ldots, \mathbf{x}_d$ form a non-degenerate simplex in $\mathbb{R}^d$, the set of vectors $\mathbf{x}_i - \mathbf{x}_m$, where $i \in \{0, 1, \ldots, d\} \setminus \{m\}$ is linearly independent. Then (15) implies that $c_i = 0$, $i \neq m, k$; that is, T is linearly independent. Thus, $\mathbf{y}$ appears to be orthogonal to the $(d - 1)$-dimensional subspace spanned by T. Then $\mathbf{y}$ belongs to the intersection of the one-dimensional subspace $T^{\perp}$ with S^{d-1}; that is there exist only two (opposite) vectors that $\mathbf{y}$ can equal to.

Let $\mathbf{u} := \frac{2\sqrt{d}}{d+1} \sum_{j \in I} \mathbf{x}_j$. Then $\mathbf{u} \in S^{d-1} \cap \overline{\omega}_{\kappa(d)}$ and equation (13) implies that, for every $i \in I \setminus \{m\}$, we have $\mathbf{u} \cdot (\mathbf{x}_i - \mathbf{x}_m) = \frac{1}{\sqrt{d}} - \frac{1}{\sqrt{d}} = 0$. Equation (14) implies that, for every $i \in J \setminus \{k\}$, we have $\mathbf{u} \cdot (\mathbf{x}_i - \mathbf{x}_k) = -\frac{1}{\sqrt{d}} + \frac{1}{\sqrt{d}} = 0$. Thus, $\mathbf{u} \in T^{\perp} \cap S^{d-1}$. This implies that $\mathbf{y} = \mathbf{u}$ or $\mathbf{y} = -\mathbf{u}$. Since $\sum_{i=0}^{d} \mathbf{x}_i = \mathbf{0}$, we have $-\mathbf{u} = \frac{2\sqrt{d}}{d+1} \sum_{j \in J} \mathbf{x}_j \in \overline{\omega}_{\kappa(d)}$. Thus, $\mathbf{y} \in \overline{\omega}_{\kappa(d)}$. □

Lemma 9 *For every $d \geq 3$ odd, the configuration $\overline{\omega}_{\kappa(d)} \subset S^{d-1}$ is a spherical 3-design.*

Proof To prove the lemma, we will use Theorem 3. Since $\overline{\omega}_{\kappa(d)}$ is centrally symmetric and Gegenbauer polynomials $P_n^{(d-1)}$, $n = 1, 3$, are odd, we have

$$M_n^{(d-1)}(\overline{\omega}_{\kappa(d)}) := \sum_{\mathbf{x}\in\overline{\omega}_{\kappa(d)}} \sum_{\mathbf{y}\in\overline{\omega}_{\kappa(d)}} P_n^{(d-1)}(\mathbf{x}\cdot\mathbf{y}) = 0 \tag{16}$$

for $n = 1, 3$. It remains to show equality (16) for $n = 2$. Let $\mathbf{x}_0, \mathbf{x}_1, \ldots, \mathbf{x}_d$ be the vertices of the regular simplex in the definition of the configuration $\overline{\omega}_{\kappa(d)}$ in (12). Let $\mathbf{x}_I := \frac{2\sqrt{d}}{d+1}\sum_{i\in I}\mathbf{x}_i$, where I is any $\ell := \frac{d+1}{2}$-element subset of $\{0, 1, \ldots, d\}$. Let $0 \le k \le \ell$ be any integer and $J \subset \{0, 1, \ldots, d\}$ be any ℓ-element subset whose intersection with I has exactly k elements. There are $\binom{\ell}{k}\binom{\ell}{\ell-k} = \binom{\ell}{k}^2$ such sets J and, for every such J,

$$\begin{aligned}
\mathbf{x}_I\cdot\mathbf{x}_J &= \frac{4d}{(d+1)^2}\left(\sum_{i\in I\cap J}\mathbf{x}_i + \sum_{i\in I\setminus J}\mathbf{x}_i\right)\cdot\sum_{j\in J}\mathbf{x}_j \\
&= \frac{d}{\ell^2}\left(\sum_{i\in I\cap J}\left(1-\frac{\ell-1}{d}\right) + \sum_{i\in I\setminus J}\left(-\frac{\ell}{d}\right)\right) = \frac{2k}{\ell} - 1,
\end{aligned}$$

$k = 0, 1, \ldots, \ell$. Then

$$\begin{aligned}
M_2^{(d-1)}(\overline{\omega}_{\kappa(d)}) &= \kappa(d)\sum_{k=0}^{\ell}\binom{\ell}{k}^2 P_2^{(d-1)}\left(\frac{2k}{\ell}-1\right) \\
&= \frac{d\kappa(d)}{d-1}\sum_{k=0}^{\ell}\binom{\ell}{k}^2\left(\left(\frac{2k}{\ell}-1\right)^2 - \frac{1}{d}\right) \\
&= \frac{d\kappa(d)}{d-1}\left(\frac{4}{\ell^2}\sum_{k=0}^{\ell}k^2\binom{\ell}{k}^2 - \frac{4}{\ell}\sum_{k=0}^{\ell}k\binom{\ell}{k}^2 + \frac{d-1}{d}\sum_{k=0}^{\ell}\binom{\ell}{k}^2\right).
\end{aligned} \tag{17}$$

Using the standard formula $\sum\limits_{k=0}^{\ell}\binom{\ell}{k}^2 = \binom{2\ell}{\ell}$ (which can be seen by counting the total number of subsets J above), we obtain that

$$\sum_{k=0}^{\ell}k^2\binom{\ell}{k}^2 = \sum_{k=1}^{\ell}\ell^2\binom{\ell-1}{k-1}^2 = \ell^2\sum_{k=0}^{\ell-1}\binom{\ell-1}{k}^2 = \ell^2\binom{2\ell-2}{\ell-1}.$$

Finally, from (16) we obtain that

$$0 = \frac{M_1^{(d-1)}(\overline{\omega}_{\kappa(d)})}{\kappa(d)} = \sum_{k=0}^{\ell} \binom{\ell}{k}^2 P_1^{(d-1)}\left(\frac{2k}{\ell} - 1\right)$$

$$= \sum_{k=0}^{\ell} \binom{\ell}{k}^2 \left(\frac{2k}{\ell} - 1\right) = \frac{2}{\ell} \sum_{k=0}^{\ell} k \binom{\ell}{k}^2 - \binom{2\ell}{\ell}.$$

Then $\sum_{k=0}^{\ell} k\binom{\ell}{k}^2 = \frac{\ell}{2}\binom{2\ell}{\ell}$ and from (17) we obtain

$$M_2^{(d-1)}(\overline{\omega}_{\kappa(d)}) = \frac{d\kappa(d)}{d-1}\left(4\binom{2\ell-2}{\ell-1} - 2\binom{2\ell}{\ell} + \frac{2\ell-2}{2\ell-1}\binom{2\ell}{\ell}\right) = 0.$$

Thus, $\overline{\omega}_{\kappa(d)}$ is a spherical 3-design. □

Lemma 10 *For every $d \geq 3$ odd, $\mathcal{D}_2(\overline{\omega}_{\kappa(d)}) = \overline{\omega}_{2d+2}$.*

Proof Relations (13) and (14) imply that $\overline{\omega}_{2d+2} \subset \mathcal{D}_2(\overline{\omega}_{\kappa(d)})$. Let $\mathbf{x}$ be any element of $\mathcal{D}_2(\overline{\omega}_{\kappa(d)})$ and $\mathbf{x}_0, \mathbf{x}_1, \ldots, \mathbf{x}_d$ be the vertices of the regular simplex in the definition of the configuration $\overline{\omega}_{\kappa(d)}$ in (12). There exist numbers $\alpha_0, \alpha_1, \ldots, \alpha_d$ such that $\sum_{i=0}^{d} \alpha_i = 1$ and $\mathbf{x} = \sum_{i=0}^{d} \alpha_i \mathbf{x}_i$. Let I be any $(d+1)/2$-element subset of the set $\{0, 1, \ldots, d\}$. Then relations (13) and (14) imply that

$$\theta := \left(\frac{2\sqrt{d}}{d+1} \sum_{i \in I} \mathbf{x}_i\right) \cdot \mathbf{x} = \left(\frac{2\sqrt{d}}{d+1} \sum_{i \in I} \mathbf{x}_i\right) \cdot \left(\sum_{j \in I} \alpha_j \mathbf{x}_j + \sum_{j \notin I} \alpha_j \mathbf{x}_j\right)$$

$$= \frac{1}{\sqrt{d}}\left(\sum_{j \in I} \alpha_j - \sum_{j \notin I} \alpha_j\right).$$

The set $\overline{\omega}_{\kappa(d)}$ is a 3-design by Lemma 9. Since $\mathcal{D}_2(\overline{\omega}_{\kappa(d)}) \neq \emptyset$ (it contains $\overline{\omega}_{2d+2}$), the set $\overline{\omega}_{\kappa(d)}$ is 2-stiff. Then in view of Proposition 1, the above dot product is $\theta = \pm\frac{1}{\sqrt{d}}$. Then $\sum_{i \in I} \alpha_i - \sum_{i \notin I} \alpha_i = \pm 1$. Since $\sum_{i=0}^{d} \alpha_i = 1$, for any I, we have $\sum_{i \in I} \alpha_i = 1$ or $\sum_{i \in I} \alpha_i = 0$.

Since $\mathbf{x}$ and each $\mathbf{x}_i$ lie on the unit sphere, we have

$$|\mathbf{x}|^2 = \left(\sum_{i=0}^{d} \alpha_i \mathbf{x}_i\right)^2 = \sum_{i=0}^{d} \alpha_i^2 - \frac{1}{d} \sum_{i \neq j} \alpha_i \alpha_j = \left(1 + \frac{1}{d}\right) \sum_{i=0}^{d} \alpha_i^2 - \frac{1}{d}\left(\sum_{i=0}^{d} \alpha_i\right)^2$$

$$= \left(1 + \frac{1}{d}\right) \sum_{i=0}^{d} \alpha_i^2 - \frac{1}{d} = 1,$$

which implies that $\sum_{i=0}^{d} \alpha_i^2 = 1$.

Choose arbitrary indices $j \neq k$ and let $J \subset \{0, 1, \ldots, d\}$ be any $(d + 1)/2$-element subset such that $k \in J$ and $j \notin J$. Let $J_1 := (J \cup \{j\}) \setminus \{k\}$. Since each sum $\sum_{i\in J} \alpha_i$ and $\sum_{i\in J_1} \alpha_i$ equals 0 or 1, the difference

$$\alpha_j - \alpha_k = \sum_{i\in J_1} \alpha_i - \sum_{i\in J} \alpha_i$$

can only equal 1, -1, or 0. Let $\alpha := \max\limits_{i=\overline{0,d}} \alpha_i$. If it were that $\alpha_i = \alpha$ for all indices i, then α_i would equal $\frac{1}{d+1}$ for all i contradicting the equality of $\sum_{i\in J} \alpha_i$ to 0 or 1. Thus, for some index ℓ, we have $\alpha_\ell < \alpha$. For every index i, $\alpha - \alpha_i$ can only equal 0 or 1 (since α is some maximal α_j). Then $\alpha_i = \alpha$ or $\alpha_i = \alpha - 1$ for every index i. Let n be the number of indices i such that $\alpha_i = \alpha$. Since the sums of α_i's and of their squares both equal 1, we obtain the system of equations

$$\begin{cases} n\alpha + (d + 1 - n)(\alpha - 1) = 1, \\ n\alpha^2 + (d + 1 - n)(\alpha - 1)^2 = 1. \end{cases}$$

Multiplying the first equation by α and subtracting it from the second equation, we obtain that $(d + 1 - n)(1 - \alpha) = 1 - \alpha$. This yields two solutions: $\alpha = 1, n = 1$ and $\alpha = \frac{2}{d+1}, n = d$. In the first case, we have $\alpha_\ell = 1$ for some index ℓ and $\alpha_i = 0$ for $i \neq \ell$. Then $\mathbf{x} = \mathbf{x}_\ell \in \overline{\omega}_{2d+2}$. In the second case, we have $\alpha_\ell = \frac{1-d}{d+1}$ for some index ℓ and $\alpha_i = \frac{2}{d+1}, i \neq \ell$. Then $\mathbf{x} = \frac{2}{d+1} \sum\limits_{i:i\neq\ell} \mathbf{x}_i + \frac{1-d}{d+1}\mathbf{x}_\ell$. Using the equality $\sum_{i=0}^{d} \mathbf{x}_i = \mathbf{0}$, we obtain that $\mathbf{x} = -\mathbf{x}_\ell \in \overline{\omega}_{2d+2}$. Thus, $\mathcal{D}_2(\overline{\omega}_{\kappa(d)}) \subset \overline{\omega}_{2d+2}$ and the assertion of the lemma follows. □

We are ready to prove the main result of this section.

Theorem 4 *Let $d \geq 3$ be odd and $f : [0, 4] \to (-\infty, \infty]$ be a function continuous on $(0, 4]$ with $f(0) = \lim\limits_{t\to 0^+} f(t)$ and a convex derivative f'' on $(0, 4)$. Then*

(i) the potential $p_f(\cdot, \overline{\omega}_{2d+2})$ of the symmetrized regular d-simplex $\overline{\omega}_{2d+2}$ attains its absolute minimum over S^{d-1} at every point of the set $\overline{\omega}_{\kappa(d)}$;

(ii) the potential $p_f(\cdot, \overline{\omega}_{\kappa(d)})$ of the configuration $\overline{\omega}_{\kappa(d)}$ attains its absolute minimum over S^{d-1} at every point of the set $\overline{\omega}_{2d+2}$.

If the convexity of f'' is strict on $(0, 4)$, then the potential $p_f(\cdot, \overline{\omega}_{2d+2})$ has no other points of absolute minimum on S^{d-1} in (i) and the potential $p_f(\cdot, \overline{\omega}_{\kappa(d)})$ has no other points of absolute minimum on S^{d-1} in (ii).

Proof of Theorem 4 The set $\overline{\omega}_{2d+2}$ is a 3-design as a symmetrization about the origin of a 2-design. In view of Lemma 8, we have $\mathcal{D}_2(\overline{\omega}_{2d+2}) = \overline{\omega}_{\kappa(d)} \neq \emptyset$; that is, $\overline{\omega}_{2d+2}$ is 2-stiff. Theorem 1 now implies that the potential $p_f(\,\cdot\,, \overline{\omega}_{2d+2})$ attains its absolute minimum on S^{d-1} at every point of $\overline{\omega}_{\kappa(d)}$.

Lemma 9 implies that $\overline{\omega}_{\kappa(d)}$ is a spherical 3-design, while Lemma 10 implies that $\mathcal{D}_2(\overline{\omega}_{\kappa(d)}) = \overline{\omega}_{2d+2} \neq \emptyset$; that is, $\overline{\omega}_{\kappa(d)}$ is 2-stiff. Theorem 1 now implies that the potential $p_f(\,\cdot\,, \overline{\omega}_{\kappa(d)})$ attains its absolute minimum on S^{d-1} at every point of $\overline{\omega}_{2d+2}$.

If the convexity of f'' is strict on (0, 4), then by Theorem 1, the set $\mathcal{D}_2(\overline{\omega}_{2d+2}) = \overline{\omega}_{\kappa(d)}$ contains all points of absolute minimum of $p_f(\cdot, \overline{\omega}_{2d+2})$ on S^{d-1}, while the set $\mathcal{D}_2(\overline{\omega}_{\kappa(d)}) = \overline{\omega}_{2d+2}$ contains all points of absolute minimum of $p_f(\,\cdot\,, \overline{\omega}_{\kappa(d)})$ on S^{d-1}. □

10 A Pair of Stiff Configurations on S^5

In this section, we consider the symmetrized Schläfli configuration $\overline{\omega}_{54}$ and the (normalized) root system of the E_6 root lattice, denoted by $\overline{\omega}_{72}$. Recall that the Schläfli configuration, denoted by $\overline{\omega}_{27}$, consists of the following $N = 27$ points (see, e.g., [13]): the point (0, 0, 0, 0, 0, 1), ten points whose sixth coordinate is $-1/2$, one of the first five coordinates is $\pm\sqrt{3}/2$, and the remaining four coordinates equal 0, and sixteen points whose sixth coordinate is 1/4 and each of the remaining five coordinates is $\pm\sqrt{3}/4$ with an even number of minus signs. The set $\overline{\omega}_{27}$ is a spherical 4-design. The symmetrized Schläfli configuration is then defined as $\overline{\omega}_{54} := \overline{\omega}_{27} \cup (-\overline{\omega}_{27})$, which is a spherical 5-design.

The configuration $\overline{\omega}_{72}$ consists of 40 vectors on S^5 with two of the first five coordinates equal $\frac{1}{\sqrt{2}}$ or $-\frac{1}{\sqrt{2}}$ and the remaining four coordinates equal zero and 32 vectors of the form $\left(\pm\frac{1}{2\sqrt{2}}, \pm\frac{1}{2\sqrt{2}}, \pm\frac{1}{2\sqrt{2}}, \pm\frac{1}{2\sqrt{2}}, \pm\frac{1}{2\sqrt{2}}, \pm\frac{\sqrt{3}}{2\sqrt{2}}\right)$ with an odd number of minus signs. This is a spherical 5-design. To show that both configurations are 3-stiff, we use the following two auxiliary statements.

Lemma 11 $\mathcal{D}_3(\overline{\omega}_{54}) = \overline{\omega}_{72}$.

Proof One can verify directly that $\overline{\omega}_{72} \subset \mathcal{D}_3(\overline{\omega}_{54})$. The set $\overline{\omega}_{54}$ is a 5-design; that is, $\overline{\omega}_{54}$ is 3-stiff. Let now $\mathbf{x} = (x_1, \ldots, x_6)$ be any vector in $\mathcal{D}_3(\overline{\omega}_{54})$. By Proposition 1, we have $D(\mathbf{x}, \overline{\omega}_{54}) = \left\{\pm\sqrt{\frac{3}{8}}, 0\right\}$. If $\mathbf{u} = (0, 0, 0, 0, 0, 1) \in \overline{\omega}_{54}$, then $x_6 = \mathbf{x} \cdot \mathbf{u} = \pm\sqrt{\frac{3}{8}}$ or 0. For every $1 \le i \le 5$, choose the vector $\mathbf{v}_i \in \overline{\omega}_{54}$ whose i-th coordinate equals $\frac{\sqrt{3}}{2}$ and the sixth coordinate equals $-\frac{1}{2}$ (all other coordinates are zero). Then

$$\mathbf{x} \cdot \mathbf{v}_i = \frac{\sqrt{3}}{2} x_i - \frac{x_6}{2} = \pm\sqrt{\frac{3}{8}} \ \text{ or } 0. \tag{18}$$

If $x_6 = 0$, then $x_i = \pm\frac{1}{\sqrt{2}}$ or 0 for $i = 1, \ldots, 5$. Since $|\mathbf{x}| = 1$, exactly two coordinates out of first five equal $\pm\frac{1}{\sqrt{2}}$ and four remaining coordinates are zero; that is, $\mathbf{x} \in \overline{\omega}_{72}$.

Assume now that $x_6 = \pm\sqrt{\frac{3}{8}}$. Then, for each $1 \leq i \leq 5$, from (18), we have $\frac{\sqrt{3}}{2}x_i = \pm\frac{3}{2}\sqrt{\frac{3}{8}}$ or $\pm\frac{1}{2}\sqrt{\frac{3}{8}}$. Then $x_i = \pm\frac{1}{\sqrt{8}}$, $i = 1, \ldots, 5$ (the values $x_i = \pm\frac{3}{\sqrt{8}}$ are impossible, since $|x_i| \leq 1$). Therefore, $\mathbf{x}=\left(\pm\frac{1}{\sqrt{8}}, \pm\frac{1}{\sqrt{8}}, \pm\frac{1}{\sqrt{8}}, \pm\frac{1}{\sqrt{8}}, \pm\frac{1}{\sqrt{8}}, \pm\sqrt{\frac{3}{8}}\right)$. Let $\mathbf{w} = \left(\pm\frac{\sqrt{3}}{4}, \pm\frac{\sqrt{3}}{4}, \pm\frac{\sqrt{3}}{4}, \pm\frac{\sqrt{3}}{4}, \pm\frac{\sqrt{3}}{4}, \frac{1}{4}\right)$ with an even number of minus signs. Assume to the contrary that $\mathbf{x}$ has an even number of negative coordinates. Then, for each $i = 1, \ldots, 6$, the product of the i-th coordinates of vectors $\mathbf{x}$ and $\mathbf{w}$ will be $\pm\frac{\sqrt{3}}{4\sqrt{8}}$ and the number of negative products will be even. This implies that possible values of dot product $\mathbf{x} \cdot \mathbf{w}$ will be $\pm\frac{3\sqrt{3}}{2\sqrt{8}}$ or $\pm\frac{\sqrt{3}}{2\sqrt{8}}$, which are not in $D(\mathbf{x}, \overline{\omega}_{54})$. Then $\mathbf{x}$ must have an odd number of negative coordinates, which makes it an element of $\overline{\omega}_{72}$. □

Lemma 12 $\mathcal{D}_3(\overline{\omega}_{72}) = \overline{\omega}_{54}$.

Proof One can verify directly that $\overline{\omega}_{54} \subset \mathcal{D}_3(\overline{\omega}_{72})$ and that $\overline{\omega}_{72}$ is a 5-design. Then $\overline{\omega}_{72}$ is 3-stiff. Let $\mathbf{x} = (x_1, \ldots, x_6)$ be any point in $\mathcal{D}_3(\overline{\omega}_{72})$. Assume to the contrary that the set $S := \{|x_i| : i = 1, \ldots, 5\}$ has at least two distinct non-zero elements. Then we have $0 < |x_k| < |x_\ell|$ for some indices $1 \leq k \neq \ell \leq 5$. Let $\mathbf{u} \in \overline{\omega}_{72}$ be one of the four vectors with the k-th and ℓ-th coordinates being $\pm\frac{1}{\sqrt{2}}$ and all other coordinates being zero. Then dot product $\mathbf{x} \cdot \mathbf{u}$ will have four distinct values $\frac{\pm|x_k|\pm|x_\ell|}{\sqrt{2}}$ contradicting the assumption that $\mathbf{x}$ is in $\mathcal{D}_3(\overline{\omega}_{72})$. Thus, $x_1 = \ldots = x_5 = 0$ or all non-zero elements of S are equal. If $x_1 = \ldots = x_5 = 0$, then $\mathbf{x} = (0, 0, 0, 0, 0, \pm 1) \in \overline{\omega}_{54}$.

Assume that all non-zero elements of S are equal and denote by a their common value. Denote also $b := |x_6|$. Let G be the set of 32 vectors

$$\left(\pm\frac{1}{2\sqrt{2}}, \pm\frac{1}{2\sqrt{2}}, \pm\frac{1}{2\sqrt{2}}, \pm\frac{1}{2\sqrt{2}}, \pm\frac{1}{2\sqrt{2}}, \pm\frac{\sqrt{3}}{2\sqrt{2}}\right)$$

with an odd number of minus signs.

Assume to the contrary that $b = 0$. Let $1 \leq k \leq 5$ be such that $|x_i| = a$ for exactly k indices i between 1 and 5 inclusive. Then $a = 1/\sqrt{k}$ and the remaining $6-k$ coordinates of $\mathbf{x}$ are zero. Since $\overline{\omega}_{72}$ is 3-stiff, by Proposition 1, all dot products of $\mathbf{x}$ with vectors from G must be in the set $\left\{\pm\frac{\sqrt{3}}{2\sqrt{2}}, 0\right\}$, in particular, there is only one positive dot product. There is a vector $\mathbf{u} \in G$ with components corresponding to non-zero x_i having the same sign as x_i. Then $\mathbf{x} \cdot \mathbf{u} = \frac{ka}{2\sqrt{2}} = \frac{\sqrt{k}}{2\sqrt{2}}$, which must equal $\frac{\sqrt{3}}{2\sqrt{2}}$. Consequently, $k = 3$. But then there is a vector $\mathbf{v} \in G$ (with all but one components corresponding to non-zero x_i having the same sign as x_i) such that $\mathbf{x} \cdot \mathbf{v} = \frac{a}{2\sqrt{2}} = \frac{1}{2\sqrt{6}}$, which is not a possible dot product. Thus, b must be positive.

If $|x_i| = a$ for exactly one index $1 \le i \le 5$, then $\mathbf{x}$, in particular, forms dot products 0 and $\pm\frac{a}{\sqrt{2}}$ with points from $\overline{\omega}_{72}$. Since $\overline{\omega}_{72}$ is 3-stiff, by Proposition 1, all these dot products must be in the set $\left\{\pm\frac{\sqrt{3}}{2\sqrt{2}}, 0\right\}$. This implies that $|x_i| = a = \frac{\sqrt{3}}{2}$ and, since $|\mathbf{x}| = 1$, we have $x_6 = \pm 1/2$. Since all other coordinates of $\mathbf{x}$ are zero in this case, we have $\mathbf{x} \in \overline{\omega}_{54}$.

Assume that $|x_i| = a$ for exactly k indices i between 1 and 5 inclusive, where $2 \le k \le 4$. Then $\mathbf{x}$ has at least dot products $0, \pm\frac{a}{\sqrt{2}}, \pm\sqrt{2}a$ with vectors from $\overline{\omega}_{72}$ contradicting the assumption that $\mathbf{x} \in \mathcal{D}_3(\overline{\omega}_{72})$.

Finally, assume that $|x_1| = \ldots = |x_5| = a$. Let $\mathbf{w}$ be any of the four vectors from $\overline{\omega}_{72}$ whose first and second coordinates equal $\pm\frac{1}{\sqrt{2}}$ and the remaining four coordinates equal 0. Then $\mathbf{x} \cdot \mathbf{w}$ has values $\pm\sqrt{2}a$ and 0. Then $\sqrt{2}a = \frac{\sqrt{3}}{2\sqrt{2}}$ or $a = \frac{\sqrt{3}}{4}$. Since $|\mathbf{x}| = 1$, we have $b = 1/4$.

Assume to the contrary that $x_i < 0$ for an odd number of indices $1 \le i \le 6$. Let H_1 be the set of all such indices. For any vector $\mathbf{u} = (u_1, \ldots, u_6) \in G$, let $H_2 := \{i : u_i < 0\}$ (by definition of the set G, $\#H_2$ is also odd). For every i, we have $|x_i u_i| = \frac{\sqrt{3}}{8\sqrt{2}}$. Then $x_i u_i = -\frac{\sqrt{3}}{8\sqrt{2}}$ for $\#H_1 + \#H_2 - 2(\#(H_1 \cap H_2))$ indices i, which is an even number. As $\mathbf{u}$ runs through the set G, dot product $\mathbf{x} \cdot \mathbf{u}$ can only assume a value from the set $\left\{\pm\frac{3\sqrt{3}}{4\sqrt{2}}, \pm\frac{\sqrt{3}}{4\sqrt{2}}\right\}$, none of which is in the set $\left\{\pm\frac{\sqrt{3}}{2\sqrt{2}}, 0\right\}$ contradicting Proposition 1. Therefore, coordinates of vector $\mathbf{x}$ must have an even number of minus signs, which puts vector $\mathbf{x}$ in $\overline{\omega}_{54}$. Thus, $\mathcal{D}_3(\overline{\omega}_{72}) \subset \overline{\omega}_{54}$. □

We are now ready to prove the main result of this section.

Theorem 5 *Let $f : [0, 4] \to (-\infty, \infty]$ be a function continuous on $(0, 4]$ with $f(0) = \lim_{t \to 0^+} f(t)$ and a convex derivative $f^{(4)}$ on $(0, 4)$. Then*

(i) the potential $p_f(\cdot, \overline{\omega}_{54})$ attains its absolute minimum over S^5 at every point of $\overline{\omega}_{72}$;
(ii) the potential $p_f(\cdot, \overline{\omega}_{72})$ attains its absolute minimum over S^5 at every point of $\overline{\omega}_{54}$.

If the convexity of $f^{(4)}$ is strict on $(0, 4)$, then the potential $p_f(\cdot, \overline{\omega}_{54})$ has no other points of absolute minimum in (i) and the potential $p_f(\cdot, \overline{\omega}_{72})$ has no other points of absolute minimum in (ii).

Proof of Theorem 5 The set $\overline{\omega}_{54}$ is a spherical 5-design. In view of Lemma 11, we have $\mathcal{D}_3(\overline{\omega}_{54}) = \overline{\omega}_{72}$; that is, $\overline{\omega}_{54}$ is 3-stiff. Theorem 1 now implies that the potential $p_f(\cdot, \overline{\omega}_{54})$ attains its absolute minimum on S^5 at every point of $\overline{\omega}_{72}$. The set $\overline{\omega}_{72}$ is a spherical 5-design. Lemma 12 implies that $\mathcal{D}_3(\overline{\omega}_{72}) = \overline{\omega}_{54}$; that is, $\overline{\omega}_{72}$ is 3-stiff. Theorem 1 now implies that the potential $p_f(\cdot, \overline{\omega}_{72})$ attains its absolute minimum on S^5 at every point of $\overline{\omega}_{54}$.

If the convexity of $f^{(4)}$ is strict on $(0, 4)$, then by Theorem 1, the set $\mathcal{D}_3(\overline{\omega}_{54}) = \overline{\omega}_{72}$ contains all points of absolute minimum of $p_f(\cdot, \overline{\omega}_{54})$ on S^5, while the set $\mathcal{D}_3(\overline{\omega}_{72}) = \overline{\omega}_{54}$ contains all points of absolute minimum of $p_f(\cdot, \overline{\omega}_{72})$ on S^5. □

11 A Pair of Stiff Configurations on S^6

On S^6, the kissing arrangement of $N = 56$ points, denoted by $\overline{\omega}_{56}$, was proved in [12] to be 3-stiff. We find that its dual is the set of $N = 126$ minimal non-zero vectors of E_7 root lattice, denoted by $\overline{\omega}_{126}$, which is also 3-stiff. The configuration $\overline{\omega}_{56}$ can be defined as the set of points on the intersection of the sphere $S^7 \subset \mathbb{R}^8$ with the seven-dimensional linear subspace $\mathcal{H}$ of $\mathbb{R}^8$ perpendicular to vector $(1, 1, 1, 1, 1, 1, 1, 1)$ that have two coordinates equal to $-\frac{3}{2\sqrt{6}}$ and six coordinates equal to $\frac{1}{2\sqrt{6}}$ or two coordinates equal to $\frac{3}{2\sqrt{6}}$ and six coordinates equal to $-\frac{1}{2\sqrt{6}}$. There are only three distinct values of the dot product between any two distinct vectors in $\overline{\omega}_{56}$, which are $\pm\frac{1}{3}$ and -1. It is a known 3-sharp configuration.

The configuration $\overline{\omega}_{126}$ can be described as the arrangement of 126 points on the intersection $S^7 \cap \mathcal{H}$ in $\mathbb{R}^8$ that have one coordinate equal to $\frac{1}{\sqrt{2}}$, one coordinate equal to $-\frac{1}{\sqrt{2}}$, and six zero coordinates or four coordinates equal to $\frac{1}{2\sqrt{2}}$ and four coordinates equal to $-\frac{1}{2\sqrt{2}}$. One can also think of $\overline{\omega}_{126}$ as the set of minimal non-zero vectors of E_8 root lattice whose coordinates sum to zero.

To prove the main result of this section, we use the following two auxiliary statements.

Lemma 13 $\mathcal{D}_3(\overline{\omega}_{56}) = \overline{\omega}_{126}$.

Proof It is not difficult to verify directly that $\overline{\omega}_{126} \subset \mathcal{D}_3(\overline{\omega}_{56})$. To show that $\mathcal{D}_3(\overline{\omega}_{56})$ is contained in $\overline{\omega}_{126}$, we will consider these configurations as subsets of the sphere

$$\widetilde{S}^6 := S^7 \cap \mathcal{H} \subset \mathbb{R}^8.$$

Let $\mathbf{x} = (x_1, \ldots, x_8) \in \widetilde{S}^6$ be arbitrary point in $\mathcal{D}_3(\overline{\omega}_{56})$. For every pair of indices $1 \leq i \neq j \leq 8$, let $\mathbf{u}$ be the vector in $\overline{\omega}_{56}$ with the i-th and the j-th coordinates equal to $-\frac{3}{2\sqrt{6}}$. Since $x_1 + \ldots + x_8 = 0$, we have

$$\mathbf{x} \cdot \mathbf{u} = \frac{1}{2\sqrt{6}} \left(-3x_i - 3x_j + \sum_{k:k\neq i,j} x_k \right) = -\frac{2}{\sqrt{6}}(x_i + x_j)$$

and $\mathbf{x} \cdot (-\mathbf{u}) = \frac{2}{\sqrt{6}}(x_i + x_j)$. Since $\overline{\omega}_{56}$ is a 5-design and $\mathcal{D}_3(\overline{\omega}_{56}) \neq \emptyset$, the set $\overline{\omega}_{56}$ is 3-stiff. Since $\mathbf{x} \in \mathcal{D}_3(\overline{\omega}_{56})$, by Proposition 1, we have

$$V := \{\pm(x_i + x_j) : 1 \leq i \neq j \leq 8\} \subset \left\{ \pm\frac{1}{\sqrt{2}}, 0 \right\}. \tag{19}$$

Assume that $\mathbf{x}$ has at least three distinct coordinates, say $x_k < x_\ell < x_m$. Then V contains the numbers $x_k + x_\ell < x_k + x_m < x_\ell + x_m$. This forces $x_k + x_\ell = -\frac{1}{\sqrt{2}}$,

$x_k + x_m = 0$, and $x_\ell + x_m = \frac{1}{\sqrt{2}}$. The only solution to this system is $x_k = -\frac{1}{\sqrt{2}}$, $x_\ell = 0$, and $x_m = \frac{1}{\sqrt{2}}$. If for some $i \neq k, \ell, m$, we had $x_i \neq 0$, then V would contain the fourth element: $x_i + x_m > \frac{1}{\sqrt{2}}$ if $x_i > 0$ or $x_i + x_k < -\frac{1}{\sqrt{2}}$ if $x_i < 0$. Thus, $x_k = -\frac{1}{\sqrt{2}}$, $x_m = \frac{1}{\sqrt{2}}$, and the remaining six coordinates are 0; i.e., $\mathbf{x} \in \overline{\omega}_{126}$.

Assume now that $\mathbf{x}$ has at most two distinct coordinates. If all coordinates of $\mathbf{x}$ were the same, then $\mathbf{x}$ would have to be a zero vector, since the sum of its coordinates is zero. Thus, $\mathbf{x}$ has exactly two distinct coordinates. Assume to the contrary that one of the two values of coordinates of $\mathbf{x}$ appears only once. Denote this value by c and let the other value equal b. Then $c = -7b \neq 0$ and V contains four distinct values: $\pm 6b, \pm 2b$. This contradicts (19).

Thus, each of the two values of coordinates of $\mathbf{x}$ appears more than once. Denote them so that $b < c$. Then V contains the numbers $2b < b + c < 2c$, which forces $2c = \frac{1}{\sqrt{2}}$ and $2b = -\frac{1}{\sqrt{2}}$; that is, all coordinates of $\mathbf{x}$ are $\pm\frac{1}{2\sqrt{2}}$. Since the sum of coordinates of $\mathbf{x}$ is zero, four of its coordinates equal $\frac{1}{2\sqrt{2}}$ and four equal $-\frac{1}{2\sqrt{2}}$; i.e., $\mathbf{x} \in \overline{\omega}_{126}$. □

Lemma 14 $\mathcal{D}_3(\overline{\omega}_{126}) = \overline{\omega}_{56}$.

Proof It is not difficult to verify directly that $\overline{\omega}_{56} \subset \mathcal{D}_3(\overline{\omega}_{126})$. To show that $\mathcal{D}_3(\overline{\omega}_{126})$ is contained in $\overline{\omega}_{56}$, we will consider these configurations as subsets of the sphere $\widetilde{S}^6$. Let $\mathbf{x} \in \widetilde{S}^6$ be an arbitrary point in $\mathcal{D}_3(\overline{\omega}_{126})$. For any $1 \leq i \neq j \leq 8$, let $\mathbf{u} \in \overline{\omega}_{126}$ be the vector whose i-th coordinate is $\frac{1}{\sqrt{2}}$, j-th coordinate equals $-\frac{1}{\sqrt{2}}$, and the remaining six coordinates equal zero. Then $\mathbf{x} \cdot \mathbf{u} = \frac{x_i - x_j}{\sqrt{2}}$.

Assume to the contrary that vector $\mathbf{x}$ has at least three distinct coordinates, which we denote by $x_k < x_\ell < x_m$. Then $\mathbf{x}\cdot\mathbf{u}$ has, in particular, four distinct values $\pm\frac{x_\ell - x_k}{\sqrt{2}}$ and $\pm\frac{x_m - x_k}{\sqrt{2}}$ contradicting the fact that $\mathbf{x} \in \mathcal{D}_3(\overline{\omega}_{126})$. Thus, there are at most two distinct numbers among coordinates of $\mathbf{x}$. If all coordinates of $\mathbf{x}$ were equal, since they sum to zero, $\mathbf{x}$ would have to be a zero vector contradicting the assumption that $\mathbf{x} \in \widetilde{S}^6$.

Thus, there are exactly two distinct numbers among coordinates of $\mathbf{x}$, which we denote by $b < c$. By choosing vector $\mathbf{u} \in \overline{\omega}_{126}$ with $\frac{1}{\sqrt{2}}$ and $-\frac{1}{\sqrt{2}}$ on appropriate positions and 0 on the remaining six positions, we get value $\mathbf{x} \cdot \mathbf{u} = \frac{c-b}{\sqrt{2}}$. Since $\mathbf{x} \in \mathcal{D}_3(\overline{\omega}_{126})$, by Proposition 1, $\mathbf{x} \cdot \mathbf{u} \in \{\pm\frac{1}{\sqrt{3}}, 0\}$. The only value the positive dot product $\frac{c-b}{\sqrt{2}}$ can have is $\frac{1}{\sqrt{3}}$; that is, $c = b + \sqrt{\frac{2}{3}}$. Denote by k the number of coordinates of $\mathbf{x}$ that equal c. Then $(8 - k)b + k\left(b + \sqrt{\frac{2}{3}}\right) = 0$ and $(8 - k)b^2 + k\left(b + \sqrt{\frac{2}{3}}\right)^2 = 1$. From the first equation, we have $b = -\frac{k}{8}\sqrt{\frac{2}{3}}$. Then the second equation implies that $k(8 - k) = 12$; that is, $k = 2$ or 6. When $k = 2$, we have $b = -\frac{1}{4}\sqrt{\frac{2}{3}} = -\frac{1}{2\sqrt{6}}$ and $c = \frac{3}{4}\sqrt{\frac{2}{3}} = \frac{3}{2\sqrt{6}}$. When $k = 6$, we have $b = -\frac{3}{4}\sqrt{\frac{2}{3}} = -\frac{3}{2\sqrt{6}}$ and $c = \frac{1}{4}\sqrt{\frac{2}{3}} = \frac{1}{2\sqrt{6}}$. In both cases, $\mathbf{x} \in \overline{\omega}_{56}$. □

We are now ready to prove the main result of this section.

Theorem 6 *Let* $f : [0, 4] \to (-\infty, \infty]$ *be a function continuous on* $(0, 4]$ *with* $f(0) = \lim\limits_{t\to 0^+} f(t)$ *and a convex derivative* $f^{(4)}$ *on* $(0, 4)$. *Then*

(i) the potential $p_f(\cdot, \overline{\omega}_{56})$ *attains its absolute minimum over* S^6 *at every point of* $\overline{\omega}_{126}$*;*
(ii) the potential $p_f(\cdot, \overline{\omega}_{126})$ *attains its absolute minimum over* S^6 *at every point of* $\overline{\omega}_{56}$.

If the convexity of $f^{(4)}$ *is strict on* $(0, 4)$, *then the potential* $p_f(\cdot, \overline{\omega}_{56})$ *has no other points of absolute minimum in (i) and the potential* $p_f(\cdot, \overline{\omega}_{126})$ *has no other points of absolute minimum in (ii).*

We remark that one universal minimum of $\overline{\omega}_{56}$ and the value of the absolute minimum of its potential on S^6 were found in [12].

Proof of Theorem 6 Since $\overline{\omega}_{56}$ is a spherical 5-design and, by Lemma 13, $\mathcal{D}_3(\overline{\omega}_{56}) = \overline{\omega}_{126} \neq \emptyset$, the configuration $\overline{\omega}_{56}$ is 3-stiff (see also [12]). Then we apply Theorem 1 to establish item (i). The set $\overline{\omega}_{126}$ is a spherical 5-design. Lemma 14 implies that $\mathcal{D}_3(\overline{\omega}_{126}) = \overline{\omega}_{56} \neq \emptyset$; that is, $\overline{\omega}_{126}$ is 3-stiff. Theorem 1 now implies that the potential $p_f(\cdot, \overline{\omega}_{126})$ attains its absolute minimum on S^6 at every point of $\overline{\omega}_{56}$.

If, in addition, $f^{(4)}$ is strictly convex on $(0, 4)$, Theorem 1 implies that the potential $p_f(\cdot, \overline{\omega}_{56})$ attains its absolute minimum over S^6 only at points of $\overline{\omega}_{126}$ and the set $\overline{\omega}_{56}$ contains all points of absolute minimum of $p_f(\cdot, \overline{\omega}_{126})$ on S^6. □

References

1. E. Bannai, R.M. Damerell, Tight spherical designs I, *J. Math. Soc. Japan* **31** (1979), no. 1, 199–207.
2. M. Bilogliadov, Equilibria of Riesz potentials generated by point charges at the roots of unity, *Comput. Methods Funct. Theory* **15** (2015), no. 4, 471–491.
3. S.V. Borodachov, Absolute minima of potentials of certain regular spherical configurations, *Journal of Approximation Theory* **294** (2023), 105930.
4. S.V. Borodachov, Odd strength spherical designs attaining the Fazekas–Levenshtein bound for covering and universal minima of potentials, *Aequationes Mathematicae*, **98** (2024), 509–533.
5. S.V. Borodachov, Min-max polarization for certain classes of sharp configurations on the sphere, *Constructive Approximation* **60** (2024), 237–252. https://doi.org/10.1007/s00365-023-09661-1.
6. S.V. Borodachov, The current state of the minimal dicrete polarization theory, *Workshop "Minimal Energy Problems with Riesz Potentials"*, American Institute of Mathematics, Zoom, May 3–7, 2021.

7. S.V. Borodachov, Min-max polarization for certain classes of sharp configurations on the sphere, *Workshop "Optimal Point Configurations on Manifolds"*, ESI, Vienna, January 17–21, 2022. https://www.youtube.com/watch?v=L-szPTFMsX8
8. S.V. Borodachov, Extreme values of potentials of spherical designs and the polarization problem, *XII Annual International Conference of the Georgian Mathematical Union*, Batumi State University, Georgia, August 29–September 3.
9. S.V. Borodachov, Polarization problem on a higher-dimensional sphere for a simplex, *Discrete and Computational Geometry* **67** (2022), 525–542.
10. S. Borodachov, D. Hardin, E. Saff, *Discrete Energy on Rectifiable Sets*. Springer, 2019.
11. P.G. Boyvalenkov, P.D. Dragnev, D.P. Hardin, E.B. Saff, M.M. Stoyanova, On polarization of spherical codes and designs, *Journal of Mathematical Analysis and Applications* **524** (2023), no. 1, 127065.
12. P.G. Boyvalenkov, P.D. Dragnev, D.P. Hardin, E.B. Saff, M.M. Stoyanova, Universal minima of discrete potentials for sharp spherical codes, to appear in *Rev. Mat. Iberoam.*, arXiv:2211.00092.
13. P. Boyvalenkov, Computing distance distributions of spherical designs, *Linear Algebra Appl.* **226/228** (1995), 277–286.
14. P. Boyvalenkov, D. Danev, P. Kazakov, Indexes of spherical codes, *Codes and association schemes* (Piscataway, NJ, 1999), 47–57. *DIMACS Ser. Discrete Math. Theoret. Comput. Sci.*, **56**, Amer. Math. Soc., Providence, RI, 2001.
15. H. Cohn, A. Kumar, Universally optimal distribution of points on spheres, *J. Amer. Math. Soc.* **20** (2007), no. 1, 99–148.
16. P.J. Davis, *Interpolation and Approximation*, Second edition, Dover Publications, New York, NY, 1975.
17. P. Delsarte, J.M. Goethals, J.J. Seidel, Spherical codes and designs, *Geometriae Dedicata*, **6** (1977), no. 3, 363–388.
18. J.F. Epperson, *An introduction to numerical methods and analysis*, Wiley, 2013.
19. G. Fazekas, V.I. Levenshtein, On upper bounds for code distance and covering radius of designs in polynomial metric spaces, *Journal of Combinatorial Theory*, Series A, **70** (1995), no. 2, 267–288.
20. G. Giorgadze, G. Khimshiashvili, Stable equilibria of three constrained unit charges, *Proc. I. Vekua Inst. Appl. Math.* **70** (2020), 25–31.
21. D. Hardin, A. Kendall, E. Saff, Polarization optimality of equally spaced points on the circle for discrete potentials. *Discrete Comput. Geom.* **50** (2013), no. 1, 236–243.
22. E. Isaacson, H. Keller, *Analysis of numerical methods*. Dover Books, 1994.
23. V.I.Levenshtein, On bounds for packings in n-dimensional Euclidean space, *Soviet Math. Dokladi*, **20** (1979), 417–421.
24. V.I. Levenshtein, Designs as maximum codes in polynomial metric spaces, *Acta Appl. Math.* **25** (1992), 1–82.
25. V.I. Levenshtein, Universal bounds for codes and designs, Chapter 6 in *Handbook of Coding Theory*, V. Pless and W.C. Huffman, Eds., Elsevier Science B.V., 1998.
26. O.R. Musin, The kissing number in four dimensions, *Annals of Mathematics*, **168** (2008), 1–32.
27. N. Nikolov, R. Rafailov, On the sum of powered distances to certain sets of points on the circle, *Pacific J. Math.* **253** (2011), no. 1, 157–168.
28. N. Nikolov, R. Rafailov, On extremums of sums of powered distances to a finite set of points. *Geom. Dedicata* **167** (2013), 69–89.
29. K. Stolarsky, The sum of the distances to certain pointsets on the unit circle, *Pacific J. Math.* **59** (1975), no. 1, 241–251.
30. K. Stolarsky, The sum of the distances to N points on a sphere, *Pacific J. Math.* **57** (1975), no. 2, 563–573.
31. G. Szegö, *Orthogonal polynomials*. Fourth edition. American Mathematical Society, Colloquium Publications, Vol. XXIII. American Mathematical Society, Providence, R.I., 1975.

Logarithmic and Riesz Energy on the Sphere: Better Bounds via Elementary Methods

Johann S. Brauchart

Dedicated to Edward B. Saff on the occasion of his 80th birthday.

1 Introduction

Let me begin with a personal note. Ed Saff and Arno Kuijlaars' paper [28] has been pivotal[1] for my study and later career and life. It was required reading for preparation of my diploma thesis [9] and an important paper for my doctoral thesis [10]. I met Ed first at one of the Bommerholz conferences and he took upon himself to be a post-doc supervisor, mentor, scientific advisor (for my Oberwolfach Fellowship), and I dare to say friend over the years. This paper is dedicated to Ed on the occasion of his 80th birthday.

A central research question discussed in [28] and revisited in this paper concerns bounds and asymptotics for the *minimal* logarithmic and Riesz s-energy

$$E_{\log}(\mathbf{x}_1, \dots, \mathbf{x}_N) := \sum_{j \neq k} \log \frac{1}{\|\mathbf{x}_j - \mathbf{x}_k\|},$$

$$E_s(\mathbf{x}_1, \dots, \mathbf{x}_N) := \sum_{j \neq k} \frac{1}{\|\mathbf{x}_j - \mathbf{x}_k\|^s}, \quad s > 0,$$

[1] As I recall, there were two topics on the announcement board near Peter Grabner's office. The other was about something in graph theory. I dithered and returned the next day. There was only one topic left .... Thus, I was set on my path.—I never looked back.

J. S. Brauchart (✉)
Institute of Analysis and Number Theory, Graz University of Technology, Kopernikusgasse, Austria
e-mail: j.brauchart@tugraz.at

D. Bilyk et al. (eds.), *Recent Advances in Approximation and Potential Theory*, Applied and Numerical Harmonic Analysis,
https://doi.org/10.1007/978-3-031-95551-8_6

of N points $\mathbf{x}_1, \dots, \mathbf{x}_N$ on the unit sphere $\mathbb{S}^d$ in the Euclidean space $\mathbb{R}^{d+1}$, $d \geq 2$, provided with the Euclidean distance $\| \cdot \|$. In case of negative Riesz parameter $-2 < s < 0$, we shall talk about maximal sum of (general) distances. The paper [15] of 2012 jointly with Doug Hardin and Ed provides a survey of such bounds and conjectures for asymptotic terms (see also the book [8] by Borodachov, Hardin, and Saff on the topic). One aim of this paper is to provide an alternative way of proving results relying on elementary arguments and simple asymptotic reasoning.

This paper is inspired by joint work with Bilyk [4]. Therein we use simple and elementary arguments to obtain surprisingly sharp lower bounds for the *spherical cap* $\mathbb{L}_2$*-discrepancy*

$$D_{\mathbb{L}_2}^{\mathrm{C}}(\mathbf{x}_1, \dots, \mathbf{x}_N) := \sqrt{\int_{-1}^{1} \int_{\mathbb{S}^d} \left| \frac{\left|\{k : \mathbf{x}_k \in C(\mathbf{x}, t)\}\right|}{N} - \sigma_d(C(\mathbf{x}, t)) \right|^2 \mathrm{d}\sigma_d(\mathbf{x}) \, \mathrm{d}t}.$$

This quantity measures the irregularity of N points $\mathbf{x}_1, \dots, \mathbf{x}_N$ on $\mathbb{S}^d$ with respect to spherical caps $C(\mathbf{x}, t) := \{\mathbf{z} \in \mathbb{S}^d : \mathbf{z} \cdot \mathbf{x} \geq t\}$ in the $\mathbb{L}_2$-sense. Here, σ_d denotes the normalized uniform surface area measure on $\mathbb{S}^d$. We want to explore how powerful those ideas are when adapted for use with logarithmic and Riesz potentials. We reprove (and even improve on) known results for logarithmic and Riesz-energy.

Outline of the Paper In Sect. 2, we summarize the approach in [4]. We give the Taylor expansion with respect to inner products of points of the spherical cap $\mathbb{L}_2$-discrepancy (10) and include Lemmas 1 and 2 about non-negativity of the terms and the asymptotic behavior of the Taylor series coefficients. This section contains frequently used quantities on the sphere like surface area ω_d given in (5) and concludes with a list of the Epstein zeta functions of the hexagonal ($d = 2$), checkerboard ($d = 4$), $\boldsymbol{E}_8$ ($d = 8$), and Leech lattice ($d = 24$).

In Sect. 3, we consider the logarithmic energy and adapt the method in [4] for use with the logarithmic kernel. We discuss the 2-dimensional case separately from the general case $d \geq 2$. We derive expansions of the logarithmic kernel which are useful in the asymptotic sense; see Lemmas 3 and 5. In particular, we reprove Steinerberger's bound [33] as the optimal bound when omitting certain non-negative quantities (Lemma 4). The section leading up to this result demonstrates how to apply the new approach. Partial summation applied to the full energy expansion gives a relation between the logarithmic and the Riesz 2-energy of minimal logarithmic energy points. An interesting result is an identity between the conjectured asymptotic constants for optimal asymptotics for these two energies that can be validated numerically to high-precision and is provably true (Theorem 1). Other connecting formulas are obtainable via a modified partial summation method and left for future work. In the general case $d \geq 2$, we improve our result in [12] with a much simpler method and thus avoid the use of the "δ-Trick" for kernel smoothing (Theorem 2). Table 2 compares the obtained bound with the conjectured one for the four cases $d = 2$, $d = 4$, $d = 8$, and $d = 24$. Figure 1 compares graphically the new improved bound with the bound in [12].

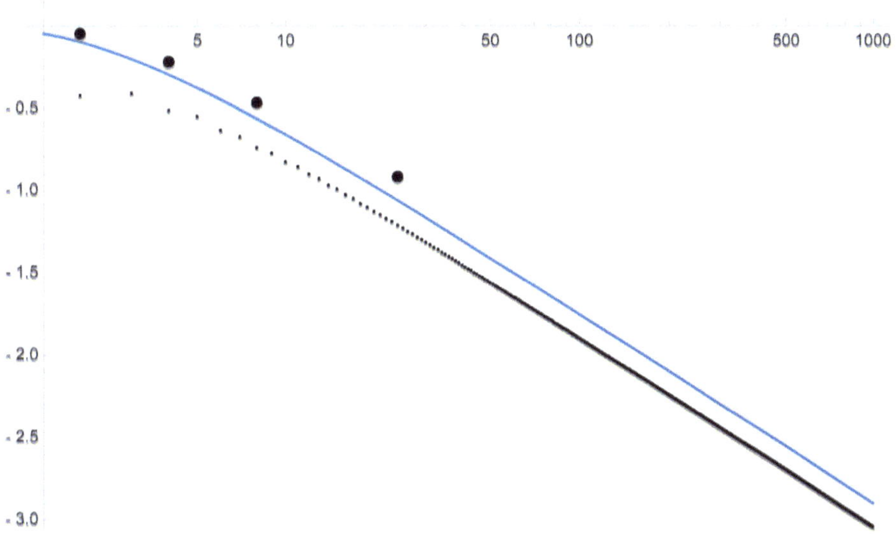

Fig. 1 The behaviour of $F(c^*)$ in (36) as a function of (*continuous*) d. Large Points indicate the values of $C_{\log,d}^{\text{conj}}$ for $d = 2, 4, 8, 16$. Small points mark the bound (32) evaluated for $d = 2, 3, 4, \ldots, 1000$

In Sect. 4, we discuss the Riesz s-energy for $d \geq 2$ and in the singular case $0 < s < d$. Lemma 6 gives the asymptotic expansion of the Riesz s-kernel. For technical reasons, we can only provide concrete results for the 2-sphere; see Theorem 3. A graphical comparison shows that our new bounds obtained by a simpler method are surprisingly close to the conjectured value (less than 3% relative error over the full range $0 < s < 2$); cf. Fig. 2. As a by-product, we observe that a lower bound for the hyper-singular case also works in the singular case.

In Sect. 5, we consider the Riesz d-energy for $d = 2, 4, 8, 24$. The asymptotic form of the Riesz kernel is given in Lemma 7. Our improved bounds are given in Theorems 4, 5, 6, and 7. Comparisons with the conjectured values and previously known bounds are summarized in remarks following the theorems.

2 The Method Applied to Spherical Cap $\mathbb{L}_2$-discrepancy

Beck [1] obtained his famous bound

$$D_{\mathbb{L}_2}^{C}(\mathbf{x}_1, \ldots, \mathbf{x}_N) \geq c_d N^{-\frac{1}{2}-\frac{1}{2d}} \tag{1}$$

by means of Fourier transform methods. Bilyk and Dai [5] give an alternate proof based on spherical harmonics and Gegenbauer expansions used in a generalised

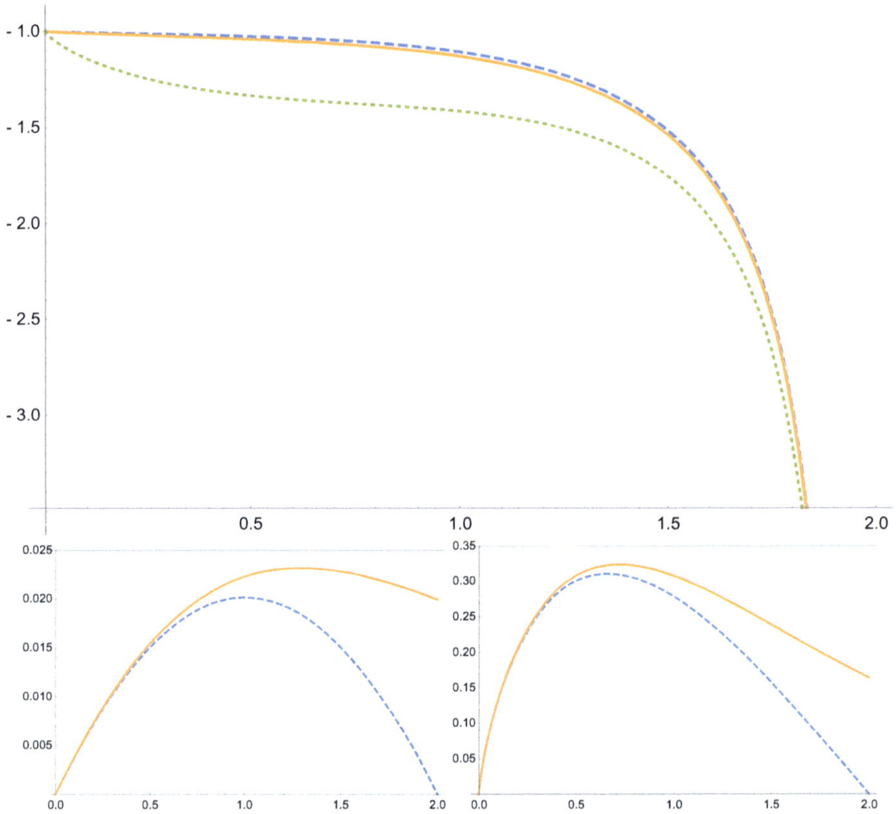

Fig. 2 Top: comparison of the bound $\tilde{c}_2(s)$ in (45) (*dotted*), $F(c^*)$ in (42) (*continuous*), and the conjectured value $\frac{C_{s,d}}{(\omega_d)^{\frac{s}{d}}}\big|_{d=2}$ in (45) (*dashed*) as a function of s. Bottom left and right: absolute (*dashed*) and relative error (*continuous*) in (46) and (47), respectively

Table 1 Comparison of conjectured c_d^{conj} and asymptotic c_d^{asymp} bound for the cases when (7) holds or is conjectured to hold

d	c_d^{conj}	c_d^{asymp}	$c_d^{\text{conj}} - c_d^{\text{asymp}}$	$\frac{c_d^{\text{conj}} - c_d^{\text{asymp}}}{c_d^{\text{conj}}}$
2	0.446...	0.433...	0.013...	3%
4	0.342...	0.328...	0.013...	4%
8	0.255...	0.243...	0.012...	5%
24	0.155...	0.145...	0.010...	7%

invariance principle. The paper [1] demonstrates that a Taylor expansion of the distance in terms of inner products substituted into Stolarsky's invariance principle and elementary asymptotic considerations suffice to give the bound in (1) with explicit constant c_d for every sphere dimension $d \geq 2$. These bounds are surprisingly tight in the asymptotic sense as $N \to \infty$: one digit of the conjectured asymptotic constant is recovered in the four cases $d = 2$, $d = 4$, $d = 8$, and $d = 24$; cf. Table 1.

Stolarsky's invariance principle (cf. [6, 14, 34])

$$\frac{1}{C_d}\left(D_{\mathbb{L}_2}^{\mathrm{C}}(\mathbf{x}_1,\dots,\mathbf{x}_N)\right)^2$$
$$= \int_{\mathbb{S}^d}\int_{\mathbb{S}^d}\|\mathbf{x}-\mathbf{y}\|\,\mathrm{d}\sigma_d(\mathbf{x})\,\mathrm{d}\sigma_d(\mathbf{y}) - \frac{1}{N^2}\sum_{j=1}^{N}\sum_{k=1}^{N}\|\mathbf{x}_j-\mathbf{x}_k\| \tag{2}$$

establishes that discrepancy is related to the second term of the asymptotics of the maximal sum of distances of N points as $N \to \infty$. Here, the Riesz parameter $s = -1$ is negative. Known bounds for the maximal sum of distance translate into bounds for the discrepancy (and vice versa[2]) giving that the order of N cannot be improved in Beck's bound. A fundamental conjecture for the Riesz s-energy is the basis of a conjecture for the leading term constant (cf. [13])

$$\lim_{N\to\infty} N^{\frac{1}{2}+\frac{1}{2d}} D_{\mathbb{L}_2}^{\mathrm{C}}(\mathbf{x}_1^*,\dots,\mathbf{x}_N^*) = c_d^{\mathrm{conj}} := \sqrt{-C_d\left[\frac{C_{s,d}}{(\mathcal{H}_d(\mathbb{S}^d))^{\frac{s}{d}}}\right]_{s=-1}}. \tag{3}$$

The conjectured constant depends on the number C_d appearing in Stolarsky's invariance principle which is given as the ratio of the normalized d-dimensional Hausdorff measure of the solid ball $\mathbb{B}^d$ in $\mathbb{R}^d$ and the surface area (denoted by ω_d) of the unit sphere $\mathbb{S}^d$ in $\mathbb{R}^{d+1}$:

$$C_2 := \frac{1}{4}, \qquad C_d := \frac{\mathcal{H}_d(\mathbb{B}^d)}{\mathcal{H}_d(\mathbb{S}^d)} = \frac{1}{d}\frac{\omega_{d-1}}{\omega_d} = \frac{1}{d}\frac{\Gamma(\frac{d+1}{2})}{\sqrt{\pi}\,\Gamma(\frac{d}{2})}, \qquad d \geq 2. \tag{4}$$

Incidentally, C_d is also the leading term constant of the asymptotics of the minimal Riesz s-energy of the boundary case $s = d$. The surface area ω_d of $\mathbb{S}^d$ is given by

$$\omega_d = \mathcal{H}_d(\mathbb{S}^d) = \frac{2\,\pi^{\frac{d+1}{2}}}{\Gamma(\frac{d+1}{2})}. \tag{5}$$

The ratio $\frac{C_{s,d}}{(\mathcal{H}_d(\mathbb{S}^d))^{\frac{s}{d}}}$ evaluated at $s = -1$ is understood to be the analytic continuation into the complex s-plane of the leading term constant of the minimal Riesz s-energy for $s > d$. The manifold independent constant $C_{s,d}$ is defined as (cf. [8])

$$C_{s,d} := \lim_{N\to\infty} \frac{\mathcal{E}_s([0,1]^d; N)}{N^{1+\frac{s}{d}}}, \qquad s > d, \tag{6}$$

[2] Indeed, Stolarsky showed bounds for the sum of distances using known bounds on the discrepancy.

where $\mathcal{E}_s([0,1]^d;N)$ denotes the minimal Riesz s-energy of N points on the unit cube $[0,1]^d$. The ground-breaking papers of Cohn and Kumar [16] and Cohn et al. [18] on universal optimality show that

$$\frac{C_{s,d}}{|\Lambda|^{\frac{s}{d}}} = \zeta_\Lambda\big(\frac{s}{2}\big) := \sum_{\mathbf{0}\neq\mathbf{x}\in\Lambda} \frac{1}{(\mathbf{x}\cdot\mathbf{x})^{\frac{s}{2}}}, \qquad s > d, \tag{7}$$

where $|\Lambda|$ is the co-volume and ζ_Λ the Epstein zeta function of the integer lattice $\mathbb{Z}$ (if $d=1$), the $\boldsymbol{E}_8$ lattice (if $d=8$), and the Leech lattice $\boldsymbol{\Lambda}_{24}$ (if $d=24$). It is conjectured that (7) is true for the hexagonal lattice $\boldsymbol{A}_2$ (if $d=2$) and the checkerboard lattice $\boldsymbol{D}_4$ (if $d=4$). Since the Epstein zeta function[3] $\zeta_\Lambda(s)$ of a full-rank lattice $\Lambda\subset\mathbb{R}^d$ has an analytic continuation to $\mathbb{C}$ except for a simple pole at $s=\frac{d}{2}$ with residue $\frac{d}{2}\frac{\mathcal{H}_d(\mathbb{B}^d)}{|\Lambda|} = \frac{\pi^{\frac{d}{2}}}{\Gamma(\frac{d}{2})}\frac{1}{|\Lambda|}$, the constant $C_{s,d}$ has a simple pole at $s=d$ with residue

$$\lim_{s\to d}(s-d)\,C_{s,d} = 2\,\frac{\pi^{\frac{d}{2}}}{\Gamma(\frac{d}{2})}$$

and is analytic elsewhere in the complex s-plane. Thus, when (7) holds, we get

$$c_d^{\mathrm{conj}} = \sqrt{C_d\,(\omega_d)^{\frac{1}{d}}\left(-|\Lambda|^{-\frac{1}{d}}\zeta_\Lambda\big(-\frac{1}{2}\big)\right)}, \tag{8}$$

where d is the dimension of the lattice $\Lambda\in\{\boldsymbol{A}_2,\boldsymbol{D}_4,\boldsymbol{E}_8,\boldsymbol{\Lambda}_{24}\}$. This conjectured bound can now be compared with the asymptotic bound (as $N\to\infty$)

$$c_d^{\mathrm{asymp}} := \sqrt{C_d\,(\omega_d)^{\frac{1}{d}}\,\frac{1}{\pi}\,\frac{1}{1+\frac{1}{d}}} \tag{9}$$

found in [4]. Explicit computations presented in [4] show that $c_d^{\mathrm{conj}} > c_d^{\mathrm{asymp}}$ in the four cases given in Table 1. The formula in (9) can be computed for every sphere dimension $d\geq 2$ whereas the conjectured constant (8) is unknown except in the four given cases.

The asymptotic bound (9) and other results are obtained in an elementary way. In general,

$$\|\mathbf{x}-\mathbf{y}\|^2 = 2\,(1-\mathbf{x}\cdot\mathbf{y}), \qquad \mathbf{x},\mathbf{y}\in\mathbb{S}^d.$$

---

[3] We use the definition of $\zeta_\Lambda(s)$ in [30] and can use the formulas for zeta function of several lattices directly. In [8, 15] the zeta function is defined as $\sum_{\mathbf{0}\neq\mathbf{x}\in\Lambda}\|\mathbf{x}\|^{-s}$.

Substituting the Taylor inner-product expansion of $\|\mathbf{x}-\mathbf{y}\|$ into (2), one gets

$$\begin{aligned}\left(D^{\mathrm{C}}_{\mathbb{L}_2}(\mathbf{x}_1,\ldots,\mathbf{x}_N)\right)^2 \\ &= \sqrt{2}\,C_d \sum_{m=1}^{\infty} \frac{-\left(-\frac{1}{2}\right)_m}{m!} \left(\frac{1}{N^2}\sum_{j=1}^{N}\sum_{k=1}^{N}\left(\mathbf{x}_j\cdot\mathbf{x}_k\right)^m\right. \\ &\left. - \int_{\mathbb{S}^d}\int_{\mathbb{S}^d} (\mathbf{x}\cdot\mathbf{y})^m \,\mathrm{d}\sigma_d(\mathbf{x})\,\mathrm{d}\sigma_d(\mathbf{y})\right).\end{aligned} \tag{10}$$

This series expansion sums up non-negative terms. We refer to the following results of [4].[4]

Lemma 1 (cf. [4]) *Let $d \geq 2$. For all $\mathbf{x}_1,\ldots,\mathbf{x}_N \in \mathbb{S}^d$ and for any $m \geq 0$,*

$$\begin{aligned} S_m(\mathbf{x}_1,\ldots,\mathbf{x}_N) {:=} &\frac{1}{N^2}\sum_{j=1}^{N}\sum_{k=1}^{N}\left(\mathbf{x}_j\cdot\mathbf{x}_k\right)^m \\ &- \int_{\mathbb{S}^d}\int_{\mathbb{S}^d} (\mathbf{x}\cdot\mathbf{y})^m \,\mathrm{d}\sigma_d(\mathbf{x})\,\mathrm{d}\sigma_d(\mathbf{y}) \geq 0.\end{aligned} \tag{11}$$

Lemma 2 (cf. [4]) *Let $d \geq 2$. If $m = 2r-1$ is odd, then*

$$a_m {:=} \int_{\mathbb{S}^d}\int_{\mathbb{S}^d} (\mathbf{x}\cdot\mathbf{y})^m \,\mathrm{d}\sigma_d(\mathbf{x})\,\mathrm{d}\sigma_d(\mathbf{y}) = 0.$$

If $m = 2r$ is even, then

$$\begin{aligned} a_m {:=} &\int_{\mathbb{S}^d}\int_{\mathbb{S}^d} (\mathbf{x}\cdot\mathbf{y})^m \,\mathrm{d}\sigma_d(\mathbf{x})\,\mathrm{d}\sigma_d(\mathbf{y}) \\ &= \frac{\left(\frac{1}{2}\right)_r}{\left(\frac{d+1}{2}\right)_r} \sim \frac{\Gamma(\frac{d+1}{2})}{\Gamma(\frac{1}{2})}\frac{1}{r^{\frac{d}{2}}} \qquad \text{as } r \to \infty \end{aligned} \tag{12}$$

and the asymptotic term is a strict upper bound for all $r \geq 1$.

The Pochhammer symbol (rising factorial) used in (12) is defined as

$$(a)_0{:=}1, \qquad (a)_{n+1} = (a)_n\,(n+a)\,, \quad n \geq 0.$$

---

[4] The first lemma is well-known in classical potential theory and it is proven in [4] using elementary methods.

Opposed to the usual approach of truncating the expansion after M terms, lower bounds are obtained by analysing and rearranging terms in

$$\left(D_{\mathbb{L}_2}^{\mathrm{C}}(\mathbf{x}_1,\ldots,\mathbf{x}_N)\right)^2 \geq \sqrt{2}\,C_d \sum_{m=M}^{\infty} \frac{-\left(-\frac{1}{2}\right)_m}{m!} S_m(\mathbf{x}_1,\ldots,\mathbf{x}_N),$$

where M is some suitably chosen index related to N by means of

$$\alpha \leq \frac{M}{N^{\frac{2}{d}}} \leq \alpha + 1, \qquad N \geq N_0, \tag{13}$$

for some sufficiently large α depending only on d and N_0.

We conclude this section by listing the Epstein zeta functions for the lattices in the four cases $d = 2, 4, 8$, and 24 using the definition of $\zeta_\Lambda(s)$ in (7); cf. [4, 30]. The hexagonal lattice $\boldsymbol{A}_2$ has co-volume $|\boldsymbol{A}_2| = \frac{\sqrt{3}}{2}$ and

$$\zeta_{\boldsymbol{A}_2}(s) = 6\zeta(s)\,3^{-s}\left(\zeta\left(s,\frac{1}{3}\right) - \zeta\left(s,\frac{2}{3}\right)\right),$$

given in terms of the Riemann zeta function and the analytically continued Hurwitz zeta function

$$\zeta(s,a) := \sum_{k=0}^{\infty} \frac{1}{(k+a)^s}, \qquad \operatorname{Re} s > 1, a \neq 0, -1, -2, \ldots. \tag{14}$$

All the other lattices are scaled to have co-volume 1. The Epstein zeta function for the checkerboard lattice $\boldsymbol{D}_4$ is

$$\zeta_{\boldsymbol{D}_4}(s) = 24\,2^{-\frac{s}{2}}\left(1 - 2^{1-s}\right)\zeta(s)\,\zeta(s-1) \tag{15}$$

and for both the $\boldsymbol{E}_8$ lattice and the Leech lattice $\boldsymbol{\Lambda}_{24}$ one has

$$\zeta_{\boldsymbol{E}_8}(s) = 240\,2^{-s}\zeta(s)\zeta(s-3) \tag{16}$$

and

$$\zeta_{\boldsymbol{\Lambda}_{24}}(s) = \frac{65520}{691}\,2^{-s}\left(\zeta(s)\zeta(s-11) - \mathrm{L}(s,\Delta)\right), \tag{17}$$

where

$$\mathrm{L}(s,\Delta) := \sum_{k=1}^{\infty} \frac{\tau(k)}{k^s}$$

denotes the Ramanujan tau Dirichlet L-function and $\tau(n)$ is defined by

$$\Delta(z):=q\prod_{n=1}^{\infty}(1-q^n)^{24}=\sum_{n=1}^{\infty}\tau(n)\,q^n, \qquad q:=e^{i2\pi z}.$$

3 The Method Applied to Logarithmic Energy

We consider the minimal logarithmic energy of N points on the unit sphere $\mathbb{S}^d$ in $\mathbb{R}^{d+1}$, $d \geq 2$. This quantity is given by

$$\mathcal{E}_{\log}(\mathbb{S}^d; N):=\min_{\mathbf{x}_1,\ldots,\mathbf{x}_N\in\mathbb{S}^d}\sum_{j\neq k}\log\frac{1}{\|\mathbf{x}_j-\mathbf{x}_k\|},$$

where $\|\cdot\|$ denotes the Euclidean distance in $\mathbb{R}^{d+1}$. The sum extends over all pairs of points excluding diagonal terms.

We are interested in the asymptotic expansion of $\mathcal{E}_{\log}(\mathbb{S}^d; N)$ as $N \to \infty$.

3.1 The 2-sphere

For the 2-sphere Bétermin and Sandier [3] prove that

$$\mathcal{E}_{\log}(\mathbb{S}^2; N)=\left(\frac{1}{2}-\log 2\right)N^2-\frac{1}{2}N\log N+C_{\log}N+o(N) \qquad \text{as } N\to\infty.$$

This result provides the most recent progress regarding asymptotic constants. It still leaves open if $o(N)$ can be replaced with $D_{\log}\log N + \mathcal{O}(1)$ as conjectured in Rakhmanov et al. [26, 27]. We recall here that Smale's 7th Problem [32] requires that the logarithmic energy of nearly optimal points may deviate at most by $c\log N$ from the optimal value.

The value of $C_{\log}$ is conjectured in [15] using an analytical continuation argument. Bétermin and Sandier show in their paper that this conjectured value is an upper bound for $C_{\log}$,

$$\begin{aligned} C_{\log} \leq C_{\mathrm{BHS}}&:=2\log 2+\frac{1}{2}\log\frac{2}{3}+3\log\frac{\sqrt{\pi}}{\Gamma(\frac{1}{3})} \\ &=-0.0556053049433925185 0\ldots. \end{aligned} \tag{18}$$

Equality in (18) seems to be closely related to optimality properties of the suitably scaled hexagonal/triangular lattice in dimension 2. Bétermin and Sandier prove that equality holds if and only if the Vortex/Abrikosov Conjecture in the Ginzburg-Landau model of superconductivity is true; cf. Sandier and Serfaty [29] and Serfaty [31]. This means that the hexagonal lattice minimises a certain renormalised energy. It is strongly believed that the hexagonal lattice is universally optimal in the sense of Cohn and Kumar [16],[5] and thus, due to Petrache and Serfaty [24], the Vortex Conjecture would be true and equality holds in (18). Using a simple explicit renormalization procedure involving the heat-kernel, Steinerberger [33] connects equality in (18) with a "sufficiently quantitative solution" of the following *Problem 1 in [33]: let* $0 < c < \infty$ *and consider for* $N \geq 2$

$$\min_{\mathbf{x}_1,\dots,\mathbf{x}_N \in \mathbb{S}^2} \sum_{j,k=1}^{N} \exp\left(- cN \left\| \mathbf{x}_j - \mathbf{x}_k \right\|^2 \right).$$

Prove that minimizing configurations behave locally like the hexagonal lattice (on most of the domain as $N \to \infty$*).*

The lower bound $C_{\log} \geq -0.223\dots$ due to Dubickas [21] is a slight improvement over the bound $C_{\log} \geq -0.225\dots$ given in Rakhmanov et al. [26, 27]; cf. the book [8]. Steinerberger [33] improves it to

$$C_{\log} \geq \log 2 - \frac{1}{2} - \frac{\gamma}{2} = -0.09546065189082112\dots,$$

where $\gamma := \lim_{n\to\infty} \left(\sum_{k=1}^{n} \frac{1}{k} - \log n \right)$ is the Euler-Mascheroni constant. The best lower bound seems to be due to Lauritsen [23]

$$C_{\log} = 2e_{\text{Jel}} + \frac{\log 4\pi}{2} \geq \log 2 - \frac{3}{4} = -0.05685281944005469\dots,$$

where e_{Jel} is the thermodynamic limit of the properly scaled jellium energy of N particles in a domain of measure N. See also Beltrán and Lizarte [2] for an alternate proof of the result of Lauritsen.

We adapt the method of [4] to reprove Steinerberger's lower bound. We use

$$\|\mathbf{x} - \mathbf{y}\|^2 = 2\,(1 - \mathbf{x} \cdot \mathbf{y})\,, \qquad \mathbf{x}, \mathbf{y} \in \mathbb{S}^d$$

to write the logarithmic potential $\log \frac{1}{\|\mathbf{x}-\mathbf{y}\|}$ as a power series of the inner product $\mathbf{x} \cdot \mathbf{y}$. Let M be a positive integer to be determined later. The series expansion

$$-\log(1 - t) = \sum_{m=1}^{\infty} \frac{1}{m} t^m$$

[5] Dimension 2 is the only remaining unresolved case of the Cohn-Kumar Conjecture. It is true for dimensions 8 and 24 by Cohn et al. [18] based on the seminal work of Viazovska [35] and Cohn et al. [17].

converges absolutely for $|t| < 1$, uniformly for $|t| \leq \rho < 1$ and at $t = -1$ by Leibniz criterion. Starting with an odd integer M, a remainder term is non-negative: if $t \geq 0$, then all terms are non-negative; if $t < 0$, then the sign of the alternating series is determined by the first term with even index of summation $M + 1$; i.e.,

$$\sum_{m=M+1}^{\infty} \frac{1}{m} t^m \geq 0, \qquad -1 \leq t < 1. \tag{19}$$

Given two different points $\mathbf{x}, \mathbf{y} \in \mathbb{S}^2$, we write

$$\begin{aligned} \log \frac{1}{\|\mathbf{x}-\mathbf{y}\|} &= -\frac{1}{2} \log(2\,(1-\mathbf{x}\cdot\mathbf{y})) = -\frac{1}{2}\log 2 - \frac{1}{2}\log(1-\mathbf{x}\cdot\mathbf{y}) \\ &= -\frac{1}{2}\log 2 + \sum_{m=1}^{M} b_m\,(\mathbf{x}\cdot\mathbf{y})^m \\ &\quad + \sum_{m=M+1}^{\infty} b_m\,(\mathbf{x}\cdot\mathbf{y})^m, \qquad b_m := \frac{1}{2}\frac{1}{m}, \quad m \geq 1. \end{aligned} \tag{20}$$

We rewrite the kernel as a balanced sum. One part does not depend on the points and determines the first few asymptotic terms as $N \to \infty$. The second part provides a non-zero contribution in the energy formula. We utilize the fact that $t \mapsto t^m, m \geq 1$, is positive definite and introduce the quantity

$$\begin{aligned} S_m(t) &:= t^m - \int_{\mathbb{S}^2}\int_{\mathbb{S}^2} (\mathbf{x}\cdot\mathbf{y})^m \,\mathrm{d}\sigma_2(\mathbf{x})\,\mathrm{d}\sigma_2(\mathbf{y}) \\ &= t^m - a_m, \qquad m \geq 1, -1 \leq t \leq 1. \end{aligned} \tag{21}$$

For $d = 2$, Lemma 2 gives $a_m = \frac{1+(-1)^m}{2}\frac{1}{m+1}$, $m \geq 0$. Thus

$$\log \frac{1}{\|\mathbf{x}-\mathbf{y}\|} = -\frac{1}{2}\log 2 + \sum_{m=1}^{M} a_m b_m + \sum_{m=1}^{M} b_m\, S_m(\mathbf{x}\cdot\mathbf{y}) + \sum_{m=M+1}^{\infty} b_m\,(\mathbf{x}\cdot\mathbf{y})^m.$$

Since $\frac{1}{m(m+1)} = \frac{1}{m} - \frac{1}{m+1}$, elementary computations give for odd integers M the relation

$$\sum_{m=1}^{M} a_m b_m = \frac{1}{2}\sum_{m=1}^{M} \frac{1+(-1)^m}{2}\frac{1}{m}\frac{1}{m+1} = \frac{1}{2} - \frac{1}{2}\sum_{m=1}^{M}\frac{(-1)^{m-1}}{m}.$$

The alternating harmonic number can be written either in terms of harmonic numbers or digamma functions $\psi(z) := \frac{\Gamma'(z)}{\Gamma(z)}$ by means of the identity $H_M := \sum_{m=1}^{M} \frac{1}{m} = \gamma + \psi(M+1)$; i.e., for odd integers $M \geq 1$,

$$\sum_{m=1}^{M} a_m b_m = \frac{1}{2} - \frac{1}{2} H_M + \frac{1}{2} H_{\frac{M-1}{2}} = \frac{1}{2} - \frac{1}{2}\psi(M+1) + \frac{1}{2}\psi\Big(\frac{M+1}{2}\Big).$$

The asymptotic expansion as $M \to \infty$ follows from the well-known asymptotic relation

$$H_M = \sum_{m=1}^{M} \frac{1}{m} = \gamma + \psi(M+1) = \log M + \gamma + \frac{1}{2}\frac{1}{M} - \frac{1}{12}\frac{1}{M^2} + \mathcal{O}\Big(\frac{1}{M^4}\Big). \tag{22}$$

We arrive at a representation of the logarithmic potential with a constant term that is the leading term coefficient in the asymptotic expansion of $\mathcal{E}_{\log}(\mathbb{S}^2; N)$ and terms that will be relevant for the second and third term of the minimal logarithmic energy asymptotics.

Lemma 3 *Let $\mathbf{x}, \mathbf{y} \in \mathbb{S}^2$ with $\mathbf{x} \neq \mathbf{y}$. Then for large odd integers M,*

$$\log \frac{1}{\|\mathbf{x} - \mathbf{y}\|} = -\frac{1}{2}\log 2 + \sum_{m=1}^{M} a_m b_m + \sum_{m=1}^{M} b_m\, S_m(\mathbf{x}\cdot\mathbf{y}) + \sum_{m=M+1}^{\infty} b_m\, (\mathbf{x}\cdot\mathbf{y})^m, \tag{23}$$

where $a_m = \frac{1+(-1)^m}{2}\frac{1}{m+1}$, $b_m = \frac{1}{2}\frac{1}{m}$, S_m is given in (21)*, and the remainder term is non-negative by* (19)*. Furthermore,*

$$\begin{aligned}-\frac{1}{2}\log 2 + \sum_{m=1}^{M} a_m b_m &= \frac{1}{2} - \log 2 + \frac{1}{2}\Big(\log 2 - \psi(M+1) + \psi\Big(\frac{M+1}{2}\Big)\Big) \\ &= \frac{1}{2} - \log 2 - \frac{1}{4}\frac{1}{M} + \frac{1}{8}\frac{1}{M^2} + \mathcal{O}\Big(\frac{1}{M^3}\Big) \text{ as} M \to \infty.\end{aligned} \tag{24}$$

Thus, for N pairwise different points $\mathbf{x}_1, \dots, \mathbf{x}_N \in \mathbb{S}^2$ and M an odd integer, we get

$$\begin{aligned}\frac{1}{N^2}\sum_{j\neq k}\log\frac{1}{\|\mathbf{x}_j - \mathbf{x}_k\|} &= -\frac{\log 2}{2}\frac{N-1}{N} + \sum_{m=1}^{M} b_m \frac{1}{N^2}\sum_{j\neq k}(\mathbf{x}_j\cdot\mathbf{x}_k)^m \\ &\quad + \frac{1}{N^2}\sum_{j\neq k}\sum_{m=M+1}^{\infty} b_m\,(\mathbf{x}_j\cdot\mathbf{x}_k)^m\end{aligned}$$

$$= -\frac{1}{2}\log 2 + \sum_{m=1}^{M} a_m b_m + \left(\frac{1}{2}\log 2 - \sum_{m=1}^{M} b_m\right)\frac{1}{N}$$
$$+ \sum_{m=1}^{M} b_m \frac{1}{N^2}\sum_{j,k} S_m(\mathbf{x}_j \cdot \mathbf{x}_k)$$
$$+ \frac{1}{N^2}\sum_{j\neq k}\sum_{m=M+1}^{\infty} b_m (\mathbf{x}_j \cdot \mathbf{x}_k)^m,$$

where by Definition (21) and Lemma 1,

$$\frac{1}{N^2}\sum_{j,k} S_m(\mathbf{x}_j \cdot \mathbf{x}_k) = S_m(\mathbf{x}_1, \ldots, \mathbf{x}_N).$$

By (24) and (22) applied to $\sum_{m=1}^{M} b_m = \frac{1}{2}\sum_{m=1}^{M}\frac{1}{m} = \frac{1}{2}H_M$, we arrive at

$$\frac{1}{N^2}\sum_{j\neq k}\log\frac{1}{\|\mathbf{x}_j - \mathbf{x}_k\|} = \frac{1}{2} - \log 2 - \frac{1}{2}\frac{\log M}{N} + \left(\frac{\log 2}{2} - \frac{\gamma}{2}\right)\frac{1}{N} - \frac{1}{4}\frac{1}{M}$$
$$- \frac{1}{4}\frac{1}{MN} + \frac{1}{8}\frac{1}{M^2} + O\Big(\frac{1}{M^2 N}\Big) + O\Big(\frac{1}{M^3}\Big)$$
$$+ \sum_{m=1}^{M} b_m S_m(\mathbf{x}_1, \ldots, \mathbf{x}_N) + \frac{1}{N^2}\sum_{j\neq k}\sum_{m=M+1}^{\infty} b_m (\mathbf{x}_j \cdot \mathbf{x}_k)^m \tag{25}$$

as $M \to \infty$ uniformly in $N \geq 2$. The second to the last term is non-negative by Lemma 1. The last term is non-negative if M is an odd integer by (19).

We consider now minimal logarithmic energy points and assume that $M = cN + \delta$ with $c \geq c_0 > 0$ (say $c_0 = 10^{-2}$) and $0 \leq \delta < 2$ such that M is an odd integer. Then we have

$$\mathcal{E}_{\log}(\mathbb{S}^2; N) \geq \left(\frac{1}{2} - \log 2\right)N^2 - \frac{1}{2}N\log(cN+\delta) + \left(\frac{\log 2}{2} - \frac{\gamma}{2}\right)N - \frac{1}{4}\frac{N}{c + \frac{\delta}{N}}$$
$$+ \left(-\frac{1}{4}\frac{N}{M} + \frac{1}{8}\frac{N^2}{M^2} + O\Big(\frac{N}{M^2}\Big) + O\Big(\frac{N^2}{M^3}\Big)\right)\Bigg|_{M=cN+\delta}.$$

Since

$$N\log(cN+\delta) = N\log N + (\log c)N + O_{c_0}(1), \qquad \frac{N}{c+\frac{\delta}{N}} = \frac{1}{c}N + O_{c_0}(1),$$

and

$$-\frac{1}{4}\frac{N}{M}+\frac{1}{8}\frac{N^2}{M^2}+\mathcal{O}\Big(\frac{N}{M^2}\Big)+\mathcal{O}\Big(\frac{N^2}{M^3}\Big)\Big|_{M=cN+\delta}=\frac{1-2c}{8c^2}+\mathcal{O}_{c_0}\Big(\frac{1}{N}\Big),$$

we arrive at the following result.

Lemma 4 *Let $d=2$. Then*

$$\mathcal{E}_{\log}(\mathbb{S}^2;N)\geq\left(\frac{1}{2}-\log 2\right)N^2-\frac{1}{2}N\log N+F(c)\,N+\mathcal{O}(1), \tag{26}$$

where the function

$$F(c):=\frac{\log 2}{2}-\frac{\gamma}{2}-\frac{\log c}{2}-\frac{1}{4c}$$

has a unique maximum at $c=c^*:=\frac{1}{2}$ in $(0,\infty)$ with value

$$F(c^*)=\log 2-\frac{1}{2}-\frac{\gamma}{2}=-0.09546065189082112\dots. \tag{27}$$

Thus, we recover Steinerberger's lower bound for $C_{\log}$ as an optimal bound in our elementary approach when ignoring the non-negative contribution of

$$\sum_{m=1}^{M}b_m\,S_m(\mathbf{x}_1^*,\dots,\mathbf{x}_N^*)+\frac{1}{N^2}\sum_{j\neq k}\sum_{m=M+1}^{\infty}b_m\,(\mathbf{x}_j^*\cdot\mathbf{x}_k^*)^m \tag{28}$$

evaluated for minimal logarithmic energy configurations $\mathbf{x}_1^*,\dots,\mathbf{x}_N^*\in\mathbb{S}^2$.

3.2 Partial Summation Attempt

For further improvements we have to find non-trivial lower bounds for (28). Summation by parts provides a connection to the minimal Riesz 2-energy. We start with

$$\sum_{m=1}^{M}b_m\,S_m(\mathbf{x}_1^*,\dots,\mathbf{x}_N^*)=\frac{1}{2M}\sum_{m=1}^{M}S_m(\mathbf{x}_1^*,\dots,\mathbf{x}_N^*)$$
$$+\frac{1}{2}\sum_{m=1}^{M-1}\frac{1}{\ell(\ell+1)}\sum_{m=1}^{\ell}S_m(\mathbf{x}_1^*,\dots,\mathbf{x}_N^*).$$

We bound the non-negative last sum from below by 0. Expanding $S_m(\mathbf{x}_1^*, \dots, \mathbf{x}_N^*)$ and splitting off diagonal terms, we get for odd M

$$\begin{aligned}
&\frac{1}{2M} \sum_{m=1}^{M} S_m(\mathbf{x}_1^*, \dots, \mathbf{x}_N^*) \\
&= \frac{1}{2M} \sum_{m=1}^{M} \left(\frac{1}{N} + \frac{1}{N^2} \sum_{j \neq k} (\mathbf{x}_j^* \cdot \mathbf{x}_k^*)^m - \frac{1 + (-1)^m}{2} \frac{1}{m+1} \right) \\
&= \frac{1}{2M} \left(\frac{M}{N} - \frac{N-1}{N} + \frac{1}{N^2} \sum_{j \neq k} \left(\frac{1}{1 - \mathbf{x}_j^* \cdot \mathbf{x}_k^*} - \frac{(\mathbf{x}_j^* \cdot \mathbf{x}_k^*)^{M+1}}{1 - \mathbf{x}_j^* \cdot \mathbf{x}_k^*} \right) - \sum_{r=1}^{\frac{M-1}{2}} \frac{1}{2r+1} \right).
\end{aligned}$$

Since

$$\sum_{j \neq k} \frac{1}{1 - \mathbf{x}_j^* \cdot \mathbf{x}_k^*} = 2E_2(\mathbf{x}_1^*, \dots, \mathbf{x}_N^*) \geq 2\mathcal{E}_2(\mathbb{S}^2; N)$$

and

$$\sum_{r=1}^{\frac{M-1}{2}} \frac{1}{2r+1} = \frac{1}{2} \psi\left(1 + \frac{M}{2}\right) - 1 + \frac{\gamma}{2} + \log 2 = \frac{1}{2} \log(2M) - 1 + \frac{\gamma}{2} + O\left(\frac{1}{2M}\right),$$

we obtain

$$\begin{aligned}
&\frac{1}{2M} \sum_{m=1}^{M} S_m(\mathbf{x}_1^*, \dots, \mathbf{x}_N^*) \\
&= \frac{1}{2M} \left(\frac{1}{2} - \frac{\gamma}{2} + \frac{2}{N^2} \left(E_2(\mathbf{x}_1^*, \dots, \mathbf{x}_N^*) - \frac{1}{4} N^2 \log N \right) \right. \\
&\quad + \frac{1}{N} - \frac{1}{N^2} \sum_{j \neq k} \frac{(\mathbf{x}_j^* \cdot \mathbf{x}_k^*)^{M+1}}{1 - \mathbf{x}_j^* \cdot \mathbf{x}_k^*} - \frac{1}{2} \log \frac{2M}{N} \\
&\quad \left. + \frac{1}{2} \frac{2M - N}{N} + O\left(\frac{1}{2M}\right) \right).
\end{aligned}$$

Using that $M = \frac{1}{2}N + \delta$, $0 \leq \delta < 2$, such that M is an odd integer, we arrive at

$$
\begin{aligned}
&\frac{N^2}{2M}\sum_{m=1}^{M} S_m(\mathbf{x}_1^*,\dots,\mathbf{x}_N^*)\\
&\quad\geq \frac{N}{N+2\delta}\Big(\frac{1}{2}-\frac{\gamma}{2}+\frac{2}{N^2}\Big(\mathcal{E}_2(\mathbb{S}^2;N)-\frac{1}{4}N^2\log N\Big)\\
&\qquad -\frac{1}{N^2}\sum_{j\neq k}\frac{(\mathbf{x}_j^*\cdot\mathbf{x}_k^*)^{M+1}}{1-\mathbf{x}_j^*\cdot\mathbf{x}_k^*}+o\Big(\frac{1}{N}\Big)\Big)N\\
&\quad=\Big(\frac{1}{2}-\frac{\gamma}{2}+\frac{2}{N^2}\Big(\mathcal{E}_2(\mathbb{S}^2;N)-\frac{1}{4}N^2\log N\Big)\Big)N-\frac{1}{2M}\sum_{j\neq k}\frac{(\mathbf{x}_j^*\cdot\mathbf{x}_k^*)^{M+1}}{1-\mathbf{x}_j^*\cdot\mathbf{x}_k^*}\\
&\qquad+O(1)+O\Big(\frac{2}{N^2}\Big(\mathcal{E}_2(\mathbb{S}^2;N)-\frac{1}{4}N^2\log N\Big)\Big).
\end{aligned}
$$

Thus, the right-hand side is a lower bound for the first part in (28). Summation by parts (for infinite series remainder) shows that

$$
\begin{aligned}
&-\frac{1}{2M}\sum_{j\neq k}\frac{(\mathbf{x}_j^*\cdot\mathbf{x}_k^*)^{M+1}}{1-\mathbf{x}_j^*\cdot\mathbf{x}_k^*}+\frac{1}{2}\sum_{m=M+1}^{\infty}\frac{1}{m}\sum_{j\neq k}(\mathbf{x}_j^*\cdot\mathbf{x}_k^*)^m\\
&\qquad\qquad=-\sum_{m=M+1}^{\infty}\frac{1}{m(m-1)}\sum_{j\neq k}\frac{(\mathbf{x}_j^*\cdot\mathbf{x}_k^*)^m}{2(1-\mathbf{x}_j^*\cdot\mathbf{x}_k^*)}.
\end{aligned}
$$

The right-hand side improves the left-most term by adding a non-negative quantity; cf. (19). It is known that minimal logarithmic energy points are well-separated; e.g., $\|\mathbf{x}_j^*-\mathbf{x}_k^*\|\geq\frac{2}{\sqrt{N-1}}$ for $j\neq k$ (cf. Dragnev [20]). A standard argument (cf., e.g., [4]) yields that

$$
\left|\sum_{j\neq k}\frac{(\mathbf{x}_j^*\cdot\mathbf{x}_k^*)^m}{2(1-\mathbf{x}_j^*\cdot\mathbf{x}_k^*)}\right|\leq\frac{1}{\frac{4}{N-1}}\sum_{j\neq k}\left|\mathbf{x}_j^*\cdot\mathbf{x}_k^*\right|^m\leq\frac{N-1}{4}\frac{\beta'+1}{N}N^2,
$$

where for at most $\beta' N$ many points holds $|\mathbf{x}_j^*\cdot\mathbf{x}_k^*|\geq 1-\frac{2}{N-1}$, provided $m\geq\frac{1}{2}N\log N$. Then

$$
\begin{aligned}
&-\sum_{m=M+1}^{\infty}\frac{1}{m(m-1)}\sum_{j\neq k}\frac{(\mathbf{x}_j^*\cdot\mathbf{x}_k^*)^m}{2(1-\mathbf{x}_j^*\cdot\mathbf{x}_k^*)}\\
&\qquad=-\sum_{M+1\leq m\leq\frac{1}{2}N\log N}\frac{1}{m(m-1)}\sum_{j\neq k}\frac{(\mathbf{x}_j^*\cdot\mathbf{x}_k^*)^m}{2(1-\mathbf{x}_j^*\cdot\mathbf{x}_k^*)}+O\Big(\frac{N}{\log N}\Big).
\end{aligned}
$$

Using (26), (27), and above bounds, we get ($2M = N + \delta$, $0 \leq \delta < 2$, and M odd integer)

$$
\begin{aligned}
\mathcal{E}_{\log}(\mathbb{S}^2; N) \geq{}& \left(\frac{1}{2} - \log 2\right) N^2 - \frac{1}{2} N \log N \\
&+ \left(\log 2 - \gamma + \frac{2}{N^2}\left(\mathcal{E}_2(\mathbb{S}^2; N) - \frac{1}{4} N^2 \log N\right)\right) N \\
&+ \frac{1}{2} \sum_{m=1}^{M-1} \frac{1}{\ell(\ell+1)} \sum_{m=1}^{\ell} S_m(\mathbf{x}_1^*, \dots, \mathbf{x}_N^*) \\
&- \sum_{m=M+1}^{\infty} \frac{1}{m(m-1)} \sum_{j \neq k} \frac{(\mathbf{x}_j^* \cdot \mathbf{x}_k^*)^m}{2(1 - \mathbf{x}_j^* \cdot \mathbf{x}_k^*)} \\
&+ O(1) + O\Big(\frac{2}{N^2}\left(\mathcal{E}_2(\mathbb{S}^2; N) - \frac{1}{4} N^2 \log N\right)\Big).
\end{aligned}
\tag{29}
$$

This relation provides a type of connection formula for two different kinds of energies. An inspection of the derivation of this inequality shows that the inequality sign is due to the use of the smaller minimal Riesz 2-energy $\mathcal{E}_2(\mathbb{S}^2; N)$. We do not know whether the third line in (29) is of order $o(N)$ and leave this as an unresolved question.

Assuming that conjectures (52) and (53) are true, we get

$$
\begin{aligned}
&\log 2 - \gamma + \frac{2}{N^2}\left(\mathcal{E}_2(\mathbb{S}^2; N) - \frac{1}{4} N^2 \log N\right) = \log 2 - \gamma + 2C_{2,2} + o(1) \\
&\quad = -0.05560530494339251850\ldots + o(1).
\end{aligned}
$$

The numerical value agrees in every given digit of C_{BHS} in (18). High-precision computations using Mathematica [11] show agreement in at least 1000 digits. In fact, we prove the following result.

Theorem 1

$$
C_{\mathrm{BHS}} - 2C_{2,2} = \log 2 - \gamma > 0,
$$

where C_{BHS} is given in (18) *and $C_{2,2}$ is given in* (53).

Proof We apply Malmsten's identity for the difference of the first generalized Stieltjes constants of rational argument and its reflected version (usually attributed to Almkvist and Meurman; cf. [7, Eq. (16)])

$$\gamma_1\Big(\frac{m}{n}\Big) - \gamma_1\Big(1-\frac{m}{n}\Big) = 2\pi\sum_{\ell=1}^{n-1}\sin\frac{2\pi m\ell}{n}\cdot\log\Gamma\Big(\frac{\ell}{n}\Big)$$
$$-\pi\,(\gamma+\log(2\pi n))\cot\frac{\pi m}{n} \tag{30}$$

with $m=2$ and $n=3$ in (53). Then after simplifications

$$C_{\mathrm{BHS}} - 2C_{2,2} = -\gamma + \frac{5}{2}\log 2 - \frac{3}{4}\log 3 + \frac{3}{2}\log\frac{\pi}{\Gamma(\frac{1}{3})\Gamma(\frac{2}{3})} = \log 2 - \gamma > 0.$$

We used that $\Gamma(\frac{1}{3})\Gamma(\frac{2}{3}) = \frac{2\pi}{\sqrt{3}}$. Positivity can be verified using Mathematica. □

3.3 The d-sphere

In general, [15] conjectures for $d \geq 2$

$$\mathcal{E}_{\log}(\mathbb{S}^d; N) = \mathbb{W}_{\log}(\mathbb{S}^d)\,N^2 - \frac{1}{d}\,N\log N + C_{\log,d}^{\mathrm{conj}}\,N + o(N) \qquad \text{as} N\to\infty.$$

Here,

$$\mathbb{W}_{\log}(\mathbb{S}^d) := \int_{\mathbb{S}^d}\int_{\mathbb{S}^d}\log\frac{1}{\|\mathbf{x}-\mathbf{y}\|}\,\mathrm{d}\sigma_d(\mathbf{x})\,\mathrm{d}\sigma_d(\mathbf{y}) = -\log 2 + \frac{1}{2}\Big(\psi(d) - \psi\Big(\frac{d}{2}\Big)\Big)$$

expressed in terms of the digamma function $\psi(z) := \frac{\Gamma'(z)}{\Gamma(z)}$. The motivation for this conjecture is an analytic approach using $\mathcal{E}_{\log}(\mathbb{S}^d; N) = \lim_{s\to 0}\frac{1}{s}\mathcal{E}_s(\mathbb{S}^d; N)$, the conjecture for the asymptotic of the Riesz s-energy, and assumptions on the behavior of remainder terms. This approach suggests that[6]

$$C_{\log,d}^{\mathrm{conj}} = \frac{1}{d}\log\frac{\omega_d}{|\Lambda_d|} + \frac{1}{2}\zeta_{\Lambda_d}'(0). \tag{31}$$

In the four cases $d = 2, 4, 8$, and 24 we compare our derived bound with the conjectured ones: For $d=2$ (hexagonal lattice), we refer to Sect. 3.1 and (cf. (18))

[6] Observe, that the factor $\frac{1}{2}$ is a consequence of the definition of the Epstein zeta function used in our paper and is not present in [8, 15]. Note, that [8] has $1+\log\frac{\omega_d}{|\Lambda_d|}$ instead of $\log\frac{\omega_d}{|\Lambda_d|}$, which would not give the correct result for $d=2$.

$$C_{\log,2}^{\text{conj}} = C_{\text{BHS}} = 2\log 2 + \frac{1}{2}\log\frac{2}{3} + 3\log\frac{\sqrt{\pi}}{\Gamma(\frac{1}{3})} = -0.0556053049433925185\,0\ldots.$$

For the checkerboard lattice $\boldsymbol{D}_4$, we get

$$C_{\log,4}^{\text{conj}} = \frac{3\log 2}{4} + \frac{1}{4}\log\left(\frac{8}{3}\right) + 6\zeta'(-1) = -0.2274591635298150340\ldots.$$

For the $\boldsymbol{E}_8$ lattice, we obtain

$$C_{\log,8}^{\text{conj}} = -60\zeta'(-3) - \frac{1}{8}\log\left(\frac{105}{32}\right) = -0.4712426373861826716\ldots.$$

For the Leech lattice $\boldsymbol{\Lambda}_{24}$, we have

$$\begin{aligned} C_{\log,24}^{\text{conj}} &= -\frac{1}{24}\log\left(\frac{316234143225}{8192}\right) + \frac{32760}{691}\log(2)\,\mathrm{L}(0) \\ &\quad - \frac{16380}{691}\zeta'(-11) - \frac{32760}{691}\,\mathrm{L}'(0) \\ &= -0.9227110093940102197\ldots, \end{aligned}$$

where we can use that $\mathrm{L}(0) = 0$. In [12], we obtain the bound

$$\mathcal{E}_{\log}(\mathbb{S}^d; N) = \mathbb{W}_{\log}(\mathbb{S}^d)\,N^2 - \frac{1}{d}\,N\log N + C'_{\log,d}\,N + O\left(N^{1-\frac{2\varepsilon}{d}}\right) \qquad \text{as} N \to \infty$$

valid for every dimension $d = 2, 3, 4, 5, \ldots$, where

$$C'_{\log,d} := -\,\mathbb{W}_{\log}(\mathbb{S}^d) - \frac{1}{d}\,\frac{\Gamma(d)\Gamma(1 + \lfloor\frac{d}{2}\rfloor - \frac{d}{2})}{2^d\Gamma(\frac{d}{2})\Gamma(1 + \lfloor\frac{d}{2}\rfloor)} - \frac{1}{2}\sum_{r=1}^{\lfloor\frac{d}{2}\rfloor}\frac{1}{r} < 0 \tag{32}$$

and $0 < \varepsilon < 1$ for even d and $0 < \varepsilon < \frac{1}{2}$ for odd d.

We apply the method in [4] to prove new bounds and compare them with the conjectured values and the bound (32). This new approach is simpler than using the "δ-trick" of smoothing the energy kernel and provides improved bounds.

We start with (20). By Lemma 2,

$$S_m(t) = t^m - \frac{1 + (-1)^m}{2}\,\frac{\left(\frac{1}{2}\right)_{\frac{m}{2}}}{\left(\frac{d+1}{2}\right)_{\frac{m}{2}}}, \qquad m \geq 1, -1 \leq t \leq 1. \tag{33}$$

Thus

$$\log \frac{1}{\|\mathbf{x}-\mathbf{y}\|} = -\frac{1}{2}\log 2 + \frac{1}{2}\sum_{m=1}^{M} \frac{1+(-1)^m}{2} \frac{1}{m} \frac{\left(\frac{1}{2}\right)_{\frac{m}{2}}}{\left(\frac{d+1}{2}\right)_{\frac{m}{2}}} + \frac{1}{2}\sum_{m=1}^{M} \frac{S_m(\mathbf{x}\cdot\mathbf{y})}{m} + \frac{1}{2}\sum_{m=M+1}^{\infty} \frac{(\mathbf{x}\cdot\mathbf{y})^m}{m}.$$

With the help of Mathematica, we get (using that M is assumed to be odd)

$$\begin{aligned}\sum_{m=1}^{M} \frac{1+(-1)^m}{2} \frac{1}{m} \frac{\left(\frac{1}{2}\right)_{\frac{m}{2}}}{\left(\frac{d+1}{2}\right)_{\frac{m}{2}}} &= \sum_{r=1}^{\frac{M-1}{2}} \frac{1}{2r} \frac{\left(\frac{1}{2}\right)_r}{\left(\frac{d+1}{2}\right)_r} \\ &= \frac{1}{2}\left(\psi\left(\frac{d+1}{2}\right) - \psi\left(\frac{d}{2}\right)\right) \\ &\quad - \frac{\Gamma(\frac{d+1}{2})\Gamma(\frac{M}{2}+1)}{\sqrt{\pi}\,(M+1)\Gamma(\frac{M+d+2}{2})}\, {}_3\mathrm{F}_2\left(\left.\begin{matrix}\frac{M+1}{2}, \frac{M}{2}+1, 1 \\ \frac{M+3}{2}, \frac{M+d+2}{2}\end{matrix}\right| 1\right).\end{aligned}$$

Since $d > 0$ and $M + 1 > 0$ for $M \geq 1$, we may apply [25, Eq. 7.4.4.1, p. 533] to the $3F2$-hypergeometric function to obtain

$${}_3\mathrm{F}_2\left(\left.\begin{matrix}\frac{M+1}{2}, \frac{M}{2}+1, 1 \\ \frac{M+3}{2}, \frac{M+d+2}{2}\end{matrix}\right| 1\right) = \frac{M+1}{d}\, {}_3\mathrm{F}_2\left(\left.\begin{matrix}\frac{d+1}{2}, \frac{d}{2}, 1 \\ \frac{d}{2}+1, \frac{M+d+2}{2}\end{matrix}\right| 1\right).$$

The right-hand side serves as a convergent generalised asymptotic expansion as $M \to \infty$. For our purpose it suffices to use the first few terms of the series. Furthermore, using the duplication formula for the digamma function, we get

$$\frac{1}{4}\left(\psi\left(\frac{d+1}{2}\right) - \psi\left(\frac{d}{2}\right)\right) = \mathbb{W}_{\log}(\mathbb{S}^d) + \frac{1}{2}\log 2.$$

Thus

$$\frac{1}{2}\sum_{m=1}^{M} \frac{1+(-1)^m}{2} \frac{1}{m} \frac{\left(\frac{1}{2}\right)_{\frac{m}{2}}}{\left(\frac{d+1}{2}\right)_{\frac{m}{2}}} = \mathbb{W}_{\log}(\mathbb{S}^d) + \frac{1}{2}\log 2 - \frac{1}{M^{\frac{d}{2}}}\left(2^{\frac{d}{2}-1}\frac{1}{d}\frac{\Gamma(\frac{d+1}{2})}{\sqrt{\pi}} + O\left(\frac{1}{M}\right)\right).$$

Putting everything together, we arrive at the analogue of Lemma 3 for general $d \geq 2$.

Lemma 5 *Let* $\mathbf{x}, \mathbf{y} \in \mathbb{S}^d$, $d \geq 2$, *with* $\mathbf{x} \neq \mathbf{y}$. *Then for large odd integers* M

$$\log \frac{1}{\|\mathbf{x}-\mathbf{y}\|} = \mathbb{W}_{\log}(\mathbb{S}^d) - \frac{1}{M^{\frac{d}{2}}}\left(2^{\frac{d}{2}-1}\frac{1}{d}\frac{\Gamma(\frac{d+1}{2})}{\sqrt{\pi}} + O\Big(\frac{1}{M}\Big)\right) + \frac{1}{2}\sum_{m=1}^{M}\frac{1}{m}S_m(\mathbf{x}\cdot\mathbf{y}) + \frac{1}{2}\sum_{m=M+1}^{\infty}\frac{1}{m}(\mathbf{x}\cdot\mathbf{y})^m, \tag{34}$$

where S_m *is given in* (33) *and the rightmost remainder term is non-negative by* (19).

For N pairwise different points $\mathbf{x}_1, \ldots, \mathbf{x}_N \in \mathbb{S}^d$, we get

$$\begin{aligned}\frac{1}{N^2}\sum_{j\neq k}\log\frac{1}{\|\mathbf{x}_j-\mathbf{x}_k\|} &= \mathbb{W}_{\log}(\mathbb{S}^d)\frac{N-1}{N} - \frac{N-1}{M^{\frac{d}{2}}N}\left(2^{\frac{d}{2}-1}\frac{1}{d}\frac{\Gamma(\frac{d+1}{2})}{\sqrt{\pi}} + O\Big(\frac{1}{M}\Big)\right)\\ &\quad - \frac{1}{2}\frac{1}{N}\sum_{m=1}^{M}\frac{1}{m}S_m(1) + \frac{1}{2}\sum_{m=1}^{M}\frac{1}{m}\frac{1}{N^2}\sum_{j,k}S_m(\mathbf{x}_j\cdot\mathbf{x}_k)\\ &\quad + \frac{1}{2}\sum_{m=M+1}^{\infty}\frac{1}{m}\frac{1}{N^2}\sum_{j\neq k}(\mathbf{x}_j\cdot\mathbf{x}_k)^m.\end{aligned}$$

The diagonal terms behave like

$$\begin{aligned}-\frac{1}{2}\frac{1}{N}\sum_{m=1}^{M}\frac{1}{m}S_m(1) &= -\frac{1}{2}\frac{1}{N}\left(\sum_{m=1}^{M}\frac{1}{m} - \sum_{m=1}^{M}\frac{1+(-1)^m}{2}\frac{1}{m}\frac{\left(\frac{1}{2}\right)_{\frac{m}{2}}}{\left(\frac{d+1}{2}\right)_{\frac{m}{2}}}\right)\\ &= -\frac{1}{2}\frac{1}{N}\left(\gamma + \psi(M+1) - \sum_{m=1}^{M}\frac{1+(-1)^m}{2}\frac{1}{m}\frac{\left(\frac{1}{2}\right)_{\frac{m}{2}}}{\left(\frac{d+1}{2}\right)_{\frac{m}{2}}}\right)\\ &= -\frac{1}{2}\frac{1}{N}\left(\log M + \gamma + \frac{1}{2M} + O\Big(\frac{1}{M^2}\Big) - 2\left(\mathbb{W}_{\log}(\mathbb{S}^d) + \frac{1}{2}\log 2\right.\right.\\ &\qquad\left.\left. - \frac{1}{M^{\frac{d}{2}}}\left(\frac{2^{\frac{d}{2}-1}}{d}\frac{\Gamma(\frac{d+1}{2})}{\sqrt{\pi}} + O\Big(\frac{1}{M}\Big)\right)\right)\right)\\ &= -\frac{1}{d}\frac{\log M^{\frac{d}{2}}}{N} + \left(\mathbb{W}_{\log}(\mathbb{S}^d) + \frac{\log 2}{2} - \frac{\gamma}{2}\right)\frac{1}{N} - \frac{1}{4}\frac{1}{MN}\end{aligned}$$

$$-2^{\frac{d}{2}-1}\frac{1}{d}\frac{\Gamma(\frac{d+1}{2})}{\sqrt{\pi}}\frac{1}{M^{\frac{d}{2}}N}+\mathcal{O}\Big(\frac{1}{M^{\frac{d}{2}+1}N}\Big).$$

Hence, we arrive at

$$\begin{aligned}\frac{1}{N^2}\sum_{j\neq k}\log\frac{1}{\|\mathbf{x}_j-\mathbf{x}_k\|} &= \mathbb{W}_{\log}(\mathbb{S}^d)-\frac{1}{d}\frac{\log M^{\frac{d}{2}}}{N}\\ &\quad+\left(\frac{\log 2}{2}-\frac{\gamma}{2}-2^{\frac{d}{2}-1}\frac{1}{d}\frac{\Gamma(\frac{d+1}{2})}{\sqrt{\pi}}\frac{N}{M^{\frac{d}{2}}}\right)\frac{1}{N}\\ &\quad-\frac{1}{4}\frac{1}{MN}+\mathcal{O}\Big(\frac{1}{M^{\frac{d}{2}+1}}\Big)+\mathcal{O}\Big(\frac{1}{M^{\frac{d}{2}+1}N}\Big)\\ &\quad+\frac{1}{2}\sum_{m=1}^{M}\frac{1}{m}\frac{1}{N^2}\sum_{j,k}S_m(\mathbf{x}_j\cdot\mathbf{x}_k)\\ &\quad+\frac{1}{2}\sum_{m=M+1}^{\infty}\frac{1}{m}\frac{1}{N^2}\sum_{j\neq k}(\mathbf{x}_j\cdot\mathbf{x}_k)^m\end{aligned}$$

as $M\to\infty$ uniformly in $N\geq 2$. As in the case of $d=2$, the second to the last term is non-negative by Lemma 1 and the last term is non-negative if M is an odd integer by (19).

We consider now minimal logarithmic energy points and assume that $M=(cN)^{\frac{2}{d}}+\delta$ with $c\geq c_0>0$ (say $c_0=10^{-2}$) and $0\leq\delta<2$ such that M is an odd integer. Then we have

$$\begin{aligned}\mathcal{E}_{\log}(\mathbb{S}^d;N) &\geq \mathbb{W}_{\log}(\mathbb{S}^d)N^2-\frac{1}{d}N\log N\\ &\quad+\left(\frac{\log 2}{2}-\frac{\gamma}{2}-2^{\frac{d}{2}-1}\frac{1}{d}\frac{\Gamma(\frac{d+1}{2})}{\sqrt{\pi}}\frac{N}{M^{\frac{d}{2}}}\right)N\\ &\quad-\frac{1}{d}N\log\frac{M^{\frac{d}{2}}}{N}+\left(-\frac{1}{4}\frac{N}{M}+\mathcal{O}\Big(\frac{N^2}{M^{\frac{d}{2}+1}}\Big)\right)\Bigg|_{M=(cN)^{\frac{2}{d}}+\delta}.\end{aligned}$$

Since

$$-\frac{1}{d}\log\frac{M^{\frac{d}{2}}}{N}=-\frac{1}{d}\log c-\frac{1}{2}\log\Big(1+\frac{\delta}{(cN)^{\frac{2}{d}}}\Big)=-\frac{1}{d}\log c+\mathcal{O}\Big(\frac{\delta}{c^{\frac{2}{d}}}\frac{1}{N^{\frac{2}{d}}}\Big)$$

and

$$\frac{N}{M^{\frac{d}{2}}} = \frac{1}{c}\left(1 + \frac{\delta}{(cN)^{\frac{2}{d}}}\right)^{\frac{d}{2}} = \frac{1}{c}\left(1 + O\left(\frac{\delta}{c^{\frac{2}{d}}}\frac{1}{N^{\frac{2}{d}}}\right)\right)$$

and

$$\left(-\frac{1}{4}\frac{N}{M} + O\left(\frac{N^2}{M^{\frac{d}{2}+1}}\right)\right)\Bigg|_{M=(cN)^{\frac{2}{d}}+\delta} = O_{c_0}\left(N^{1-\frac{2}{d}}\right),$$

we arrive at the following result.

Theorem 2 *Let $d \geq 2$. Then*

$$\mathcal{E}_{\log}(\mathbb{S}^d; N) \geq \mathbb{W}_{\log}(\mathbb{S}^d) N^2 - \frac{1}{d} N \log N + F(c)\, N + O_{c_0}\left(N^{1-\frac{2}{d}}\right), \tag{35}$$

where the function

$$F(c) := \frac{\log 2}{2} - \frac{\gamma}{2} - \frac{\log c}{d} - 2^{\frac{d}{2}-1}\frac{1}{d}\frac{\Gamma(\frac{d+1}{2})}{\sqrt{\pi}}\frac{1}{c}$$

has a unique maximum at $c = c^* := 2^{\frac{d}{2}-1}\frac{\Gamma(\frac{d+1}{2})}{\sqrt{\pi}}$ in $(0, \infty)$ with value

$$F(c^*) = \frac{\log 2}{d} - \frac{1}{d}\log\frac{\Gamma(\frac{d+1}{2})}{\sqrt{\pi}} - \frac{1}{d} - \frac{\gamma}{2}. \tag{36}$$

We compare this bound with the conjectured values; see Table 2. and with the general bound (32). We use Mathematica to compute the values in the table. The relative error is rounded up. Interestingly, the discrepancy in the 2-dimensional case is significantly larger than in the other cases $d = 4, 8, 24$.

4 The Method Applied to the Riesz s-energy (Singular Case $0 < s < d$)

We consider the minimal Riesz s-energy of N points on the unit sphere $\mathbb{S}^d$ in $\mathbb{R}^{d+1}$, $d \geq 2$, in the singular case $0 < s < d$. This quantity is given by

$$\mathcal{E}_s(\mathbb{S}^d; N) := \min_{\mathbf{x}_1,\ldots,\mathbf{x}_N \in \mathbb{S}^d} \sum_{j \neq k} \frac{1}{\|\mathbf{x}_j - \mathbf{x}_k\|^s}.$$

Table 2 Comparison of conjectured $C_{\log,d}^{\mathrm{conj}}$ and asymptotic $F(c^*)$ bound given in (36) for the cases when (31) is conjectured to hold

d	$C_{\log,d}^{\mathrm{conj}}$	$F(c^*)$	$C_{\log,d}^{\mathrm{conj}} - F(c^*)$	$\left\lvert \frac{C_{\log,d}^{\mathrm{conj}} - F(c^*)}{C_{\log,d}^{\mathrm{conj}}} \right\rvert$	Lattice
2	$-0.055605304943392\dots$	$-0.095460651890821\dots$	$0.039\dots$	72%	$\boldsymbol{A}_2$
4	$-0.227459163529815\dots$	$-0.293400519197834\dots$	$0.065\dots$	29%	$\boldsymbol{D}_4$
8	$-0.471242637386182\dots$	$-0.562135888370491\dots$	$0.090\dots$	20%	$\boldsymbol{E}_8$
24	$-0.922711009394010\dots$	$-1.058142640302924\dots$	$0.135\dots$	15%	$\boldsymbol{\Lambda}_{24}$

A standard argument from classical potential theory yields

$$\lim_{N\to\infty} \frac{\mathcal{E}_s(\mathbb{S}^d, N)}{N^2} = \mathbb{W}_s(\mathbb{S}^d),$$

where

$$\begin{aligned} \mathbb{W}_s(\mathbb{S}^d) &= \int_{\mathbb{S}^d}\int_{\mathbb{S}^d} \frac{1}{\|\mathbf{x}-\mathbf{y}\|^s}\, \mathrm{d}\sigma_d(\mathbf{x})\, \mathrm{d}\sigma_d(\mathbf{x}) \\ &= 2^{d-s-1}\frac{\Gamma(\frac{d+1}{2})\Gamma(\frac{d-s}{2})}{\sqrt{\pi}\,\Gamma(d-\frac{s}{2})}, \qquad 0 < s < d. \end{aligned} \tag{37}$$

Combined work of several authors and papers, for details we refer to [8, 15], yield bounds of the correct order in N and unspecified constants: There exist two negative constants c_1 and c_2 such that for every N sufficiently large,

$$c_1\, N^{1+\frac{s}{d}} \leq \mathcal{E}_s(\mathbb{S}^d; N) - \mathbb{W}_s(\mathbb{S}^d)\, N^2 \leq c_2\, N^{1+\frac{s}{d}}.$$

In [15] it is conjectured that

$$\mathcal{E}_s(\mathbb{S}^d; N) = \mathbb{W}_s(\mathbb{S}^d)\, N^2 + \frac{C_{s,d}}{(\omega_d)^{\frac{s}{d}}}\, N^{1+\frac{s}{d}} + o\Big(N^{1+\frac{s}{d}}\Big),$$

where $C_{s,d}$ is the constant in (6) and takes the form (7) for $d = 2, 4, 8$, and $d = 24$ (for details, see Sect. 2).

We use our approach to obtain a lower bound for the Riesz s-energy and compare it to the conjecture.

The Riesz s-kernel has the following expansion

$$\frac{1}{\|\mathbf{x}-\mathbf{y}\|^s} = 2^{-\frac{s}{2}}\,(1-\mathbf{x}\cdot\mathbf{y})^{-\frac{s}{2}} = \sum_{m=0}^{\infty} 2^{-\frac{s}{2}} \frac{(\frac{s}{2})_m}{m!} (\mathbf{x}\cdot\mathbf{y})^m,$$

where the coefficients have the asymptotic behavior

$$b_m := 2^{-\frac{s}{2}}\frac{(\frac{s}{2})_m}{m!} = 2^{-\frac{s}{2}}\frac{1}{\Gamma(\frac{s}{2})}\frac{\Gamma(m+\frac{s}{2})}{\Gamma(m+1)} \sim 2^{-\frac{s}{2}}\frac{1}{\Gamma(\frac{s}{2})}\, m^{\frac{s}{2}-1} \qquad \text{as} m \to \infty.$$

Using (33), we write

$$\frac{1}{\|\mathbf{x}-\mathbf{y}\|^s} = A_{s,d}(M) + \sum_{m=1}^{M} b_m\, S_m(\mathbf{x}\cdot\mathbf{y}) + \sum_{m=M+1}^{\infty} b_m\, (\mathbf{x}\cdot\mathbf{y})^m,$$

where we set

$$A_{s,d}(M) := \sum_{m=0}^{M} \frac{1+(-1)^m}{2} \, b_m \, \frac{\left(\frac{1}{2}\right)_{\frac{m}{2}}}{\left(\frac{d+1}{2}\right)_{\frac{m}{2}}} = \sum_{r=0}^{\frac{M-1}{2}} b_{2r} \, \frac{\left(\frac{1}{2}\right)_r}{\left(\frac{d+1}{2}\right)_r}. \tag{38}$$

In the case $d = 2$ we are able to compute this quantity with the help of Mathematica. It is assumed that M is an odd integer:

$$A_{s,2}(M) = \mathbb{W}_s(\mathbb{S}^2) - \frac{2^{-\frac{s}{2}}}{\Gamma(\frac{s}{2})} \frac{\Gamma(M+1+\frac{s}{2})}{\Gamma(M+1)} \, {}_3\mathrm{F}_2\left(\begin{matrix} \frac{1}{2}+\frac{M}{2}+\frac{s}{4}, 1+\frac{M}{2}+\frac{s}{4}, 1 \\ \frac{3}{2}+\frac{M}{2}, 2+\frac{M}{2} \end{matrix} \middle| 1 \right).$$

The transformation [25, Eq. 7.4.4.1, p. 533] yields

$$\begin{aligned} A_{s,2}(M) =& \mathbb{W}_s(\mathbb{S}^2) \\ & - \frac{2^{-\frac{s}{2}}}{(2-s)\Gamma(\frac{s}{2})} \frac{(M+1)\Gamma(M+1+\frac{s}{2})}{\Gamma(M+1)} \, {}_3\mathrm{F}_2\left(\begin{matrix} \frac{3}{2}-\frac{s}{4}, 1-\frac{s}{4}, 1 \\ 2-\frac{s}{2}, 2+\frac{M}{2} \end{matrix} \middle| 1 \right). \end{aligned} \tag{39}$$

The series expansion of the $3F2$-hypergeometric function serves as a convergent generalised asymptotic expansion for large M. We get (using that $\mathbb{W}_s(\mathbb{S}^2) = \frac{2^{1-s}}{2-s}$),

$$A_{s,2}(M) = \frac{2^{1-s}}{2-s} - \frac{2^{-\frac{s}{2}}}{(2-s)\Gamma(\frac{s}{2})} M^{-1+\frac{s}{2}} + O\left(M^{-2+\frac{s}{2}}\right). \tag{40}$$

For the remainder term, we obtain with the help of Mathematica and application of the linear transformation [19, Last formula of 15.8.1] the expression

$$\sum_{m=M+1}^{\infty} b_m \, (\mathbf{x}\cdot\mathbf{y})^m = 2^{-\frac{s}{2}} \frac{\left(\frac{s}{2}\right)_{M+1}}{(M+1)!} \frac{t^{M+1}}{(1-t)^{\frac{s}{2}}} \, {}_2\mathrm{F}_1\left(\begin{matrix} 1-\frac{s}{2}, M+1 \\ M+2 \end{matrix} \middle| t \right).$$

The series expansion of the $2F1$-hypergeometric function is absolutely convergent for all $-1 \leq \mathbf{x}\cdot\mathbf{y} \leq 1$ and has positive strictly decreasing power series coefficients a_ℓ, $\ell \geq 0$, if $0 < s < 2$. Thus we may rewrite for odd integers M,

$$\begin{aligned} {}_2\mathrm{F}_1\left(\begin{matrix} 1-\frac{s}{2}, M+1 \\ M+2 \end{matrix} \middle| t \right) &= \sum_{r=\frac{M+1}{2}}^{\infty} (a_{2r} - a_{2r+1}) \, (\mathbf{x}\cdot\mathbf{y})^{2r} \\ &\quad + \sum_{r=\frac{M+1}{2}}^{\infty} a_{2r+1} \, (\mathbf{x}\cdot\mathbf{y})^{2r} (1+\mathbf{x}\cdot\mathbf{y}). \end{aligned}$$

Thus

$$\sum_{m=M+1}^{\infty} b_m\,(\mathbf{x}\cdot\mathbf{y})^m \geq 0, \qquad 0 < s < 2.$$

Lemma 6 *Let $d \geq 2$, $0 < s < d$, and $\mathbf{x}, \mathbf{y} \in \mathbb{S}^d$ with $\mathbf{x} \neq \mathbf{y}$. Then for large odd integers M,*

$$\frac{1}{\|\mathbf{x}-\mathbf{y}\|^s} = A_{s,d}(M) + \sum_{m=1}^{M} 2^{-\frac{s}{2}} \frac{\left(\frac{s}{2}\right)_m}{m!}\, S_m(\mathbf{x}\cdot\mathbf{y}) + \sum_{m=M+1}^{\infty} 2^{-\frac{s}{2}} \frac{\left(\frac{s}{2}\right)_m}{m!}\,(\mathbf{x}\cdot\mathbf{y})^m, \tag{41}$$

where $A_{s,d}(M)$ is defined in (38). *For $0 < s < 2$, the remainder term is non-negative. For $d = 2$, the coefficient $A_{s,2}(M)$ and its asymptotic form are given in* (39) *and* (40).

For N pairwise different points $\mathbf{x}_1, \ldots, \mathbf{x}_N \in \mathbb{S}^d$, we get

$$\begin{aligned}\frac{1}{N^2}\sum_{j\neq k}\frac{1}{\|\mathbf{x}_j-\mathbf{x}_k\|^{\frac{s}{2}}} &= A_{s,d}(M)\,\frac{N-1}{N} - \frac{1}{N}\sum_{m=1}^{M} b_m\,S_m(1)\\ &\quad + \sum_{m=1}^{M} b_m\,\frac{1}{N^2}\sum_{j,k} S_m(\mathbf{x}_j\cdot\mathbf{x}_k)\\ &\quad + \sum_{m=M+1}^{\infty} b_m\,\frac{1}{N^2}\sum_{j\neq k}(\mathbf{x}_j\cdot\mathbf{x}_k)^m.\end{aligned}$$

By (33) and (38),

$$\sum_{m=1}^{M} b_m\,S_m(1) = \sum_{m=1}^{M} b_m - \big(A_{s,d}(M) - b_0\big) = \frac{2^{-\frac{s}{2}}}{\Gamma(1+\frac{s}{2})}\,\frac{\Gamma(M+1+\frac{s}{2})}{\Gamma(M+1)} - A_d(M).$$

Thus for large odd integers M,

$$\begin{aligned}\frac{1}{N^2}\sum_{j\neq k}\frac{1}{\|\mathbf{x}_j-\mathbf{x}_k\|^{\frac{s}{2}}} &= A_{s,d}(M) - 2^{-\frac{s}{2}}\frac{\left(\frac{s}{2}+1\right)_M}{M!}\,\frac{1}{N}\\ &\quad + \sum_{m=1}^{M} b_m\,S_m(\mathbf{x}_1,\ldots,\mathbf{x}_N) + \sum_{m=M+1}^{\infty} b_m\,\frac{1}{N^2}\sum_{j\neq k}(\mathbf{x}_j\cdot\mathbf{x}_k)^m.\end{aligned}$$

This relations holds for all $d \geq 2$ and $0 < s < d$. The second to the last sum is non-negative by construction and the last sum is non-negative for $0 < s < 2$.

We now consider $d = 2$. We set $M = (cN)^{\frac{2}{d}} + \delta = cN + \delta$ for $c \geq c_0 > 0$ and $0 \leq \delta < 2$ such that M is an odd integer. Then

$$\mathcal{E}_s(\mathbb{S}^2; N) \geq \frac{2^{1-s}}{2-s} N^2 + F(c)\, N^{1+\frac{s}{2}} + \frac{2^{-\frac{s}{2}-3} c^{\frac{s}{2}-2}(2c-1)(4\delta+s+2)}{\Gamma(\frac{s}{2})} N^{\frac{s}{2}} + O_{c_0}\Big(N^{-1+\frac{s}{2}}\Big),$$

where the function

$$F(c) := \frac{2^{-1-\frac{s}{2}}}{(2-s)\Gamma(1+\frac{s}{2})} c^{\frac{s}{2}-1} (2cs - 4c - s)$$

has a unique maximum in $(0, \infty)$ at $c = c^*$ with $c^* = \frac{1}{2}$ with value

$$F(c^*) = -\frac{2^{1-s}}{(2-s)\Gamma(1+\frac{s}{2})}. \tag{42}$$

Note that the $N^{\frac{s}{2}}$-term disappears because of $c = \frac{1}{2}$ for the optimal constant.

Theorem 3 *Let $d = 2$ and $0 < s < d$. Then*

$$\mathcal{E}_s(\mathbb{S}^2; N) \geq \frac{2^{1-s}}{2-s} N^2 + F(c^*)\, N^{1+\frac{s}{2}} + O\Big(N^{-1+\frac{s}{2}}\Big), \tag{43}$$

where $F(c^*)$ is given in (42).

Remark 1 Since the bounds are a function of s, we do a graphical comparison of the conjectured constant

$$\frac{C_{s,d}}{(\omega_d)^{\frac{s}{d}}}\Bigg|_{d=2} = -2^{1-\frac{3s}{2}} 3^{1-\frac{s}{4}} \pi^{-\frac{s}{2}} \zeta\Big(\frac{s}{2}\Big) \Big(\zeta\Big(\frac{s}{2}, \frac{2}{3}\Big) - \zeta\Big(\frac{s}{2}, \frac{1}{3}\Big)\Big) \tag{44}$$

and $F(c^*)$ given in (42). Surprisingly, the lower bound

$$\tilde{c}_2(s) := -\frac{2^{-\frac{s}{2}} s^{-\frac{s}{2}}}{1-\frac{s}{2}}. \tag{45}$$

for the hypersingular case $s > d = 2$ given in [15] also works for $0 < s < d$ and is improved by $F(c^*)$; see Fig. 2. This figure also shows absolute and relative errors

$$\frac{C_{s,d}}{(\omega_d)^{\frac{s}{d}}}\Big|_{d=2} - F(c^*), \qquad \left| \frac{\frac{C_{s,d}}{(\omega_d)^{\frac{s}{d}}}\big|_{d=2} - F(c^*)}{\frac{C_{s,d}}{(\omega_d)^{\frac{s}{d}}}\big|_{d=2}} \right| \tag{46}$$

and

$$\frac{C_{s,d}}{(\omega_d)^{\frac{s}{d}}}\Big|_{d=2} - \tilde{c}_2(s), \qquad \left| \frac{\frac{C_{s,d}}{(\omega_d)^{\frac{s}{d}}}\big|_{d=2} - \tilde{c}_2(s)}{\frac{C_{s,d}}{(\omega_d)^{\frac{s}{d}}}\big|_{d=2}} \right|. \tag{47}$$

The new bound is significantly closer (relative error less than 3%) to the conjectured value than to the bound $\tilde{c}_2(s)$. Moreover, the relative errors in the logarithmic and the boundary case are much larger.

5 The Method Applied to the Riesz s-energy (Boundary Case $s = d$)

We consider the minimal Riesz d-energy of N points on the unit sphere $\mathbb{S}^d$ in $\mathbb{R}^{d+1}$, $d \geq 2$. This quantity is given by

$$\mathcal{E}_d(\mathbb{S}^d; N) := \min_{\mathbf{x}_1,\ldots,\mathbf{x}_N \in \mathbb{S}^d} \sum_{j \neq k} \frac{1}{\|\mathbf{x}_j - \mathbf{x}_k\|^d}.$$

Kuijlaars and Saff [22] proved that

$$\lim_{N\to\infty} \frac{\mathcal{E}_d(\mathbb{S}^d; N)}{N^2 \log N} = C_d = \frac{1}{d} \frac{\Gamma(\frac{d+1}{2})}{\sqrt{\pi}\,\Gamma(\frac{d}{2})}$$

with C_d the constant in (4). A careful inspection of the proof in [22] (cf. [15]) yields the lower bound

$$\mathcal{E}_d(\mathbb{S}^d; N) \geq C_d\, N^2 \log N + \tilde{c}_d N^2 + O\Big(N^{2-\frac{2}{d}}\Big), \tag{48}$$

where

$$\tilde{c}_d := -\,C_d \left(1 - \log C_d + d\left(\psi\Big(\frac{d}{2}\Big) + \gamma - \log 2\right)\right) < 0. \tag{49}$$

For $d = 2$, one has $\tilde{c}_d = -\frac{1}{4}$ and the remainder term is in fact $O(N)$. The following conjecture is obtained by means of a limit process as $s \to d$ applied to the conjecture for the Riesz s-energy asymptotics ([15])

$$\mathcal{E}_d(\mathbb{S}^d; N) = C_d \, N^2 \log N + C_{d,d} \, N^2 + O(1),$$

where

$$C_{d,d} := \lim_{s \to d} \left(\mathbb{W}_s(\mathbb{S}^d) + \frac{C_{s,d}}{(\omega_d)^{\frac{s}{d}}} \right) \tag{50}$$

and both

$$\mathbb{W}_s(\mathbb{S}^d) := 2^{d-s-1} \frac{\Gamma(\frac{d+1}{2}) \Gamma(\frac{d-s}{2})}{\sqrt{\pi} \Gamma(d - \frac{s}{2})} \tag{51}$$

and $C_{s,d}$ have poles at $s = d$ that cancel in the limit as $s \to d$. The constant $C_{s,d}$ is understood as the analytic continuation of (6) to the complex s-plane and assumes the form of (7) for $d = 8$ and $d = 24$ and is conjectured to have that form for $d = 2$ and $d = 4$; cf. discussion in Sect. 2. For $d = 2$, the conjecture reduces to

$$\mathcal{E}_2(\mathbb{S}^2; N) = \frac{1}{4} N^2 \log N + C_{2,2} \, N^2 + o(N^2), \tag{52}$$

where

$$\begin{aligned} C_{2,2} &= \frac{1}{4} \left(\gamma - \log(2\sqrt{3}\pi) \right) + \frac{\sqrt{3}}{4\pi} \left(\gamma_1\left(\frac{2}{3}\right) - \gamma_1\left(\frac{1}{3}\right) \right) \\ &= -0.0857684103009024836 5\ldots \end{aligned} \tag{53}$$

and $\gamma_n(\alpha)$ is the generalized Stieltjes constant appearing as the coefficient $\frac{\gamma_n(\alpha)}{n!}$ of $(1 - s)^n$ in the Laurent series expansion of the Hurwitz zeta function $\zeta(s, a)$ about $s = 1$.

We use our approach to obtain a lower bound for the Riesz d-energy and compare it with the known bound and the conjecture.

The Riesz d-energy kernel has the following expansion

$$\frac{1}{\|\mathbf{x} - \mathbf{y}\|^d} = 2^{-\frac{d}{2}} \left(1 - \mathbf{x} \cdot \mathbf{y}\right)^{-\frac{d}{2}} = \sum_{m=0}^{\infty} 2^{-\frac{d}{2}} \frac{\left(\frac{d}{2}\right)_m}{m!} (\mathbf{x} \cdot \mathbf{y})^m.$$

The coefficients have the asymptotic behavior

$$b_m := 2^{-\frac{d}{2}} \frac{\left(\frac{d}{2}\right)_m}{m!} = 2^{-\frac{d}{2}} \frac{1}{\Gamma(\frac{d}{2})} \frac{\Gamma(m + \frac{d}{2})}{\Gamma(m + 1)} \sim 2^{-\frac{d}{2}} \frac{1}{\Gamma(\frac{d}{2})} m^{\frac{d}{2} - 1} \qquad \text{as} m \to \infty.$$

Using (33), we write

$$\frac{1}{\|\mathbf{x}-\mathbf{y}\|^d} = A_d(M) + \sum_{m=1}^{M} b_m\, S_m(\mathbf{x}\cdot\mathbf{y}) + \sum_{m=M+1}^{\infty} b_m\,(\mathbf{x}\cdot\mathbf{y})^m,$$

where we set

$$A_d(M) := \sum_{m=0}^{M} \frac{1+(-1)^m}{2}\, b_m\, \frac{\left(\frac{1}{2}\right)_{\frac{m}{2}}}{\left(\frac{d+1}{2}\right)_{\frac{m}{2}}} = \sum_{r=0}^{\frac{M-1}{2}} b_{2r}\, \frac{\left(\frac{1}{2}\right)_r}{\left(\frac{d+1}{2}\right)_r}. \tag{54}$$

We did not find a closed form representation valid for all $d \geq 2$. Using Mathematica, it is possible to compute such formulas for selected even dimensions d including the cases 2, 4, 8 and 24:

$$\begin{aligned}
A_2(M) &= \frac{1}{4}\left(\psi\left(\frac{M}{2}+1\right)+\gamma+2\log 2\right)\\
&= \frac{1}{4}\log(M) + \frac{1}{4}(\log 2+\gamma) + \frac{1}{4}\frac{1}{M} + O\left(\frac{1}{M^2}\right), \qquad (55)\\
A_4(M) &= \frac{3}{8}\left(\psi\left(\frac{M}{2}+2\right)+\gamma+2\log 2-2\right)\\
&= \frac{3}{8}\log M + \frac{3}{8}(\log 2+\gamma-2) + \frac{9}{8}\frac{1}{M} + O\left(\frac{1}{M^2}\right), \qquad (56)\\
A_8(M) &= \frac{35}{64}\left(\psi\left(\frac{M}{2}+4\right)+\gamma+2\log 2-\frac{11}{3}+\frac{3}{M+6}\right)\\
&= \frac{35}{64}\log M + \frac{35}{64}\left(\log 2+\gamma-\frac{11}{3}\right) + \frac{175}{32}\frac{1}{M} + O\left(\frac{1}{M^2}\right), \qquad (57)\\
A_{24}(M) &= \frac{2028117}{2097152}\left(\psi\left(\frac{M}{2}+12\right)+\gamma+2\log 2-\frac{83711}{13860}+\frac{693}{128(M+14)}\right.\\
&\qquad \left. -\frac{7161}{64(M+16)}+\frac{9471}{16(M+18)}-\frac{71511}{64(M+20)}+\frac{87923}{128(M+22)}\right)\\
&= \frac{2028117}{2097152}\log M + \frac{2028117}{2097152}\left(\log 2+\gamma-\frac{83711}{13860}\right)\\
&\qquad + \frac{79096563}{1048576}\frac{1}{M} + O\left(\frac{1}{M^2}\right). \qquad (58)
\end{aligned}$$

Next we consider the remainder term of the expansion of the Riesz d-energy kernel. Mathematica gives

$$\sum_{m=M+1}^{\infty} b_m\,(\mathbf{x}\cdot\mathbf{y})^m = 2^{-\frac{d}{2}}\,\frac{\left(\frac{d}{2}\right)_{M+1}}{(M+1)!}\,t^{M+1}\;{}_2F_1\left(\left.\begin{matrix}1,\,1+\frac{d}{2}+M\\ M+2\end{matrix}\,\right|\,t\right).$$

The transformation [19, Last formula of 15.8.1] of the argument of the $2F1$-hypergeometric function gives

$$\sum_{m=M+1}^{\infty} b_m \frac{(\mathbf{x}\cdot\mathbf{y})^m}{(1-t)^{\frac{d}{2}}} = 2^{-\frac{d}{2}} \frac{\left(\frac{d}{2}\right)_{M+1}}{(M+1)!} \frac{t^{M+1}}{(1-t)^{\frac{d}{2}}} {}_2\mathrm{F}_1\left(\begin{matrix} 1-\frac{d}{2}, 1+M \\ M+2 \end{matrix} \middle| t\right).$$

We consider only even dimensions d; i.e., above hypergeometric functions reduces to a polynomial of degree $\frac{d}{2}-1$. We apply the transformation [19, First formula of 15.8.7] to get

$$\sum_{m=M+1}^{\infty} b_m \frac{(\mathbf{x}\cdot\mathbf{y})^m}{(1-t)^{\frac{d}{2}}} = 2^{-\frac{d}{2}} \frac{\left(\frac{d}{2}\right)_{M+1}}{(M+1)!} \frac{\Gamma(\frac{d}{2})}{(M+2)!} \frac{t^{M+1}}{(1-t)^{\frac{d}{2}}}$$
$$\times {}_2\mathrm{F}_1\left(\begin{matrix} 1-\frac{d}{2}, 1+M \\ 1-\frac{d}{2} \end{matrix} \middle| 1-t\right).$$

Since the hypergeometric polynomial is a sum of non-negative terms, the remainder term is non-negative. Thus, we obtain the following result.

Lemma 7 *Let* $\mathbf{x}, \mathbf{y} \in \mathbb{S}^d$, $d \geq 2$, *with* $\mathbf{x} \neq \mathbf{y}$. *Then for large integers* M,

$$\frac{1}{\|\mathbf{x}-\mathbf{y}\|^d} = A_d(M) + \sum_{m=1}^{M} 2^{-\frac{d}{2}} \frac{\left(\frac{d}{2}\right)_m}{m!} S_m(\mathbf{x}\cdot\mathbf{y}) + \sum_{m=M+1}^{\infty} 2^{-\frac{d}{2}} \frac{\left(\frac{d}{2}\right)_m}{m!} (\mathbf{x}\cdot\mathbf{y})^m, \tag{59}$$

where $A_d(M)$ *is defined in* (54) *and* S_m *is given in* (33). *For even dimensions* d, *the right-most remainder term is non-negative. In the cases* $d \in \{2, 4, 8, 24\}$, *the coefficient* $A_d(M)$ *and its asymptotic expansion is given in* (55)–(57).

For N pairwise different points $\mathbf{x}_1, \dots, \mathbf{x}_N \in \mathbb{S}^d$, we get

$$\frac{1}{N^2} \sum_{j \neq k} \frac{1}{\|\mathbf{x}_j - \mathbf{x}_k\|^{\frac{d}{2}}} = A_d(M) \frac{N-1}{N} - \frac{1}{N} \sum_{m=1}^{M} b_m S_m(1)$$
$$+ \sum_{m=1}^{M} b_m \frac{1}{N^2} \sum_{j,k} S_m(\mathbf{x}_j \cdot \mathbf{x}_k)$$
$$+ \sum_{m=M+1}^{\infty} b_m \frac{1}{N^2} \sum_{j \neq k} (\mathbf{x}_j \cdot \mathbf{x}_k)^m.$$

By (33) and (54),

$$\sum_{m=1}^{M} b_m S_m(1) = \sum_{m=1}^{M} b_m - (A_d(M) - b_0) = 2^{-\frac{d}{2}} \frac{\left(\frac{d}{2}+1\right)_M}{M!} - A_d(M).$$

Thus

$$\frac{1}{N^2}\sum_{j\neq k}\frac{1}{\|\mathbf{x}_j-\mathbf{x}_k\|^{\frac{d}{2}}}=A_d(M)-2^{-\frac{d}{2}}\frac{\left(\frac{d}{2}+1\right)_M}{M!}\,\frac{1}{N}$$

$$+\sum_{m=1}^{M} b_m\,S_m(\mathbf{x}_1,\dots,\mathbf{x}_N)+\sum_{m=M+1}^{\infty} b_m\,\frac{1}{N^2}\sum_{j\neq k}(\mathbf{x}_j\cdot\mathbf{x}_k)^m.$$

Similar as before, we determine the asymptotic form for the minimal Riesz d-energy. We set $M=(cN)^{\frac{2}{d}}+\delta$ with $0\leq\delta<1$ such that M is an integer.[7] For $d=2$, we get

$$\mathcal{E}_2(\mathbb{S}^2;N)\geq\frac{1}{4}N^2\log N+F_2(c)\,N^2-\frac{(2c-1)(1+\delta)}{4c}\,N+O_{c_0}(1),$$

where the function

$$F_2(c):=\frac{1}{4}\,(\gamma+\log 2+\log c-2c)$$

has a unique maximum at $c=c_2^*$ with $c_2^*=\frac{1}{2}$ with value

$$F_2(c_2^*)=-\frac{1}{4}+\frac{\gamma}{4}=-0.105696083774616784\dots. \tag{60}$$

Note that the N-term thus disappears in the lower bound above.

Theorem 4 *Let $d=2$. Then*

$$\mathcal{E}_2(\mathbb{S}^2;N)\geq\frac{1}{4}N^2\log N+F_2(c_2^*)\,N^2+O(1), \tag{61}$$

where $F_2(c_2^*)$ is given in (60).

Remark 2 We note that the new bound improves the bound $\tilde{c}_2$ given in (49) by $\frac{\gamma}{4}$. Compared with the conjectured value (53), we have

$$C_{2,2}-F_2(c_2^*)\approx 0.0199\dots,\qquad\left|\frac{C_{2,2}-F_2(c_2^*)}{C_{2,2}}\right|\approx 19\%.$$

For $d=4$, we obtain

[7] Adding an even and an odd term to get non-negativity for the remainder term of the kernel expansion does not work here.

$$\mathcal{E}_4(\mathbb{S}^4; N) \geq \frac{3}{16} N^2 \log N + F_4(c)\, N^2 + \frac{9 - 3c - (2c - 3)\delta}{8\sqrt{c}} N^{\frac{3}{2}} + O_{c_0}(N),$$

where the function

$$F_4(c) := \frac{3}{8}\left(\gamma + \log 2 - 2 + \frac{1}{2}\log c - \frac{1}{3}c\right)$$

has a unique maximum at $c = c_4^*$ with $c_4^* = \frac{3}{2}$ with value

$$F_4(c_4^*) = \frac{3}{8}\left(\gamma + \frac{1}{2}\log 2 + \frac{1}{2}\log 3 - \frac{5}{2}\right) = -0.385089225181664864\ldots. \tag{62}$$

The dependence on δ vanishes in the $N^{\frac{3}{2}}$-term.

Theorem 5 *Let $d = 4$. Then*

$$\mathcal{E}_4(\mathbb{S}^4; N) \geq \frac{3}{16} N^2 \log N + F_4(c)\, N^2 + \frac{3\sqrt{3}}{8\sqrt{2}} N^{\frac{3}{2}} + O(N), \tag{63}$$

where $F_4(c_4^*)$ is given in (62).

Remark 3 The new bound improves the known bound given in (49) of value

$$\tilde{c}_4 = -0.731510195874729432\ldots.$$

Compared with the conjectured value

$$\begin{aligned} C_{4,4} &= \frac{3}{8}\left(\gamma + \log 2 + \frac{1}{2}\log 3 - 1 + \frac{6}{\pi^2}\zeta'(2) - \log \pi\right) \\ &= -0.335633208430649986\ldots, \end{aligned}$$

we have

$$C_{4,4} - F_4(c_4^*) \approx 0.0494, \qquad \left|\frac{C_{4,4} - F_4(c_4^*)}{C_{4,4}}\right| \approx 13\%.$$

For $d = 8$, we get

$$\mathcal{E}_8(\mathbb{S}^8; N) \geq \frac{35}{256} N^2 \log N + F_8(c)\, N^2 + \frac{1050 - 5c - (2c - 105)\delta}{192\sqrt{4c}} N^{\frac{7}{4}} + O_{c_0}\left(N^{\frac{3}{2}}\right),$$

where the function

$$F_8(c) := \frac{35}{64}\left(\gamma + \log 2 - \frac{11}{3} + \frac{1}{4}\log c - \frac{1}{210}c\right)$$

has a unique maximum at $c = c_8^*$ with $c_8^* = \frac{105}{2}$ with value

$$F_8(c_8^*) = \frac{35}{64}\left(\gamma + \frac{3}{4}\log 2 + \frac{1}{4}\log 3 + \frac{1}{4}\log 5 + \frac{1}{4}\log 7 - \frac{47}{12}\right) \\ = -0.905679976690668583\ldots. \tag{64}$$

The dependence on δ vanishes in the $N^{\frac{7}{4}}$-term.

Theorem 6 *Let* $d = 8$. *Then*

$$\mathcal{E}_8(\mathbb{S}^8; N) \geq \frac{35}{256} N^2 \log N + F_8(c)\, N^2 + \frac{5}{64}\left(\frac{105}{2}\right)^{\frac{3}{4}} N^{\frac{7}{4}} + O\!\left(N^{\frac{3}{2}}\right), \tag{65}$$

where $F_8(c_8^*)$ *is given in* (64).

Remark 4 The new bound improves the known bound given in (49) of value

$$\tilde{c}_8 = -1.65584434055157755\ldots.$$

Compared with the conjectured value

$$C_{8,8} = \frac{35}{64}\left(\gamma - \frac{1}{4}\log 2 + \frac{1}{4}\log 3 + \frac{1}{4}\log 5 + \frac{1}{4}\log 7 - \frac{11}{6} + \frac{90}{\pi^4}\zeta'(4) - \log \pi\right) \\ = -0.806265483426581312\ldots,$$

we have

$$C_{8,8} - F_8(c_8^*) \approx 0.0994, \qquad \left|\frac{C_{8,8} - F_8(c_8^*)}{C_{8,8}}\right| \approx 11\%.$$

For $d = 24$, we proceed in a similar way to obtain the following result. The computations are done using Mathematica.

Theorem 7 *Let* $d = 24$. *Then*

$$\mathcal{E}_{24}(\mathbb{S}^{24}; N) \geq \frac{676039}{8388608} N^2 \log N + F_{24}(c_{24}^*)\, N^2 \\ + \frac{13 \times 3^{\frac{7}{12}} 7^{\frac{5}{6}} 1062347^{\frac{11}{12}}}{2097152 \times 2^{\frac{11}{12}} \sqrt{65}} N^{\frac{23}{12}} + O\!\left(N^{\frac{11}{6}}\right), \tag{66}$$

where

$$F_{24}(c_{24}^*) = \frac{2028117}{2097152}\left(\gamma + \frac{11}{12}\log 2 + \frac{5}{12}\log 3 + \frac{1}{6}\log 5 + \frac{1}{6}\log 7 + \frac{1}{12}\log 11 \right.$$
$$\left. + \frac{1}{12}\log 13 + \frac{1}{12}\log 17 + \frac{1}{12}\log 19 + \frac{1}{12}\log 23 - \frac{42433}{6930}\right)$$
$$= -2.61483542770544488\dots.$$

Remark 5 The new bound improves the known bound given in (49) of value

$$\tilde{c}_{24} = -4.78382223827582489\dots.$$

Compared with the conjectured value

$$C_{24,24} = \frac{2028117}{2097152}\left(\gamma - \frac{1}{12}\log 2 + \frac{5}{12}\log 3 + \frac{1}{6}\log 5 + \frac{1}{6}\log 7 + \frac{1}{12}\log 11 \right.$$
$$+ \frac{1}{12}\log 13 + \frac{1}{12}\log 17 + \frac{1}{12}\log 19 + \frac{1}{12}\log 23 - \frac{83711}{27720}$$
$$\left. + \frac{638512875}{691\pi^{12}}\zeta'(12) - \frac{638512875}{691\pi^{12}}\mathrm{L}(12, \Delta) - \log\pi\right)$$
$$= -2.35288856549984884\dots,$$

we have

$$C_{24,24} - F_{24}(c_{24}^*) \approx 0.262, \qquad \left|\frac{C_{24,24} - F_{24}(c_{24}^*)}{C_{24,24}}\right| \approx 11\%.$$

Acknowledgments The author would like to thank the anonymous referee who provided useful comments.

References

1. J. Beck. Sums of distances between points on a sphere—an application of the theory of irregularities of distribution to discrete geometry. *Mathematika*, 31(1):33–41, 1984.
2. C. Beltrán and F. Lizarte. A lower bound for the logarithmic energy on $\mathbb{S}^2$ and for the green energy on $\mathbb{S}^n$. *Constr. Approx.*, 58(3):565–587, 2023.
3. L. Bétermin and E. Sandier. Renormalized energy and asymptotic expansion of optimal logarithmic energy on the sphere. *Constr. Approx.*, 47(1):39–74, 2018.
4. D. Bilyk and J. S. Brauchart. On the lower bounds for the spherical cap discrepancy. https://arxiv.org/abs/2502.15984, 2025.
5. D. Bilyk and F. Dai. Geodesic distance Riesz energy on the sphere. *Trans. Amer. Math. Soc.*, 372(5):3141–3166, 2019.

6. D. Bilyk, F. Dai, and R. Matzke. The Stolarsky principle and energy optimization on the sphere. *Constr. Approx.*, 48(1):31–60, 2018.
7. B. V. Blagouchine. A theorem for the closed-form evaluation of the first generalized stieltjes constant at rational arguments and some related summations. *Journal of Number Theory*, 148:537–592, 2015.
8. S. V. Borodachov, D. P. Hardin, and E. B. Saff. *Discrete energy on rectifiable sets*. Springer Monographs in Mathematics. Springer, New York, [2019] ©2019.
9. J. Brauchart. Punktverteilungen extremaler diskreter Energie auf Sphären. Master's thesis, Graz University of Technology, 2001.
10. J. Brauchart. *Points on an Unit Sphere in* $\mathbb{R}^{d+1}$, *Riesz Energy, Discrepancy and Numerical Integration*. PhD thesis, Graz University of Technology, 2005.
11. J. Brauchart. *Mathematica Calculations for "Logarithmic and Riesz Energy on the Sphere: Better Bounds via elementary Methods"*. Graz University of Technology, 2005. https://doi.org/10.3217/552dw-y3x91
12. J. S. Brauchart. Optimal logarithmic energy points on the unit sphere. *Math. Comp.*, 77(263):1599–1613, 2008.
13. J. S. Brauchart. Optimal discrete Riesz energy and discrepancy. *Unif. Distrib. Theory*, 6(2):207–220, 2011.
14. J. S. Brauchart and J. Dick. A simple proof of Stolarsky's invariance principle. *Proc. Amer. Math. Soc.*, 141(6):2085–2096, 2013.
15. J. S. Brauchart, D. P. Hardin, and E. B. Saff. The next-order term for optimal Riesz and logarithmic energy asymptotics on the sphere. In *Recent advances in orthogonal polynomials, special functions, and their applications*, volume 578 of *Contemp. Math.*, pages 31–61. Amer. Math. Soc., Providence, RI, 2012.
16. H. Cohn and A. Kumar. Universally optimal distribution of points on spheres. *J. Amer. Math. Soc.*, 20(1):99–148, 2007.
17. H. Cohn, A. Kumar, S. D. Miller, D. Radchenko, and M. Viazovska. The sphere packing problem in dimension 24. *Ann. of Math. (2)*, 185(3):1017–1033, 2017.
18. H. Cohn, A. Kumar, S. D. Miller, D. Radchenko, and M. Viazovska. Universal optimality of the E_8 and Leech lattices and interpolation formulas. *Ann. of Math. (2)*, 196(3):983–1082, 2022.
19. *NIST Digital Library of Mathematical Functions*. https://dlmf.nist.gov/, Release 1.2.0 of 2024-03-15. F. W. J. Olver, A. B. Olde Daalhuis, D. W. Lozier, B. I. Schneider, R. F. Boisvert, C. W. Clark, B. R. Miller, B. V. Saunders, H. S. Cohl, and M. A. McClain, eds.
20. P. D. Dragnev. On the separation of logarithmic points on the sphere. In *Approximation Theory X: Abstract and Classical Analysis*, pages 137–144. Vanderbilt University Press, Nashville, TN, 2002.
21. A. Dubickas. On the maximal product of distances between points on a sphere. *Liet. Mat. Rink.*, 36(3):303–312, 1996.
22. A. B. J. Kuijlaars and E. B. Saff. Asymptotics for minimal discrete energy on the sphere. *Trans. Amer. Math. Soc.*, 350(2):523–538, 1998.
23. A. B. Lauritsen. Floating Wigner crystal and periodic jellium configurations. *J. Math. Phys.*, 62(8):Paper No. 083305, 17, 2021.
24. M. Petrache and S. Serfaty. Crystallization for Coulomb and Riesz interactions as a consequence of the Cohn-Kumar conjecture. *Proc. Amer. Math. Soc.*, 148(7):3047–3057, 2020.
25. A. P. Prudnikov, Y. A. Brychkov, and O. I. Marichev. *Integrals and series. Vol. 3*. Gordon and Breach Science Publishers, New York, 1990. More special functions, Translated from the Russian by G. G. Gould.
26. E. A. Rakhmanov, E. B. Saff, and Y. M. Zhou. Minimal discrete energy on the sphere. *Math. Res. Lett.*, 1(6):647–662, 1994.
27. E. A. Rakhmanov, E. B. Saff, and Y. M. Zhou. Electrons on the sphere. In *Computational methods and function theory 1994 (Penang)*, volume 5 of *Ser. Approx. Decompos.*, pages 293–309. World Sci. Publ., River Edge, NJ, 1995.
28. E. B. Saff and A. B. J. Kuijlaars. Distributing many points on a sphere. *Math. Intelligencer*, 19(1):5–11, 1997.

29. E. Sandier and S. Serfaty. From the Ginzburg-Landau model to vortex lattice problems. *Comm. Math. Phys.*, 313(3):635–743, 2012.
30. P. Sarnak and A. Strömbergsson. Minima of Epstein's zeta function and heights of flat tori. *Invent. Math.*, 165(1):115–151, 2006.
31. S. Serfaty. Ginzburg-Landau vortices, Coulomb gases, and renormalized energies. *J. Stat. Phys.*, 154(3):660–680, 2014.
32. S. Smale. Mathematical problems for the next century. *Math. Intelligencer*, 20(2):7–15, 1998.
33. S. Steinerberger. On the logarithmic energy of points on $\mathbb{S}^2$. *J. Anal. Math.*, 148(1):187–211, 2022.
34. K. B. Stolarsky. Sums of distances between points on a sphere. II. *Proc. Amer. Math. Soc.*, 41:575–582, 1973.
35. M. S. Viazovska. The sphere packing problem in dimension 8. *Ann. of Math. (2)*, 185(3):991–1015, 2017.

Energy of Codes with Forbidden Distances in 48 Dimensions

Peter G. Boyvalenkov and Peter D. Dragnev

Dedicated to Edward B. Saff on the occasion of his 80th birthday.

1 Introduction

Let $\mathbb{S}^{n-1} = \{x = (x_1, \ldots, x_n) : x_1^2 + \cdots + x_n^2 = 1\}$ be the unit Euclidean sphere in n dimensions. A finite nonempty set $C \subset \mathbb{S}^{n-1}$ is a *spherical (n,N,s) code* if $|C| = N \geq 2$ and $s = s(C) = \max\{x \cdot y : x, y \in C, x \neq y\}$; $s(C)$ is called *maximal cosine* of C. Denote by

$$I(C) := \{x \cdot y : x, y \in C, x \neq y\}$$

the set of all inner products of distinct points of C. Note that $s(C) = \max I(C)$.

Definition 1 Let $T \subset [-1, 1)$. A spherical code $C \subset \mathbb{S}^{n-1}$ is called T-avoiding if $I(C) \cap T = \phi$.

As is often the case in the study of good spherical codes, the notion of *spherical designs* plays a significant role. Spherical designs were introduced in 1977 by Delsarte, Goethals, and Seidel with several equivalent definitions, one of them being the following.

Definition 2 A spherical τ-design is a spherical code $C \subset \mathbb{S}^{n-1}$ such that

P. G. Boyvalenkov (✉)
Institute of Mathematics and Informatics, Bulgarian Academy of Sciences, Sofia, Bulgaria
e-mail: peter@math.bas.bg

P. D. Dragnev
Department of Mathematical Sciences, Purdue University, Fort Wayne, IN, USA
e-mail: dragnevp@pfw.edu

D. Bilyk et al. (eds.), *Recent Advances in Approximation and Potential Theory*, Applied and Numerical Harmonic Analysis,
https://doi.org/10.1007/978-3-031-95551-8_7

$$\int_{\mathbb{S}^{n-1}} p(x)d\sigma_n(x) = \frac{1}{|C|}\sum_{x\in C} p(x)$$

holds for all polynomials $p(x) = p(x_1, x_2, \dots, x_n)$ of total degree at most τ. Here σ_n is the normalized (unit) Lebesgue surface measure.

Given a function $h : [-1, 1] \to (-\infty, +\infty]$, continuous on $[-1, 1)$, we consider the *discrete h -energy* of C

$$\mathcal{E}^h(C) := \frac{1}{N} \cdot \sum_{x,y\in C, x\neq y} h(x \cdot y).$$

For fixed $T \subset [-1, 1)$ we define corresponding minimum and maximum quantities

$$\mathcal{P}_{n,h}(N, T) := \min\{\mathcal{E}^h(C) : C \subset \mathbb{S}^{n-1},\ |C| = N,\ C \text{ is } T\text{-avoiding}\},$$

$$\mathcal{P}_{n,h}(\tau, N, T) :=$$
$$\min\{\mathcal{E}^h(C) : C \subset \mathbb{S}^{n-1},\ |C| = N,\ C \text{ is a } T\text{-avoiding spherical } \tau\text{-design}\},$$

$$Q^{n,h}(N, T) := \max\{\mathcal{E}^h(C) : C \subset \mathbb{S}^{n-1},\ |C| = N,\ C \text{ is } T\text{-avoiding}\},$$

$$Q^{n,h}(\tau, N, T) :=$$
$$\max\{\mathcal{E}^h(C) : C \subset \mathbb{S}^{n-1},\ |C| = N,\ C \text{ is a } T\text{-avoiding spherical } \tau\text{-design}\},$$

where in the latter two cases we additionally assume that h is continuous and finite at 1.

In this paper we consider T-avoiding spherical codes on $\mathbb{S}^{47}$ for the special choice of

$$T_1 := \left(-\frac{1}{3}, -\frac{1}{6}\right) \cup \left(\frac{1}{6}, \frac{1}{3}\right), \quad T_2 := \left(-\frac{1}{2}, -\frac{1}{3}\right) \cup \left(\frac{1}{3}, \frac{1}{2}\right). \tag{1}$$

We prove that among all codes on $\mathbb{S}^{47}$ with cardinality 52 416 000, the four exceptional codes, namely the sets of the minimal vectors (normalized to be unit) of the four known even unimodular extremal lattices P_{48p}, P_{48q}, P_{48m}, and P_{48n} in $\mathbb{R}^{48}$ (cf. [9] for the first three and [13] for the fourth) have optimal h-energy for every absolutely monotone h among all T_2-avoiding codes and among all T_1-avoiding codes that are antipodal or spherical 3-designs. Moreover, a code C in either class that attains our universal lower bound is an 11-design with prescribed inner product set $I(C)$ and frequency distribution of those inner products. Such designs necessarily have 52 416 000 points and this is the minimum possible their cardinality by Boyvalenkov and Cherkashin [3].

Our main tool is the linear programming (LP). We derive LP bounds for the quantities $\mathcal{P}_{48,h}(N, T_i)$, $i = 1, 2$, and $\mathcal{Q}_{48,h}(11, N, T_1)$, which turn out to coincide with the actual h-energy of the four 11-designs under consideration.

In [8] the authors consider lattices in $\mathbb{R}^{48}$ which avoid certain distances which are, after reformulating into inner products, leading exactly to the set T_1. The main result in [8] is that among these, say T_1-avoiding lattices, the four exceptional lattices mentioned above have maximal possible density. This is a result in the spirit of the celebrated proofs of packing optimality of the E_8 [17] and the Leech [7] lattices. The optimality of the cardinality of the kissing configurations of these lattices among the T_1-avoiding $(48, N, 1/2)$ codes and T_1-avoiding and T_2-avoiding spherical 11-designs was established in [3] (see Theorems 1 and 2 below), resembling the Levenshtein's finding [12] of the kissing numbers in dimensions 8 and 24 (see also [14]).

We present general linear programming bound for T-avoiding spherical codes and designs in Sect. 2. In Sect. 3 the four remarkable spherical 11-designs are presented with some of their properties including their (common) distance distribution. The main results are presented in Sect. 4. In Theorem 3 we prove the universal optimality of the four codes among all T_1-avoiding antipodal codes or 3-designs. In Theorem 4 the universal optimality is among all T_2-avoiding codes. We also discuss the optimality of these codes with respect to linear programming upper bounds.

2 General Linear Programming Bounds for the h-Energy of T-Avoiding Codes and Designs

2.1 Some Preliminaries

We derive (more or less folklore) general LP bounds on the minimum and maximum possible h-energy of T-avoiding spherical codes and designs. The two main theorems below are true for general $T \subset [-1, 1)$. We follow the LP framework developed in [4] for spherical designs and in [5] for general spherical codes.

With each real one-variable polynomial f we associate its (unique) expansion in terms of the *Gegenbauer polynomials*[1] $\{P_i^{(n)}(t)\}_{i=0}^{\infty}$,

$$f(t) = \sum_{i=0}^{\deg f} f_i P_i^{(n)}(t),$$

[1] We use the version of the Gegenbauer polynomials $\{P_i^{(n)}(t)\}_{i=0}^{\infty}$ in which they are orthogonal with respect to the weight $w(t) := (1 - t^2)^{(n-3)/2}$ and are normalized by $P_i^{(n)}(1) = 1$.

where, with $w(t) = (1-t^2)^{(n-3)/2}$, the coefficients f_i, $i = 0, 1, \dots, \deg f$, are uniquely determined by

$$f_i = \int_{-1}^{1} f(t) P_i^{(n)}(t)\, w(t)\, dt \Big/ \int_{-1}^{1} \left(P_i^{(n)}(t)\right)^2 w(t)\, dt. \tag{2}$$

In particular,

$$f_0 = c_n \int_{-1}^{1} f(t) \cdot w(t) dt,$$

where $c_n := \Gamma(n/2)/\sqrt{\pi}\Gamma((n-1)/2)$.

Note that the Gegenbauer polynomials are positive definite (and so are their nonnegative linear combinations). This implies, in particular, that

$$M_i(C) := \sum_{x,y \in C} P_i^{(n)}(x \cdot y) \geq 0$$

for every code $C \subset \mathbb{S}^{n-1}$ and every positive integer i (cf. [15], see also [2, Chapter 5] for comprehensive discussion). We call $M_i(C)$ moments of C. The moments serve for an equivalent (and very useful) definition of spherical designs.

Definition 3 Let τ be a positive integer. A spherical code $C \subset \mathbb{S}^{n-1}$ is a spherical τ-design if and only if $M_1(C) = \cdots = M_\tau(C) = 0$.

Note also that $M_i(C) = 0$ for all odd i if and only if the code C is antipodal, i.e. $C = -C$.

2.2 Lower LP Bounds

Definition 4 Let h be a potential function. Denote by $\mathcal{L}(n, h; T)$ the class of *lower admissible polynomials* $f(t)$ such that

(A1) $f(t) \leq h(t)$ for every $t \in [-1, 1] \setminus T$;
(A2) $f_i \geq 0$ for all $i \geq 1$.

Definition 5 Let τ be a positive integer and h be a potential function. Denote by $\mathcal{L}(n, \tau, h; T)$ the class of *lower τ-admissible polynomials* $f(t)$ such that

(A1) $f(t) \leq h(t)$ for every $t \in [-1, 1] \setminus T$,
(A2)′ $\deg(f) \leq \tau$.

Utilizing these definitions one derives the following Delsarte-Yudin type (cf. [10, 18], Chapter 5 in [2]) LP lower bounds on the minimum possible energy of any T-avoiding spherical codes and designs on $\mathbb{S}^{n-1}$.

Proposition 1 (Lower LP Bound) *Let $h(t)$ be a potential function and $f \in \mathcal{L}(n, h; T)$ (respectively, $f \in \mathcal{L}(n, \tau, h; T)$). Then*

$$\mathcal{E}^h(C) \geq f_0 N - f(1) \tag{3}$$

for every T-avoiding spherical code (respectively, τ-design) $C \subset \mathbb{S}^{n-1}$. Consequently,

$$\mathcal{P}_{n,h}(N, T) \geq \max_{f \in \mathcal{L}(n,h;T)} \{f_0 N - f(1)\}, \tag{4}$$

and

$$\mathcal{P}_{n,h}(\tau, N, T) \geq \max_{f \in \mathcal{L}(n,\tau,h;T)} \{f_0 N - f(1)\}, \tag{5}$$

respectively.

If the bound (4) *is attained by some polynomial $f \in \mathcal{L}(n, h; T)$ and some code $C \subset \mathbb{S}^{n-1}$, then $f_i M_i(C) = 0$ for each $i \geq 1$ and $I(C)$ is a subset of the set of zeros of $f - h$. If the bound* (5) *is attained by some polynomial $f \in \mathcal{L}(n, \tau, h; T)$ and some τ-design $C \subset \mathbb{S}^{n-1}$, then $I(C)$ is a subset of the set of zeros of $f - h$.*

Proof This is immediate from the identity

$$f(1)N + \sum_{x,y \in C, x \neq y} f(x \cdot y) = f_0 N^2 + \sum_{i=1}^{\deg(f)} f_i M_i(C) \tag{6}$$

(see, e.g. Equation (1.9) in [11]; also [10, Corollary 3.8]), Definition 4 (Definition 5, respectively), and the inequalities $M_i(C) \geq 0$. Indeed, the LHS of (6) is at most $N(f(1) + \mathcal{E}^h(C))$, in both cases because of the condition (A1). The RHS is at least $f_0 N^2$ because of (A2) in the case of Definition 4 and equal to $f_0 N^2$ in the case of Definition 5. □

In fact, we will use two modifications of Proposition 1 which impose requirements only either on f_i for odd i or on f_i for $i \geq 4$.

2.3 Upper LP Bounds

Similarly, we define the upper admissible polynomials and derive the corresponding upper linear programming bounds.

Definition 6 Let h be a potential function. Denote by $\mathcal{U}(n, h; T)$ the class of *upper admissible polynomials* $g(t)$ such that

(B1) $g(t) \geq h(t)$ for every $t \in [-1, 1] \setminus T$;

(B2) $g_i \leq 0$ for all $i \geq 1$.

Definition 7 Let τ be a positive integer and h be a potential function. Denote by $\mathcal{U}(n, \tau, h; T)$ the class of *upper τ-admissible polynomials* $g(t)$ such that

(B1) $g(t) \geq h(t)$ for every $t \in [-1, 1] \setminus T$,
(B2)′ $\deg(g) \leq \tau$.

The Delsarte-Yudin type upper bounds follow.

Proposition 2 (Upper LP Bound) *Let $h(t)$ be a potential function which is continuous and finite at* 1 *and $g \in \mathcal{U}(n, h; T)$ (respectively, $g \in \mathcal{U}(n, \tau, h; T)$). Then*

$$\mathcal{E}^h(C) \leq g_0 N - g(1) \tag{7}$$

for every T-avoiding spherical code (respectively, τ-design) $C \subset \mathbb{S}^{n-1}$. Consequently,

$$\mathcal{Q}_{n,h}(N, T) \leq \min_{g \in \mathcal{U}(n,h;T)} \{g_0 N - g(1)\}, \tag{8}$$

respectively,

$$\mathcal{Q}_{n,h}(\tau, N, T) \leq \min_{g \in \mathcal{U}(n,\tau,h;T)} \{g_0 N - g(1)\}. \tag{9}$$

If the bound (8) *is attained by some polynomial $g \in \mathcal{U}(n, h; T)$ and some code $C \subset \mathbb{S}^{n-1}$, then $g_i M_i(C) = 0$ for each $i \geq 1$ and $I(C)$ is a subset of the set of zeros of $g - h$. If the bound* (9) *is attained by some polynomial $g \in \mathcal{U}(n, \tau, h; T)$ and some τ-design $C \subset \mathbb{S}^{n-1}$, then $I(C)$ is a subset of the set of zeros of $g - h$.*

Proof Similarly to Proposition 1 this follows from the identity (6), Definition 6 (Definition 7, respectively), and the inequalities $M_i(C) \geq 0$. □

3 A Class of Remarkable Spherical 11-Designs on $\mathbb{S}^{47}$

Motivated by Gonçalves and Vedana [8] and Boyvalenkov and Cherkashin [3], we will apply the framework from the previous section in the particular case $n = 48$, $N = 52\,416\,000$, and avoiding sets T_i, $i = 1, 2$ (see (1)). In addition, when spherical designs are in consideration, we shall assume $\tau = 11$ (in fact, these are spherical 11½-designs, see Venkov [16]; in other words, all they have zeroth fourteenth moment $M_{14}(.) = 0$. In what follows, these values of n, τ, N, and T_i, $i = 1, 2$, will be fixed.

We recall the notion of distance distribution of a (spherical) code. For any $x \in C$ and $t \in I(C)$, one denotes by

$$A_t(x) := |\{y \in C : x \cdot y = t\}|,$$

the number of the points of C with inner product t with x. The system of nonnegative integers

$$(A_t(x) : t \in I(C))$$

is called *distance distribution of C with respect to x*. If all $A_t(x)$ do not depend on the choice of x, the code C is called *distance invariant* (cf. [10, Definition 7.2]) and one omits x in the notation. We remark that $A_{-1} = 1$ means that C is antipodal. In antipodal codes one has $A_t(x) = A_{-t}(x)$ for every $t \in I(C) \setminus \{-1\}$ and every $x \in C$.

There are at least four non-isomorphic spherical 11-designs on $\mathbb{S}^{47}$, formed as the sets of minimal vectors of the even unimodular extremal lattices P_{48p}, P_{48q}, P_{48m}, and P_{48n} in $\mathbb{R}^{48}$, normalized on the unit sphere. All these codes are distance invariant with 8 distinct distances (cf. [10, Theorem 7.4]) and all they have the same distance distribution (cf. the calculation in [3] via equation (1.10) from [11]):

$$\begin{aligned} A_{-1} &= 1, \\ A_{1/2} = A_{-1/2} &= 36\,848, \\ A_{1/3} = A_{-1/3} &= 1\,678\,887, \\ A_{1/6} = A_{-1/6} &= 12\,608\,784, \\ A_0 &= 23\,766\,960. \end{aligned} \tag{10}$$

Remark 1 We note that the set of 23 766 960 points from the distance distribution (10) defines 47-dimensional kissing configuration which is far superior than what is given (9 741 412) in the webpage of Henry Cohn [6].

The above information is enough for calculation of the h-energy of our target codes. Thus, the h-energy of each of the above four codes (say, C) is the same, given by

$$\begin{aligned} \mathcal{E}^h(C) = 36848\left(h\left(-\frac{1}{2}\right) + h\left(\frac{1}{2}\right)\right) + 1678887\left(h\left(-\frac{1}{3}\right) + h\left(\frac{1}{3}\right)\right) \\ +12608784\left(h\left(-\frac{1}{6}\right) + h\left(\frac{1}{6}\right)\right) + 23766960h(0) + h(-1). \end{aligned}$$

It was shown in [3] that the four codes above have optimal cardinality among the T_1-avoiding antipodal codes with maximal cosine $1/2$ and among the T_1-avoiding spherical 3-designs on $\mathbb{S}^{47}$.

Theorem 1 (Theorem 5.1 in [3]) *Let $C \subset \mathbb{S}^{47}$ be a T_1-avoiding spherical $(48, N, 1/2)$ code which is either antipodal or a spherical 3-design. Then $N \leq$*

52 416 000. *If the equality is attained, then C is an antipodal spherical 11-design and, moreover, it is distance invariant and its distance distribution is as given in* (10).

Also in [3], the minimal cardinality of T_1-avoiding and T_2-avoiding 11-designs on $\mathbb{S}^{47}$ was established.

Theorem 2 (Theorem 6.1 and 6.2 in [3]) *Let $C \subset \mathbb{S}^{47}$ be a T_i-avoiding spherical 11-design, with T_i, $i = 1, 2$, as in* (1). *Then $|C| \geq 52\,416\,000$. If the equality is attained in either case, then C is a* $(48\,52\,416\,000, 1/2)$ *antipodal spherical code which is distance invariant with distance distribution as in* (10).

The existence of spherical 11-designs with 52 416 000 points with distance distribution (10) implies the existence of a quadrature formula (see, e.g., Equation (1.10) in [11]) as follows.

Proposition 3 *For every polynomial f of degree at most 11, it follows that*

$$f_0 = \frac{1}{52\,416\,000}\left(36848\left(f\left(-\frac{1}{2}\right)+f\left(\frac{1}{2}\right)\right)+1678887\left(f\left(-\frac{1}{3}\right)+f\left(\frac{1}{3}\right)\right)\right.$$
$$\left.+12608784\left(f\left(-\frac{1}{6}\right)+f\left(\frac{1}{6}\right)\right)+23766960f(0)+f(-1)+f(1)\right). \quad (11)$$

Proof This follows by the existence of spherical 11-designs on $\mathbb{S}^{47}$ of cardinality 52 416 000 and, consequently, with distance distribution (10). We apply the definition of spherical designs via Equation (1.10) in [11] with f of degree at most 11 and the point y in that equation belonging to the design.

Another proof follows by applying the identity (6) for a spherical 11-designs on $\mathbb{S}^{47}$ of cardinality 52 416 000 and using the fact that such a design is distance regular (since $|I(C)| = 8 < 11$, the strength of the design [10, Theorem 7.4]). □

4 Energy Bounds for T-Avoiding Codes on $\mathbb{S}^{47}$

In this section we assume that the function h has either positive twelfth derivative or is absolutely monotone, i.e. $h^{(i)} \geq 0$ for every $i \geq 0$. In fact, only finitely many positive derivatives (up to twelfth) are enough as we apply Propositions 1 and 2 with polynomials of degree at most 11.

4.1 Lower Bounds for T_1-Avoiding Sets

We shall apply (slightly modified, see in the proof of Theorem 3) Proposition 1 with the polynomial f that interpolates the potential function h as follows:

$$f(a) = h(a),\ \ f'(a) = h'(a)$$

for $a = -1,\ \pm 1/2$, and 0 (four times double interpolation),

$$f(b) = h(b)$$

for $b = \pm 1/3$ and $\pm 1/6$ (four times single interpolation). Then f is an 11-degree polynomial to be used in the theorem below. Recall that $n = 48$, $\tau = 11$, $N = 52\,416\,000$, and $T_1 = (-1/3, -1/6) \cup (1/6, 1/3)$.

Theorem 3 *Let h be absolutely monotone with $h^{(12)} > 0$ in $(-1, 1)$. Let $C \subset \mathbb{S}^{47}$ be a T_1-avoiding spherical code with $|C| = 52\,416\,000$ which is either antipodal or a 3-design. Then*

$$\begin{aligned}\mathcal{E}^h(C) \geq 36848\left(h\left(-\frac{1}{2}\right)+h\left(\frac{1}{2}\right)\right) + 1678887\left(h\left(-\frac{1}{3}\right)+h\left(\frac{1}{3}\right)\right)\\ + 12608784\left(h\left(-\frac{1}{6}\right)+h\left(\frac{1}{6}\right)\right) + 23766960h(0) + h(-1).\end{aligned} \tag{12}$$

The equality is attained when C is an antipodal spherical 11-design that is distance invariant and its distance distribution is as given in (10). *In particular, the four codes formed by the minimum norm vectors in the even unimodular extremal lattices P_{48p}, P_{48q}, P_{48m}, and P_{48n} in $\mathbb{R}^{48}$, respectively, attain the bound* (12) *and hence, are universally optimal among the considered class of codes.*

Remark 2 The proof shows that it suffices to have potentials h with nonnegative (strictly positive) derivatives up to order 12. Note also that if $h^{(12)}(t_0) = 0$ for some $t_0 \in (-1, 1)$, then the absolute monotonicity implies that $h^{(12)}(t) \equiv 0$ on $(-1, t_0]$ and that h is a polynomial of degree at most 11.

Proof We first establish that $f(t) \leq h(t)$ for every $t \in [-1, 1) \setminus T_1$, i.e. condition (A1) in both Definitions 4 and 5 is satisfied. Indeed, the Hermite interpolation error formula implies that

$$h(t) - f(t) =$$

$$\frac{h^{(12)}(\xi)}{12!} t^2 (t+1)^2 \left(t+\frac{1}{2}\right)^2 \left(t-\frac{1}{2}\right)^2 \left(t+\frac{1}{3}\right)\left(t+\frac{1}{6}\right)\left(t-\frac{1}{6}\right)\left(t-\frac{1}{3}\right) \geq 0,$$

where $\xi \in (-1, 1)$, for every $t \in [1, 1) \setminus T_1$.

While the interpolant f does not necessarily satisfy condition $(A2)$ of Definition 4, we will prove a modified condition which will allow us to conclude (12) for codes that are antipodal or (at least) 3-designs. Utilizing Newton interpolation formula, we have (see, for example, [1])

$$f(t) = h(t_1) + \sum_{r=1}^{11} h[t_1, \ldots, t_{r+1}] \prod_{j=1}^{r} (t - t_j),$$

where

$$(t_1, t_2, t_3, t_4, t_5, t_6, t_7, t_8, t_9, t_{10}, t_{11}, t_{12}) = \left(-1, -1, -\frac{1}{2}, -\frac{1}{2}, -\frac{1}{3}, -\frac{1}{6}, 0, 0, \frac{1}{6}, \frac{1}{3}, \frac{1}{2}, \frac{1}{2}\right)$$

is the interpolation nodes multiset and $h[t_1, \ldots, t_i]$ are the corresponding divided differences.

Since h is absolutely monotone, the divided differences $h[t_1, \ldots, t_{r+1}]$ are nonnegative (actually, only non-negativity of the first 12 derivatives of h suffices). It remains to consider the Gegenbauer expansions of the partial products $PP_r := (t - t_1) \ldots (t - t_r)$ for $r = 1, \ldots, 11$. Each of the factors $t - t_i$ with $t_i \leq 0$ is positive definite, and so are the partial products PP_i, $i = 1, \ldots, 8$. Thus, we need to examine only the partial products PP_9, PP_{10}, and PP_{11}. We present the explicit Gegenbauer expansions of these.

The Gegenbauer expansion of

$$PP_9(t) = \prod_{i=1}^{9} (t - t_i) = \sum_{i=0}^{9} g_{i,9} P_i^{(48)}(t)$$

has

$$g_{0,9} = \frac{107}{336960}, \quad g_{1,9} = \frac{9559}{1684800}, \quad g_{2,9} = \frac{1457}{31104}, \quad g_{3,9} = \frac{662371}{2818800},$$

$$g_{4,9} = \frac{793877}{1002240}, \quad g_{5,9} = \frac{109123049}{58631040}, \quad g_{6,9} = \frac{2444141}{808704}, \quad g_{7,9} = \frac{1873655}{582552},$$

$$g_{8,9} = \frac{296429}{150336}, \quad g_{9,9} = \frac{296429}{575360}.$$

Further,

$$PP_{10}(t) = \prod_{i=1}^{10} (t - t_i) = \sum_{i=0}^{10} g_{i,10} P_i^{(48)}(t)$$

has

$$g_{0,10} = \frac{37}{2995200}, \quad g_{1,10} = \frac{337}{1123200}, \quad g_{2,10} = \frac{984415}{281428992}, \quad g_{3,10} = \frac{1712633}{67651200},$$

$$g_{4,10} = \frac{96599993}{781747200},\ g_{5,10} = \frac{584962}{1374165},\ g_{6,10} = \frac{1598663969}{1504189440},\ g_{7,10} = \frac{5878243}{3106944},$$

$$g_{8,10} = \frac{651254513}{288645120},\ g_{9,10} = \frac{889287}{575360},\ g_{10,10} = \frac{3260719}{7364608},$$

and

$$PP_{11}(t) = \prod_{i=1}^{11}(t - t_i) = \sum_{i=0}^{11} g_{i,11} P_i^{(48)}(t)$$

(this is the polynomial utilized in Theorem 5.1 of [3]) has

$$g_{0,11} = \frac{1}{13478400},\ g_{1,11} = \frac{3961}{1758931200},\ g_{2,11} = \frac{47}{8794656},$$

$$g_{3,11} = -\frac{118957}{811814400},\ g_{4,11} = \frac{122059}{1563494400},\ g_{5,11} = \frac{376856011}{32716120320},$$

$$g_{6,11} = \frac{231656467}{3008378880},\ g_{7,11} = \frac{399983395}{1342199808},\ g_{8,11} = \frac{439011349}{577290240},$$

$$g_{9,11} = \frac{3260719}{2589120},\ g_{10,11} = \frac{16303595}{14729216},\ g_{11,11} = \frac{2075003}{5523456}.$$

Thus, the partial products PP_9 and PP_{10} are positive definite. Furthermore, PP_{11} has only one negative Gegenbauer coefficient, namely $g_{3,11}$, and all others are (strictly) positive. Therefore, all Gegenbauer coefficients of $f(t)$ are (strictly) positive, with the possible exception of f_3, because $h^{(12)}(t) > 0$ on $(-1, 1)$.

Note that for all antipodal codes or spherical 3-designs, we have $M_3(C) = 0$, and thus $f_3 M_3(C) = 0$ in the identity (6). Therefore, for such codes

$$\mathcal{E}^h(C) \geq f_0 N - f(1)$$

for our interpolation polynomial f. Using (11) and the interpolation conditions we conclude

$$f_0 N - f(1) = 36848\left(h\left(-\frac{1}{2}\right) + h\left(\frac{1}{2}\right)\right) + 1678887\left(h\left(-\frac{1}{3}\right) + h\left(\frac{1}{3}\right)\right)$$
$$+12608784\left(h\left(-\frac{1}{6}\right) + h\left(\frac{1}{6}\right)\right) + 23766960h(0) + h(-1),$$

which proves (12). If equality holds for some potential h, as $f_i > 0$, $i = 1, \ldots, 11, i \neq 3$, we shall have $M_i(C) = 0, i = 1, \ldots, 11, i \neq 3$, which along with $M_3(C) = 0$ implies that C is an 11-design. Since $|C| = 52\,416\,000$, we complete the proof of the theorem by using the part of Theorem 2 about T_1 (i.e., [3, Theorem 6.1]). □

Remark 3 We note that we actually prove the universal optimality of the sets of minimal vectors of the even unimodular extremal lattices P_{48p}, P_{48q}, P_{48m}, and P_{48n} in $\mathbb{R}^{48}$ in the class of T_1-avoiding codes with $M_3(C) = 0$.

4.2 *Lower Bounds for T_2-Avoiding Sets*

In this case we modify f to be the polynomial that interpolates the potential function h as follows:

$$f(a) = h(a), \;\; f'(a) = h'(a)$$

for $a = -1, \pm 1/6$, and 0 (four times double interpolation),

$$f(b) = h(b)$$

for $b = \pm 1/2$ and $\pm 1/3$ (four times single interpolation). It is still uniquely determined 11-degree polynomial. We still have $n = 48$, $\tau = 11$, $N = 52\,416\,000$, but our avoiding set is $T_2 = (-1/2, -1/3) \cup (1/3, 1/2)$.

Theorem 4 *Let h be absolutely monotone with $h^{(12)} > 0$ in $(-1, 1)$. Let $C \subset \mathbb{S}^{47}$ be any T_2-avoiding spherical code with $|C| = 52\,416\,000$. Then*

$$\begin{aligned} \mathcal{E}^h(C) \geq\; & 36848\left(h\left(-\frac{1}{2}\right) + h\left(\frac{1}{2}\right)\right) + 1678887\left(h\left(-\frac{1}{3}\right) + h\left(\frac{1}{3}\right)\right) \\ & + 12608784\left(h\left(-\frac{1}{6}\right) + h\left(\frac{1}{6}\right)\right) + 23766960h(0) + h(-1). \end{aligned} \tag{13}$$

The equality is attained when C is an antipodal spherical 11-design, which is distance invariant and its distance distribution is as given in (10). *In particular, the four codes formed by the minimum norm vectors in the even unimodular extremal lattices P_{48p}, P_{48q}, P_{48m}, and P_{48n} in $\mathbb{R}^{48}$, respectively, attain the bound* (13) *and hence, are universally optimal among any T_2 -avoiding codes.*

Remark 4 As in the previous subsection it suffices to have potentials h with nonnegative (strictly positive) derivatives up to order 12.

Proof That $f(t) \leq h(t)$ for every $t \in [1, 1) \setminus T_2$ is derived similarly. The Hermite interpolation error formula in this case yields that for any such t we have with some $\xi \in (-1, 1)$

$h(t) - f(t) =$

$$\frac{h^{(12)}(\xi)}{12!} t^2(t+1)^2 \left(t+\frac{1}{6}\right)^2 \left(t-\frac{1}{6}\right)^2 \left(t+\frac{1}{2}\right) \left(t+\frac{1}{3}\right) \left(t-\frac{1}{3}\right) \left(t-\frac{1}{2}\right) \geq 0.$$

The Newton interpolation formula for the multi-set

$$(t_1, t_2, t_3, t_4, t_5, t_6, t_7, t_8, t_9, t_{10}, t_{11}, t_{12}) = \left(-1, -1, -\frac{1}{2}, -\frac{1}{3}, -\frac{1}{6}, -\frac{1}{6}, 0, 0, \frac{1}{6}, \frac{1}{6}, \frac{1}{3}, \frac{1}{2}\right)$$

yields

$$f(t) = h(t_1) + \sum_{r=1}^{11} h[t_1, \ldots, t_{r+1}] \prod_{j=1}^{r} (t - t_j),$$

where $h[t_1, \ldots, t_i]$ are the relevant divided differences, which similarly are nonnegative by the absolute monotonicity of h.

The partial products $PP_r := (t - t_1) \ldots (t - t_r)$ for $r = 1, \ldots, 11$ are all positive definite as we establish below (as in Sect. 4.1 we only need to consider $r = 9, 10, 11$).

The Gegenbauer expansion of

$$PP_9(t) = \prod_{i=1}^{9} (t - t_i) = \sum_{i=0}^{9} g_{i,9} P_i^{(48)}(t)$$

has

$$g_{0,9} = \frac{7903}{40435200}, \; g_{1,9} = \frac{371}{105300}, \; g_{2,9} = \frac{47705}{1617408}, \; g_{3,9} = \frac{15341599}{101476800},$$

$$g_{4,9} = \frac{51317749}{97718400}, \; g_{5,9} = \frac{677167211}{527679360}, \; g_{6,9} = \frac{743869}{336960}, \; g_{7,9} = \frac{12041729}{4660416},$$

$$g_{8,9} = \frac{296429}{167040}, \; g_{9,9} = \frac{296429}{575360}.$$

Similarly, we compute

$$PP_{10}(t) = \prod_{i=1}^{10} (t - t_i) = \sum_{i=0}^{10} g_{i,10} P_i^{(48)}(t)$$

with

$$g_{0,10} = \frac{1981}{48522240},\ g_{1,10} = \frac{3983}{5054400},\ g_{2,10} = \frac{680701}{93809664},\ g_{3,10} = \frac{25583369}{608860800},$$

$$g_{4,10}=\frac{79510417}{469048320},\ g_{5,10}=\frac{1585405927}{3166076160},\ g_{6,10}=\frac{331592191}{300837888},\ g_{7,10}=\frac{49704991}{27962496},$$

$$g_{8,10} = \frac{118868029}{57729024},\ g_{9,10} = \frac{5039293}{3452160},\ g_{10,10} = \frac{3260719}{7364608}.$$

Finally,

$$PP_{11}(t) = \prod_{i=1}^{11}(t - t_i) = \sum_{i=0}^{11} g_{i,11} P_i^{(48)}(t)$$

where

$$g_{0,11} = \frac{511}{181958400},\ g_{1,11} = \frac{40103}{586310400},\ g_{2,11} = \frac{328013}{422143488},$$

$$g_{3,11} = \frac{40304803}{7306329600},\ g_{4,11} = \frac{391174091}{14071449600},\ g_{5,11} = \frac{30641555483}{294445082880},$$

$$g_{6,11} = \frac{32981921}{111421440},\ g_{7,11} = \frac{98632555}{149133312},\ g_{8,11} = \frac{209575303}{192430080},$$

$$g_{9,11} = \frac{296429}{215760},\ g_{10,11} = \frac{16303595}{14729216},\ g_{11,11} = \frac{2075003}{5523456}.$$

Hence, $f(t)$ is positive definite and since $h^{(12)}(t) > 0$ on $(-1, 1)$ we obtain that $f_i > 0$, $i \in \{1, \ldots, 11\}$. We now conclude that for T_2-avoiding codes

$$\mathcal{E}^h(C) \geq f_0 N - f(1).$$

From (11) and the interpolation conditions we derive

$$f_0N - f(1) = 36848\left(h\left(-\frac{1}{2}\right) + h\left(\frac{1}{2}\right)\right) + 1678887\left(h\left(-\frac{1}{3}\right) + h\left(\frac{1}{3}\right)\right)$$
$$+12608784\left(h\left(-\frac{1}{6}\right) + h\left(\frac{1}{6}\right)\right) + 23766960h(0) + h(-1),$$

which proves (13). Should equality hold for some C and any potential h with $h^{(12)}(t) > 0$, $t \in (-1, 1)$, the strict positivity of f_i, $i = 1, \ldots, 11$, implies $M_i(C) = 0$, which shows C is an 11-design. From the part of Theorem 2 about

T_2 (i.e., [3, Theorem 6.2]) and $|C| = 52\,416\,000$ we conclude that C is distance invariant and its distance distribution is as given in (10). □

4.3 Upper Bounds for T_1-Avoiding Sets

It follows from Theorem 2 that the four codes under consideration will have maximal h-energy among all T_i-avoiding, $i = 1, 2$, 11-designs on $\mathbb{S}^{47}$. We add to the picture by constructing a upper LP bound for Proposition 2 with $T = T_1$, which is valid for all absolutely monotone h.

We apply Proposition 2 with the polynomial g that interpolates the potential function h as follows:

$$g(a) = h(a),\ g'(a) = h'(a)$$

for $a = \pm 1/2$ and 0 (three times double interpolation),

$$g(b) = h(b)$$

for $b = \pm 1, \pm 1/3$, and $\pm 1/6$ (six times single interpolation). Then g is an 11-degree polynomial,

Theorem 5 *Let h have nonnegative twelfth derivative and be finite at* 1. *The four codes formed by the minimum norm vectors in the even unimodular extremal lattices P_{48p}, P_{48q}, P_{48m}, and P_{48n} in $\mathbb{R}^{48}$, respectively, are universally optimal (with respect to the maximum h-energy) in the class of T_1-avoiding spherical 11-designs on $\mathbb{S}^{47}$.*

Proof The error formula now gives

$$h(t) - g(t) =$$

$$\frac{h^{(12)}(\xi)}{12!}t^2(t+1)(t-1)\left(t+\frac{1}{2}\right)^2\left(t-\frac{1}{2}\right)^2\left(t+\frac{1}{3}\right)\left(t+\frac{1}{6}\right)\left(t-\frac{1}{3}\right)\left(t-\frac{1}{6}\right) \leq 0$$

for every $t \in [1, 1) \setminus T$, where $\xi \in (-1, 1)$. Therefore the condition (B1) of Definition 7 is satisfied and $f \in \mathcal{U}(48, 11, h; T)$. We conclude that

$$Q_{48,h}(11, N, T) \leq g_0 N - g(1).$$

The bound $g_0 N - g(1)$ is computed as in Theorem 3 via the quadrature formula (11). Taking into account the interpolation equalities $g(t) = h(t)$ for each $t \in \{-1, \pm 1/2, \pm 1/3, \pm 1/6, 0\}$ we complete the proof. □

Acknowledgments The research of the first author is supported by Bulgarian NSF grant KP-06-N72/6-2023. The research of the second author was supported, in part, by the Lilly Endowment and by the Bulgarian Ministry of Education and Science, Scientific Programme "Enhancing the Research Capacity in Mathematical Sciences (PIKOM)", No. DO1-241/15.08.2023.

References

1. C de Boor, Divided differences, Surveys in Approximation Theory **1**, 46–69 (2005).
2. Borodachov, S. V., Hardin, D. P., Saff, E. B., Discrete Energy on Rectifiable Sets, Springer Monographs in Mathematics, Springer, 2019.
3. P. Boyvalenkov, D. Cherkashin, The kissing number in 48 dimensions for codes with certain forbidden distances is 52 416 000, Results in Mathematics, 80 (2025), no. 1, art. 3.
4. P. Boyvalenkov, P. Dragnev, D. Hardin, E. Saff, M. Stoyanova, Universal upper and lower bounds on energy of spherical designs, *Dolomites Res. Notes Approx.* **8** (2015) 51–65.
5. P. Boyvalenkov, P. Dragnev, D. Hardin, E. Saff, M. Stoyanova, Universal lower bounds for potential energy of spherical codes, *Constr. Approx.* **44** (2016) 385–415.
6. https://cohn.mit.edu/kissing-numbers/, Upper and lower bounds for the kissing numbers in dimensions 1 through 48 and 72.
7. H. Cohn, A. Kumar, S. D. Miller, D. Radchenko, M. Viazovska, The sphere packing problem in dimension 24, *Ann. of Math.* **185** (2017) 1017–1033.
8. F. Gonçalves, G. Vedana, Sphere packings in Euclidean space with forbidden distances, to appear in *Forum Math. Sigma*, **13** (2025) art. e49.
9. J. H. Conway, N. J. A. Sloane, *Sphere Packings, Lattices and Groups*, Springer-Verlag, New York, Third edition, 1999.
10. P. Delsarte, J.-M. Goethals, J. J. Seidel, Spherical codes and designs, *Geom. Dedicata* **6** (1977) 363–388.
11. G. Fazekas, V. I. Levenshtein, On upper bounds for code distance and covering radius of designs in polynomial metric spaces, *J. Comb. Theory Ser. A* **70** (1995) 267–288.
12. V. I. Levenshtein, On bounds for packings in n-dimensional Euclidean space, *Soviet Math. Dokl.* **20** (1979) 417–421.
13. G. Nebe, A fourth extremal even unimodular lattice of dimension 48, *Discr. Math.* **331** (2014) 133–136.
14. A.M. Odlyzko, N.J.A. Sloane, New bounds on the number of unit spheres that can touch a unit sphere in n dimensions, J. Combin. Theory, Ser. A, **26** (1979) 210–214.
15. I. J. Schoenberg, Positive definite functions on spheres, *Duke Math. J.* **9** (1942) 96–107.
16. B. B. Venkov, Even unimodular extremal lattices, *Trudy Mat. Inst. Steklov* **165** (1984) 43–48 (in Russian); *Proc. Steklov Inst. Math.* **165** (1985) 47–52.
17. M. Viazovska, The sphere packing problem in dimension 8, *Ann. of Math.* **185** (2017) 991–1015.
18. V. A. Yudin, Minimum potential energy of a point system of charges, *Discret. Mat.* **4** (1992) 115–121 (in Russian); English translation: *Discr. Math. Appl.* **3** (1993) 75–81.

Frequency-dependent Stability for Gabor Frames

Laura De Carli, Luis Rodriguez, and Pierluigi Vellucci

1 Introduction

A *Gabor system* in $L^2(\mathbb{R})$ is a collection of functions $\mathcal{G} = \{e^{2\pi i b_n x} g(x - a_k)\}_{n,k\in\mathbb{Z}}$, where g (the *window function*) is in $L^2(\mathbb{R})$ and $\{a_k,\ b_n\}_{k,n\in\mathbb{Z}} \subset \mathbb{R}$.

If $(a_k, b_n) = (ak, bn)$ for some $a, b > 0$ and all $n,\ k \in \mathbb{Z}$, we say that $\mathcal{G}$ is *regular* and we let $\mathcal{G}(g; a, b) = \{e^{2\pi binx} g(x - ak)\}_{n,k\in\mathbb{Z}}$.

An important problem is to determine general and verifiable conditions on the window function g, the *time sampling* $\{a_k\}_{k\in\mathbb{Z}}$ and the *frequency sampling* $\{b_n\}_{n\in\mathbb{Z}}$ which imply that a Gabor system is a frame. In the regular case many necessary and sufficient conditions on g, a and b are known (see e.g. [6, 11] and also [3, Chapt. 11], [14, Chapt. 11] and the references cited in these textbooks).

Given a regular Gabor frame $\mathcal{G}(g, a, b)$, it is important to determine *stability bounds* $\omega,\ \delta > 0$ so that each set

$$\mathcal{F} = \{e^{2\pi i b(n+\omega_{n,k})x} g(x - a(k + \delta_{n,k}))\}_{n,k\in\mathbb{Z}} \tag{1}$$

is a frame whenever $|\omega_{n,k}| < \omega$ and $|\delta_{n,k}| < \delta$. The terms $\delta_{n,k}$ models the so-called *timing jitter*, an unwelcome phenomenon of electronic systems. The stability of regular Gabor frames under perturbations of the window function and of the time-shift has been investigated in [5, 10, 24, 25]. In [18] the stability of irregular frames is investigated.

L. De Carli (✉) · L. Rodriguez
Department of Mathematics, Florida International University, Miami, FL, USA
e-mail: decarlil@fiu.edu; LURODRIG@fiu.edu

P. Vellucci
Department of Economics, Roma Tre University, Rome, Italy
e-mail: pierluigi.vellucci@uniroma3.it

D. Bilyk et al. (eds.), *Recent Advances in Approximation and Potential Theory*, Applied and Numerical Harmonic Analysis,
https://doi.org/10.1007/978-3-031-95551-8_8

When $\delta_{n,k} = \delta_k$ and $\omega_{n,k} = \omega_n$, the frame $\mathcal{F}$ in (1) can be viewed as an example of non-stationary Gabor frame (NSGF).

A NSGF is a set $\{e^{2\pi i \omega_n t} g_k(t)\}_{n,k\in\mathbb{Z}}$ where the window function g and the frequency shifts ω vary adaptively, depending on time and/or frequency. This adaptability makes NSGF particularly useful for analyzing non-stationary signals (e.g., signals whose spectral content changes over time, like speech or music).

In [8] two authors of this paper investigated the stability of Gabor frames $\mathcal{G}(\mathrm{rect}^{(p)}, a, b)$ where $\mathrm{rect}(x) = \chi_{[-\frac{1}{2},\frac{1}{2}]}(x)$ is the characteristic function of the interval $[-\frac{1}{2}, \frac{1}{2}]$ and $\mathrm{rect}^{(p)}(x) = \mathrm{rect} * \ldots * \mathrm{rect}(x)$ is the $p-$times iterated convolution of $\mathrm{rect}(x)$.

In [8, Theorem 1.1] it is shown that the set $\mathcal{F} = \{e^{2\pi inbx}\,\mathrm{rect}(x - a(k + \delta_{n,k}))\}_{n,k\in\mathbb{Z}}$ is a frame in $L^2(\mathbb{R})$ provided that $0 < a \le 1$ and $4ab \sum_n \sup_k |\delta_{n,k}| < 1$.

The timing jitter in the aforementioned theorem also depends on the frequencies n, and a simple example (and also Corollary 1 in the next section) demonstrates that stability cannot be achieved with timing jitter that solely depends on k when $a = 1$.

These observations prompted further investigation into the stability of Gabor frames $G(h; a, b)$ under frequency-dependent timing jitter. These frames model more realistic scenarios in communication and sampling systems, where jitter varies with frequency. Understanding how this affects frame stability ensures that such systems remain robust under these conditions.

Furthermore, frames in the form of (1) can be viewed as a nontrivial generalizations of the NSGF, with potentially significant applications.

In this paper we prove various stability results for Gabor frames, $\mathcal{G}(h; a, b)$ with frequency-dependent jitter under various conditions on the window function h, among which the following generalization of [8, Theorem 1.1]. Here, $\|\ \|_2$ denotes the standard norm in $L^2(\mathbb{R})$.

Theorem 1 *Let $\mathcal{G}(h; a, b) = \{e^{2\pi ibnx} h(x - ak)\}_{n,k\in\mathbb{Z}}$ be a frame with bounds $A,\ B$. Assume that $h \in L^2(\mathbb{R})$ is supported in the interval $(-\frac{c}{2}, \frac{c}{2})$ with $0 < a \le c$.*

Let $\{\delta_{n,k}\} \subset (-\frac{1}{2}, \frac{1}{2})$ be such that

$$\lambda =: 4 \sup_k \sum_n \|h - h(. - a\delta_{n,k})\|_2^2 < A. \tag{2}$$

Then, the set $\tilde{\mathcal{B}} = \{e^{2\pi ibnx} h(x - a(k + \delta_{n,k}))\}_{n,k\in\mathbb{Z}}$ is a frame with bounds

$$A' = \left(1 - \sqrt{\frac{\lambda}{A}}\right) A, \quad B' = \left(1 + \sqrt{\frac{\lambda}{A}}\right) B.$$

Corollary 1 in the next section shows that if h is not continuous in $\mathbb{R}$ and $a = b = c = 1$, the stability of the frame $\mathcal{G}(a, b, h)$ cannot be achieved with $\delta_{n,k}$ independent of n.

When $h = \mathrm{rect}$, the stability condition (2) improves that in [8, Theorem 1.1].

The condition (2) for the frame $\mathcal{G}(a, b, \text{rect})$ in our Theorem 1is

$$4A^{-1}\sup_k \sum_n \|\,\text{rect} - \text{rect}(. - a\delta_{n,k})\|_2^2 = 4aA^{-1}\sup_k \sum_n |\delta_{n,k}| < 1$$

The stability condition in [8, Theorem 1.1] is

$$4ab\sum_n \sup_k |\delta_{n,k}| < 1. \tag{3}$$

By [8, Lemma 2.3], the frame bounds of $\mathcal{G}(\text{rect}, a, b)$ are $A \geq \frac{1}{b}$ and $B \leq \frac{1}{b}[\frac{1}{a}]$, where [ ] denotes the integer part. Thus,

$$4aA^{-1}\sup_k \sum_n |\delta_{n,k}| \leq 4ab\sum_n \sup_k |\delta_{n,k}|$$

which show that the stability condition (2) in our Theorem (1) improves that in (3).

We also prove a stability result for Gabor frames $\mathcal{G}(h; a, b)$ with frequency-dependent jitter when $\hat{h}$ has compact support (Theorem 4 in Sect. 4) and when h is differentiable in $\mathbb{R}$ (with the possible exception of a set of measure zero) and h' is in the Wiener Amalgam space $W(L^\infty, \ell^1)$ (Theorem 3 in Sect. 4). We recall that a function $g \in L^p(\mathbb{R})$ belongs to the *Wiener space* $W(L^p, \ell^q)$, if

$$\|g\|_{W(L^p,\ell^q)} =: \sum_m \|g(. + m)\|^q_{L^p(-\frac{1}{2},\frac{1}{2})} < \infty.$$

Here, $1 \leq p \leq \infty$ and $1 \leq q < \infty$. See e.g. [14, Section 11.4] for an introduction to Wiener amalgam spaces.

Wiener amalgam spaces appear naturally in Gabor frame theory. See [13], and see also [18, Theorem 5], which presents some similarities with our Theorem 1. Stability conditions that depend on the derivative of the window functions appear in [25, Theorem 2.1].

Our paper is organized as follows. In Sect. 2 we have collected some preliminaries and results that are needed for our proofs. In Sect. 3 we prove Theorem 1 and some corollaries that generalize stability results for B-splines proved in [8]. In Sect. 4 we state and prove Theorems 3 and 4, and some corollaries.

After our paper was submitted, we became aware of a recent preprint by I. Krishtal and B. Miller. Theorem 1.1 in [17] can be used to prove an analogue of our Theorem 4.2.

2 Preliminaries

We denote with $\langle\ ,\ \rangle_2$ and $\|\ \|_2$ the standard inner product and norm in $L^2(\mathbb{R})$.

Let $h \in L^2(\mathbb{R})$ and $a, b > 0$; a Gabor frame in $L^2(\mathbb{R})$ is a Gabor system $\mathcal{G} = \{e^{2\pi i b_n x} g(x - a_k)\}_{n,k\in\mathbb{Z}}$ for which there exists constants $A, B > 0$ such the following inequality holds for every $\psi \in L^2(\mathbb{R})$.

$$A\|\psi\|_2^2 \leq \sum_{n,k} |\left\langle \psi,\ e^{2\pi i b_n t} h(t - a_k) \right\rangle_2 |^2 \leq B|\psi\|_2^2. \tag{4}$$

The constants $A,\ B$ are the *frame bounds* of $\mathcal{G}$.

When $\mathcal{G} = \mathcal{G}(h; a, b) = \{e^{2\pi i b n x} g(x - ak)\}_{n,k\in\mathbb{Z}}$, the condition $0 < ab \leq 1$ is necessary for $\mathcal{G}(g, a, b)$ to be a frame. Since

$$\left\langle e^{2\pi i b_n t} h\,(a_k - t)\,,\ \psi \right\rangle_2 = \left\langle e^{2\pi i a_k s} \hat{h}\,(b_n - s)\,,\ \hat{\psi} \right\rangle_2$$

where $\hat{f}(s) = \int_{-\infty}^{\infty} f(t) e^{2\pi i s t} dt$ is the Fourier transform, we can see at once that $\mathcal{G}$ is a frame if and only if $\hat{\mathcal{G}} = \{e^{2\pi i a_k s} \hat{h}(s - b_n)\}_{n,k\in\mathbb{Z}}$ is a frame, and $\mathcal{G}$ and $\hat{\mathcal{G}}$ have the same frame bounds. We refer the reader to [14] or to [3] for an introduction to Gabor frame

2.1 Non-stationary Gabor Frames

To the best of our knowledge, non-stationary Gabor frames (NSGF)

$$\mathcal{F} = \{e^{2\pi i \omega_n x} h_k(x)\}_{k,\, n\in\mathbb{Z}}$$

were first introduced by F. Jaillet, P. Balazs, M. Dörfler and N. Engelputzeder in [1] (see also an early example in [23]). Other references on the NSGFs are [2, 9, 19].

When the h_n have compact support, the following result proved by P. Balazs and al. in [2] generalizes the classical painless non-orthogonal expansions for regular Gabor frames:

Proposition 1 *Let $\mathcal{F} = \{e^{2\pi i \omega_n x} h_k(x)\}_{k,\, n\in\mathbb{Z}}$ be a frame; if $supp(h_k) \subset [c_k,\ d_k]$ for every $k \in \mathbb{Z}$, and b_k is chosen so that $d_k - c_k \leq \frac{1}{b_k}$, the frame operator $S(f) = \sum_{n,k} \left\langle f, g_{k,n} \right\rangle g_{k,n}$ of $\mathcal{F}$ is given by*

$$Sf(t) = f(t) \sum_k \frac{1}{b_k} |h_k(t)|^2.$$

This proposition shows that the set $\mathcal{F}$ is a frame if and only if there exist constant $A,\ B > 0$ for which

$$A \leq \sum_k \frac{1}{b_k} |h_k(t)|^2 \leq B.$$

If the supports of the h_k do not cover $\mathbb{R}$, then $\mathcal{F}$ is not a frame. So if $h_k = h(t - a(k+\delta_k))$ with supp $(h) = [-\frac{c}{2}, \frac{c}{2}]$ and $0 < a \leq 1$, the support of h_k intersects that of h_{k+1} if

$$\frac{c}{2} + a(k+\delta_k) \geq -\frac{c}{2} + a(k+1+\delta_{k+1}).$$

Thus, if $a(\delta_{k+1} - \delta_k) \leq c - a$ for every $k \in \mathbb{Z}$, the support of the f_k overlap. This observations prove the following corollary of Proposition 1

Corollary 1 *Let $\mathcal{G} = \{e^{2\pi i n \omega_n x} h(x - ak)\}_{k,n\in\mathbb{Z}}$ be a frame. Assume $g \in L^2(\mathbb{R})$ with support in $[-\frac{c}{2}, \frac{c}{2}]$ and that $0 < a \leq c$.*

If $\{\delta_k\}_k \subset \mathbb{R}$ is such that $\delta_{k+1} - \delta_k \leq \frac{c-a}{a}$ for every $k \in \mathbb{Z}$, the set $\{e^{2\pi i n \omega_n x} f(x - a(k+\delta_k))\}_{k,n\in\mathbb{Z}}$ is a frame.

2.2 Paley-Wiener and Kadec Theorems

One of the fundamental stability criteria for bases in Banach spaces, and historically the first, is due to R. Paley and N. Wiener (see [22]). The following generalized Paley-Wiener theorem is in [4].

Theorem 2 (Paley–Wiener Theorem for Frames) *Let $\{x_n\}_n$ be a frame in a Hilbert space H with bounds A and B. Let $\{y_n\}_n \subset H$; suppose that there exists $\lambda, \mu \geq 0$, with $\lambda + \frac{\mu}{\sqrt{A}} < 1$, for which*

$$\lambda + \frac{\mu}{\sqrt{A}} < 1 \quad \text{and} \quad \left\| \sum_n a_n(x_n - y_n) \right\| \leq \lambda \| \sum_n a_n x_n \| + \mu \left(\sum_n |a_n|^2 \right)^{\frac{1}{2}} \tag{5}$$

for every finite set of scalars $\{a_n\} \subset \mathbb{C}$. Then, $\{y_n\}_n$ is a frame with bounds $\left(1 - (\lambda + \frac{\mu}{\sqrt{A}})\right) A$ and $\left(1 + \lambda + \frac{\mu}{\sqrt{A}})\right) B$

If $\mu = 0$ and $\{x_n\}_n$ is a Riesz basis, then Theorem 2 reduces to the classical Paley-Wiener theorem and proves that also $\{y_n\}_n$ is a basis.

A celebrated stability result for the standard orthogonal basis $\{e^{inx}\}_{n\in\mathbb{Z}}$ on $L^2(-\pi, \pi)$ is the classical Kadec-$\frac{1}{4}$ theorem. It states that the set $\{e^{i(n+\delta_n)x}\}_{n\in\mathbb{Z}}$ is a basis on $L^2(-\pi, \pi)$ whenever $\sup_n |\delta_n| < \frac{1}{4}$. An example by Levinson shows that the constant $\frac{1}{4}$ cannot be replaced by any larger constant. Kadec's theorem is proved in [16] and also in [17, Chaper 1, Theorem 14, page 42]. A generalization of Kadec's theorem is in [15, Theorem 1].

The proof of Kadec's theorem in [26] uses the fact that the sequence

$$S(t) = A_0(\delta) + \sum_{m=1}^{\infty} A_m(\delta)\cos(mt) + \sum_{m=1}^{N} B_m(\delta)\sin(t(m-\frac{1}{2})),$$

where $A_0(\delta) = 1 - \frac{\sin(\pi\delta)}{\pi\delta}$, $A_m(\delta) = \frac{2(-1)^m\delta\sin(\pi\delta)}{\pi(m^2-\delta^2)}$, $B_m(\delta) = i\frac{2(-1)^m\delta\cos(\pi\delta)}{\pi\left((m-\frac{1}{2})^2-\delta^2\right)}$ converges pointwise to $f(t) = 1 - e^{i\delta t}$ in the interval $[-\pi, \pi]$. This important fact is used to show that for every sequence $\{\delta_n\}_n \subset (-\frac{1}{4}, \frac{1}{4})$ and every finite set $\{a_k\} \subset \mathbb{C}$, the following inequality holds with $\delta = \sup_n |\delta_n|$.

$$\|\sum_n a_n\left(e^{int} - e^{i(n+\delta_n)t}\right)\|_{L^2(-\pi\,\pi)} \leq (1-\cos(\pi\delta)+\sin(\pi\delta))\left(\sum_n |a_n|^2\right)^{\frac{1}{2}}. \tag{6}$$

Since $1-\cos(\pi\delta)+\sin(\pi\delta) < 1$ when $0 \leq \delta < \frac{1}{4}$, we can apply Theorem 2 with $\lambda = 0$, $\mu = 1-\cos(\pi\delta)+\sin(\pi\delta)$ and $A = 1$ and conclude that the set $\{e^{i(n+\delta_n)t}\}_{n\in\mathbb{Z}}$ is a Riesz basis of $L^2(-\pi,\pi)$.

We prove the following

Lemma 1 *Let $M > 0$ be an integer; let $\{\delta_n\}_n \subset \mathbb{R}$, with $\delta =: \sup_n |\delta_n| < \frac{1}{4M}$. Then, for every finite sequence $\{a_n\} \subset \mathbb{C}$, we have that*

$$\|\sum_n a_n e^{2\pi int}(1-e^{2\pi i\delta_n t})\|_{L^2(-\frac{M}{2}\,\frac{M}{2})} \tag{7}$$

$$\leq \sqrt{\frac{M}{2\pi}}(1-\cos(\pi\delta M)+\sin(\pi\delta M))\sum_n |a_n|^2.$$

Proof Letting $t = \frac{M}{2\pi}s$ in the L^2 norm in the inequality (7), we obtain

$$S =: \|\sum_n a_n e^{2\pi int}(1-e^{2\pi i\delta_n t})\|_{L^2(-\frac{M}{2}\,\frac{M}{2})} =$$

$$\sqrt{\frac{M}{2\pi}}\|\sum_n a_n e^{iMns}(1-e^{iM\delta_n s})\|_{L^2(-\pi\,\pi)}.$$

We let $b_m = \begin{cases} 0 & \text{if } m \neq Mn \\ a_n & \text{if } m = Mn \end{cases}$ and $\delta_m = \delta_n$ whenever $m \in [nM, (n+1)M)$. In view of (6), we obtain

$$S = \sqrt{\frac{M}{2\pi}}\|\sum_m b_m e^{ims}(1-e^{iM\delta_m s})\|_{L^2(-\pi\,\pi)}$$

$$\leq \sqrt{\frac{M}{2\pi}}(1-\cos(\pi M\delta)+\sin(\pi M\delta))\left(\sum_n |b_n|^2\right)^{\frac{1}{2}}$$

$$= \sqrt{\frac{M}{2\pi}}(1-\cos(\pi M\delta)+\sin(\pi M\delta))\left(\sum_n |a_n|^2\right)^{\frac{1}{2}}$$

as required. □

Remark We recall the elementary inequalities $\sin t \leq t$ and $0 \leq 1-\cos t \leq \frac{t^2}{2}$. When $0 \leq t \leq \frac{\pi}{2}$, we also have $\frac{2t}{\pi} \leq \sin t$. Thus, when $M\delta < \frac{1}{4}$ we have

$$2\delta M \leq (1-\cos(\pi\delta M)+\sin(\pi\delta M)) \leq \pi\delta M + \frac{(\pi\delta M)^2}{2} \tag{8}$$

2.3 Poisson Summation Formula

Let $f \in C(\mathbb{R}) \cap L^2(\mathbb{R})$. The basic Poisson summation formula states that

$$\sum_n f(x+n) = \sum_k \hat{f}(k)e^{2\pi ikx}, \quad x \in \mathbb{R}.$$

Given the parameters $P > 0$, we can let $s_P f(x) = \sum_n f(x+Pn) = \sum_n f_P(x/P+n)$ where $f_P(t) = f(Pt)$. Recalling that $\widehat{f_P}(y) = \frac{1}{P}\hat{f}(y/P)$ and using the basic Poisson formula above, we obtain

$$s_P f(x) = \sum_n f(x+Pn) = \sum_n f_P(x/P+n) = \frac{1}{P}\sum_k \hat{f}(k/P)e^{2\pi ikx/P}$$

Letting $f = \hat{F}$ and recalling that $\hat{\hat{F}}(t) = F(-t)$, we obtain

$$\sum_n \hat{F}(y+Pn) = \frac{1}{P}\sum_k F(-k/P)e^{2\pi iky/P} = \frac{1}{P}\sum_k F(k/P)e^{-2\pi iky/P}. \tag{9}$$

We prove the following

Lemma 2

(a) Let $g \in L^2(\mathbb{R})$. Then, for every $P > 0$ and a.e. $y \in \mathbb{R}$,

$$\sum_n |\hat{g}(y+Pn)|^2 = \frac{1}{P}\sum_k e^{-2\pi iky/P}\int_{-\infty}^{\infty} g(s)\overline{g}(s-k/P)dt \tag{10}$$

where $\overline{g}$ denotes the complex conjugate of g.

(b) *If g is supported in $(t-\frac{1}{2}, t+\frac{1}{2})$ for some $t \in \mathbb{R}$ and if $P \leq 1$, we have that*

$$\sum_n |\hat{g}(y+Pn)|^2 = \frac{1}{P}\|g\|_2^2. \tag{11}$$

Proof Let us show first that for a given $\psi \in L^2(\mathbb{R})$ and for a.e. $y \in \mathbb{R}$, we have that $|\hat{\psi}(y)|^2 = \widehat{\psi * \psi_-}(y)$, where $\psi_-(x) = \overline{\psi}(-x)$. Indeed,

$$|\hat{\psi}(y)|^2 = \hat{\psi}(y)\overline{\hat{\psi}(y)} = \hat{\psi}(y)\int_{-\infty}^{\infty} \overline{\psi}(t)e^{2\pi i y t}dt,$$

and with the change of variable $t \to -t$ in the integral, we obtain

$$|\hat{\psi}(y)|^2 = \hat{\psi}(y)\int_{-\infty}^{\infty} \overline{\psi}(-t)e^{-2\pi i y t}dt = \hat{\psi}(y)\hat{\psi}_-(y) = \widehat{\psi * \psi_-}(y).$$

Using this identity and (9), we obtain

$$\sum_n |\hat{g}(y+Pn)|^2 = \sum_n \widehat{g * g_-}(y+Pn) = \frac{1}{P}\sum_k g * g_-(k/P)e^{-2\pi i k y/P}$$

$$= \frac{1}{P}\sum_k e^{-2\pi i k y/P}\int_{-\infty}^{\infty} g(s)\overline{g}(s-k/P)ds$$

which is (10).

If g is supported in $(t-\frac{1}{2},\ t+\frac{1}{2})$ and $P \leq 1$, the supports of $g(s)$ and $g(s-k/P)$ are disjoint whenever $k \neq 0$. Thus, the sum above reduces to $\frac{1}{P}\int_{-\infty}^{\infty}|g(s)|^2 ds = \frac{1}{p}\|g\|_2^2$ and (11) is proved.

□

2.4 The Short-term Fourier Transform

Let $g \in L^2(\mathbb{R})$ with $\|g\|_2 \neq 0$. The *short-time Fourier transform* with respect to g, also called *Gabor transform*, is the operator $\mathcal{F}^g : L^2(\mathbb{R}) \to L^2(\mathbb{R}\times\mathbb{R})$

$$(\mathcal{F}^g f)(\xi, t) = \int_{-\infty}^{\infty} e^{-2\pi i \xi s} f(s)\, g(s-t)\, ds, \quad t, \xi \in \mathbb{R} \tag{12}$$

For a given $t \in \mathbb{R}$, we can let $g_t(x) = g(x-t)$ and observe that $\mathcal{F}^g f$ is the Fourier transform of $f g_t$. This windowed Fourier transform provide a tool for time-frequency localization, and shares many properties with the Fourier transform. For example, it is proved [3, Section 11.1] that for any $f_1, f_2, g_1, g_2 \in L^2(\mathbb{R})$, we have

$$\int_{-\infty}^{\infty}\int_{-\infty}^{\infty}(\mathcal{F}^{g_1} f_1)(\xi, t)\overline{(\mathcal{F}^{g_2} f_2)(\xi, t)}dtd\xi = \langle f_1, f_2\rangle_2 \langle g_2, g_1\rangle_2 . \tag{13}$$

From this formula we obtain a generalized Plancherel identity

$$\|\mathcal{F}^g f\|^2_{L^2(\mathbb{R}^2)} = \|f\|_2^2 \|g\|_2^2. \tag{14}$$

Other properties of the short-term Fourier transform are described e.g in [12]. We will need the following

Proposition 2 *Let $g \in L^2(\mathbb{R})$ with support in $(-\frac{1}{2}, \frac{1}{2})$. For every $f \in L^2(\mathbb{R})$, every $t \in \mathbb{R}$ and every $P \leq 1$, we have that*

$$\sum_n |\mathcal{F}^g f(Pn, t)|^2 = \frac{1}{P}\int_{-\infty}^{\infty} |f(s)|^2|g(t-s)|^2 ds$$

Proof Fix $t \in \mathbb{R}$. By Lemma 2 with $y = 0$,

$$\begin{aligned}\sum_n |\mathcal{F}^g (f(Pn, t)|^2 &= \frac{1}{P}\sum_n |\widehat{g_t f}(Pn)|^2 \\ &= \int_{-\infty}^{\infty} |fg_t|^2(s)ds = \frac{1}{P}\int_{-\infty}^{\infty} |f(s)|^2|g(t-s)|^2 ds\end{aligned}$$

as required. □

3 Proof of Theorem 1 and Corollaries

In this section we prove Theorem 1, but first we recall some basic facts about the stability of Gabor frames and about Gabor frames with compactly supported window.

Given the frame $\mathcal{G}(h; a, b) = \{e^{2\pi ibnx}h(x - ak)\}_{n,k\in\mathbb{Z}}$, we find sufficient conditions that ensure that the set

$$\tilde{\mathcal{G}} = \{e^{2\pi ibnx}h(x - a(k + \delta_{n,k}))\}_{n,k\in\mathbb{Z}}$$

is a frame. As usual, the frame bounds of $\mathcal{B}$ will be denoted with A and B. By Theorem 2, if we show that, for every finite set $\{a_{n,k}\} \subset \mathbb{C}$, we have that

$$\left\| \sum_{n,k} a_{n,k}\, e^{2\pi i bnx} \left(h\left(x-ak\right) - h\left(x - a(k+\delta_{n,k})\right)\right) \right\|_2^2 \leq \mu A \sum_{n,k} |a_{n,k}|^2 \tag{15}$$

for some $0 < \mu < 1$, we have proved that $\tilde{\mathcal{G}}$ is a frame with bounds $A' = (1 - \frac{\mu}{\sqrt{A}})A, \quad B' = (1 + \frac{\mu}{\sqrt{A}})B$.

A standard application of Plancherel theorem and of the properties of the Fourier transform show that (15) is equivalent to

$$\| \sum_{n,k} a_{n,k} \left(e^{2\pi i ak(x-bn)} - e^{2\pi i a(k+\delta_{k,n})(x-bn)}\right) \hat{h}(x-bn) \|_2^2 \leq \mu A \sum_{n,k} |a_{n,k}|^2. \tag{16}$$

Both inequalities (15) and (16) will be used in our proofs.

In the seminal paper [6] it is proved that when h is supported in an interval $[-\frac{c}{2}, \frac{c}{2}]$, with $c > 0$, then $\mathcal{G}(h; a, b)$ is a frame if and only if $a \leq c \leq b^{-1}$ and the inequality

$$Ab \leq \sum_{k \in \mathbb{Z}} |h(x-ak)|^2 \leq Bb \tag{17}$$

holds for a.e. $x \in \mathbb{R}$ with constants $A, \ B > 0$ See also [14, Theorem 11.6]. The optimal constants $A, \ B$ in the inequality (17) are the optimal frame bounds of $\mathcal{G}(h; a, b)$. It is proved in [14, Thm 11.8] that

$$A = \frac{1}{a} \inf_x \sum_k |\widehat{h}(x-bk)|^2, \quad B = \frac{1}{a} \sup_x \sum_k |\widehat{h}(x-bk)|^2.$$

In view of the definition of frames (4), a change of variables in the inner product yields

$$A\|\psi\|_2^2 \leq a^2 \sum_{n,k} |\left\langle \psi_a, e^{2\pi i banx} h_a(x-k)\right\rangle_2|^2 \leq B\|\psi_a\|_2^2 \tag{18}$$

where we have let $f_a(t) = f(at)$. Since $a^{-1}\|\psi\|_2^2 = \|\psi_a\|_2^2$, the inequality (18) can be written as

$$a^{-1}A\|\psi_a\|_2^2 \leq \sum_{n,k} |\left\langle \psi_a, e^{2\pi i banx} h_a(x-k)\right\rangle_2|^2 \leq a^{-1}B\|\psi_a\|_2^2,$$

and we can conclude that $\mathcal{G}(h; a, b)$ is a frame with bounds A and B if and only if the set

$$\mathcal{G}(h_a; 1, ab) = \{e^{2\pi i abnx} h_a(x-k)\}_{n,k \in \mathbb{Z}}$$

is a frame with bounds $a^{-1}A$ and $a^{-1}B$. Note that the support of h_a is in the interval $[-\frac{c}{2a}, \frac{c}{2a}] \supset [-\frac{1}{2}, \frac{1}{2}]$.

3.1 Proof of Theorem 1

By the previous observations, we can reduce matters to proving that if the assumptions of the theorem are satisfied, the set

$$\tilde{\mathcal{B}} =: \{e^{2\pi iabnx} h_a(x - (k + \delta_{n,k}))\}_{n,k \in \mathbb{Z}}$$

is a frame whenever the set $\mathcal{B} = \mathcal{G}(h_a;\, 1, ab)$ is a frame.

For simplicity we assume that h_a is supported in $[-\frac{1}{2}, \frac{1}{2}]$, but the proof is similar also in the other case. To simplify the notation in the proof, we will replace $k + \delta_{n,k}$ with $\mu_{n,k}$ when there is no risk of confusion.

By Theorem 2, the set $\tilde{\mathcal{B}}$ is a frame if we prove that for every finite sequence $\{a_{n,k}\} \subset \mathbb{C}$, the following inequality holds with constant $\lambda < 1$.

$$\left\| \sum_{n,k} a_{n,k}\, e^{2\pi iabnx} \left(h_a\left(x - \mu_{n,k}\right) - h_a\left(x - k\right)\right) \right\|_2^2 \leq \lambda a^{-1} A \sum_{n,k} |a_{n,k}|^2. \tag{19}$$

Fix $k \in \mathbb{Z}$ and let

$$f_k(x) = \sum_n e^{2\pi iabnx} a_{n,k} \left(h_a\left(x - \mu_{n,k}\right) - h_a\left(x - k\right)\right).$$

The inequality (19) follows if we prove that for every finite sequence $\{a_{n,k}\} \subset \mathbb{C}$, with $\sum_{n,k} |a_{n,k}|^2 = 1$, the inequality

$$\|\sum_k f_k\|_2^2 \leq \lambda a^{-1} A \tag{20}$$

holds with constant $\lambda < 1$. We let $d_j = \sup_n |\delta_{n,j}|$, and

$$I_j^{prime} = [j - d_j - \frac{1}{2},\ j + d_j - \frac{1}{2}), \quad I_j = [j + d_j - \frac{1}{2},\ j - d_j + \frac{1}{2}).$$

Clearly, $\mathbb{R} = \bigcup_j I_j \cup I'_j$, and since $|d_k| < \frac{1}{2}$, all intervals are disjoint.

We show that each f_k is supported in the interval $[k - \frac{1}{2} - d_k,\ k + \frac{1}{2} + d_k) = I'_k \cup I_k \cup I'_{k+1}$.

Indeed, for every $n \in \mathbb{Z}$, the support of the function $x \to h_a\left(x - k - \delta_{n,k}\right) - h_a\left(x - k\right)$ is in the interval $(k - \frac{1}{2},\ k + \delta_{n,k} + \frac{1}{2}) \subset (k - \frac{1}{2} - d_k,\ k + d_k + \frac{1}{2})$ whenever

$\delta_{n,k} \geq 0$, and in the interval $(k + \delta_{n,k} - \frac{1}{2},\ k + \frac{1}{2}) \subset (k - d_k - \frac{1}{2},\ k + \frac{1}{2} + d_k)$ whenever $\delta_{n,k} \leq 0$. Note that only f_k is supported in I_k but both f_k and f_{k-1} can be supported in I'_k.

In view of the elementary inequality $(a+b)^2 \leq 2(a^2+b^2)$, we can write

$$\|\sum_k f_k\|_2^2 = \|\sum_k \chi_{I_k^{prime}} f_k + \sum_k \chi_{I_k} f_k\|_2^2 \leq 2\|\sum_k \chi_{I_k} f_k\|_2^2 + 2\|\sum_k \chi_{I_k^{prime}} f_k\|_2^2. \tag{21}$$

Let us estimate the first sum in (21). Since $I_k \cup I_j = 0$ whenever $k \neq j$ and only f_k is supported in I_k, we can write

$$\sum_k \|\chi_{I_k} f_k\|_2^2 = \sum_k \int_{k-\frac{1}{2}+d_k}^{k+\frac{1}{2}-d_k} \Big|\sum_n e^{2\pi iabnx} a_{n,k}(h_a(x-k) - h_a(x-\mu_{n,k}))\Big|^2 dx.$$

After applying the Cauchy-Schwartz inequality to the sum in n and a change of variable in the integrals, we obtain

$$\|\sum_k \chi_{I_k} f_k\|_2^2 \leq \sum_k \sum_n |a_{n,k}|^2 \sum_n \int_{k-\frac{1}{2}+d_k}^{k+\frac{1}{2}-d_k} |(h_a(x-k) - h_a(x-k-\delta_{n,k}))|^2 dx$$

$$= \sum_k \sum_n |a_{n,k}|^2 \sum_n \int_{-\frac{1}{2}+d_k}^{\frac{1}{2}-d_k} |h_a(x) - h_a(x-\delta_{n,k}))|^2 dx. \tag{22}$$

We now estimate the second sum in (21). We have observed that the I'_k are disjoint and both f_j and f_{j-1} can be supported in I'_j. Using again the elementary inequality $(a+b)^2 \leq 2(a^2+b^2)$, we can write

$$\|\sum_k \chi_{I'_k} f_k\|_2^2 = \sum_k \|\chi_{I'_k}(f_k + f_{k-1})\|_2^2 \leq 2\sum_k \|\chi_{I'_k} f_k\|_2^2 + 2\sum_k \|\chi_{I'_k} f_{k-1}\|_2^2. \tag{23}$$

Recall that $I_j^{prime} = [j - d_j - \frac{1}{2},\ j + d_j - \frac{1}{2})$. Arguing as in the proof of (22), we obtain

$$\sum_k \|\chi_{I_k^{prime}} f_k\|_2^2 \leq \sum_k \sum_n |a_{n,k}|^2 \sum_n \int_{-\frac{1}{2}-d_k}^{-\frac{1}{2}+d_k} |h_a(x) - h_a(x-\delta_{n,k})|^2 dx,$$

and

$$\sum_k \|\chi_{I_k^{prime}} f_{k-1}\|_2^2 \leq \sum_k \sum_n |a_{n,k}|^2 \sum_n \int_{\frac{1}{2}-d_k}^{\frac{1}{2}+d_k} |h_a(x) - h_a(x-\delta_{n,k})|^2 dx.$$

From these inequalities and (23), we obtain

$$\|\sum_k \chi_{I_k^{prime}} f_k\|_2^2 \le 2\sum_k\sum_n |a_{n,k}|^2 \sum_n \left(\int_{\frac{1}{2}-d_k}^{\frac{1}{2}+d_k} |h_a(x) - h_a(x-\delta_{n,k})|^2 dx \right.$$
$$\left. + \int_{-\frac{1}{2}-d_k}^{-\frac{1}{2}+d_k} |h_a(x) - h_a(x-\delta_{n,k})|^2 dx \right).$$

This inequality and (22), combined with (21), yield

$$\sum_k \|f_k\|_2^2 \le 4\sum_k\sum_n |a_{n,k}|^2 \sum_n \int_{-\frac{1}{2}-d_k}^{\frac{1}{2}+d_k} |h_a(x) - h_a(x-\delta_{n,k})|^2 dx. \tag{24}$$

Recall that the support of the function $x \to h_a(x) - h_a(x-\delta_{n,k})$ is in the interval $(-\frac{1}{2}-d_k,\ \frac{1}{2}+d_k)$; letting $m(t) =: \|h - h(.-t)\|_2$, we have that

$$\int_{-\frac{1}{2}-d_k}^{\frac{1}{2}+d_k} |h_a(x) - h_a(x-\delta_{n,k})|^2 dx = \int_{-\infty}^{\infty} |h(ax) - h(ax - a\delta_{n,k})|^2 dx$$
$$= a^{-1}\|h - h(\cdot - a\delta_{n,k})\|_2^2 = a^{-1} m(a\delta_{n,k})^2.$$

We have used the change of variables $ax \to x$ in the integral. We can write the inequality (24) as

$$\sum_k \|f_k\|_2^2 \le 4a^{-1}\sum_k \left(\sum_n |a_{n,k}|^2 \sum_n m(a\delta_{n,k})^2\right)$$

from which follows that

$$\sum_k \|f_k\|_2^2 \le 4a^{-1} \sup_k \sum_n m(a\delta_{n,k})^2 \sum_{n,k} |a_{n,k}|^2 = 4a^{-1} \sup_k \sum_n m(a\delta_{n,k})^2.$$

The inequality (20) is satisfied if

$$\lambda =: 4\sup_k \sum_n m(a\delta_{n,k})^2 < \sqrt{A}.$$

which is (2). The theorem is proved. □

3.2 Corollaries

Let $h \in L^2(\mathbb{R})$ with support in the interval $\left(-\frac{c}{2}, \frac{c}{2}\right)$. As observed at the beginning of this section, if $\mathcal{G}(h; a, b)$ is a frame, necessarily $a \le c \le \frac{1}{b}$; we have also observed that if $\mathcal{G}(h; a, b)$ is a frame with bounds A and B, then $\mathcal{G}(h_a; 1, ab)$ is a frame with bounds $a^{-1}A$ and $a^{-1}B$. Here and throughout this subsection, $f_a(x) = f(ax)$.

For a given $p \in \mathbb{N}$, we consider the p-times iterated convolution $h^{(p)}(x) =: \underbrace{h * \cdots * h}_{p\ times}(x)$.

Using the properties of the convolution, it is easy to verify that the $h^{(p)}(x)$ are supported in the interval $\left(-\frac{cp}{2}, \frac{cp}{2}\right)$ and are continuous for $p \ge 2$. When $h = \text{rect}$, the functions $\text{rect}^{(p)}(x)$ are piecewise polynomial of degree $p - 1$ and a prime example of a B-spline of order $p - 1$ (see e.g [7, 21] and other classical references on B-splines).

We consider the set $\mathcal{G}(h^{(p)}; pa, b/p)$. By the previous consideration, this set is a frame with bounds A_p and B_p if and only if the set $\mathcal{G}(h^{(p)}_{ap}; 1, ab)$ is a frame with bounds$(ap)^{-1}A_p$ and $(ap)^{-1}B_p$. Note that the assumption $a \le c \le \frac{1}{b}$ yields the necessary condition $pa \le pc \le \frac{1}{b/p}$ that $\mathcal{G}(h^{(p)}; pa, b/p)$ needs to satisfy in order to be a frame.

The following easy lemma is perhaps already known, but we prove it here for the convenience of the reader.

Lemma 3 *Assume that $\mathcal{G}(h; a, b)$ is a frame with bounds A, B and h is supported in $(-\frac{c}{2}, \frac{c}{2})$. Assume also that*

$$N(h) =: \sup_x \sum_k |\hat{h}(x + kb)|^{\frac{4}{3}} < \infty.$$

Then for every $p \in \mathbb{N}$, the set $\mathcal{G}(h^{(p)}; pa,\ b/p)$ is a frame with bounds A_p, B_p that satisfy the relations

$$B_p = a^{p-1}B^p,$$

$$A_1 = A; \quad A_{p+1} = a^{\frac{1}{p}+1}\left(\frac{A_p^{\frac{1}{p}}}{N(h)}\right)^{2(p+1)} \tag{25}$$

Proof By [14, Theorems 11.4, 11.8] and the previous considerations, the set $\mathcal{G}(h^{(p)}; pa, b/p)$ is a frame if

$$A_p =: \frac{1}{ap} \inf_x \sum_k |\widehat{h^{(p)}}(x - bk/p)|^2 > 0$$

$$B_p =: \frac{1}{ap} \sup_x \sum_k |\widehat{h^{(p)}}(x - bk/p)|^2 < \infty.$$

Recall that $A_1 = A$ and $B_1 = B$ are the frame bounds of $\mathcal{G}(h; a, b)$.

The Fourier transform of the p-fold convolution of $h(x)$ is the Fourier transform of $h(x)$ raised to the power p, and so

$$A_p = \frac{1}{ap} \inf_x \sum_k |\hat{h}(x - bk/p)|^{2p}, \quad B_p = \frac{1}{ap} \sup_x \sum_k |\hat{h}(x - bk/p)|^{2p}. \tag{26}$$

Every $k \in \mathbb{Z}$ can be written as $k = k'p + j$ with $0 \le j \le p - 1$. Thus,

$$\begin{aligned} A_p =& \frac{1}{ap} \inf_x \sum_{j=0}^{p-1} \sum_{k'} |\hat{h}(x - bk' - jb/p)|^{2p} \\ \ge& \frac{1}{ap} \sum_{j=0}^{p-1} \inf_x \sum_{k'} |\hat{h}(x - bk' - bj/p)|^{2p} = \frac{1}{a} \inf_x \sum_{k'} |\hat{h}(x - bk')|^{2p}. \end{aligned}$$

$$\begin{aligned} B_p =& \frac{1}{ap} \sup_x \sum_{j=0}^{p-1} \sum_{k'} |\hat{h}(x - bk/p - bj/p)|^{2p} \\ \le& \frac{1}{ap} \sum_{j=0}^{p-1} \sup_x \sum_{k'} |\hat{h}(x - bk' - bj/p)|^{2p} = \frac{1}{a} \sup_x \sum_{k'} |\hat{h}(x - bk')|^{2p}. \end{aligned}$$

We let $B'_p =: \sup_x \sum_{k'} |\hat{h}(x - bk')|^{2p}$ and $A'_p =: \inf_x \sum_{k'} |\hat{h}(x - bk')|^{2p}$. Thus,

$$A_p \ge \frac{1}{a} A'_p \quad \text{and} \quad B_p \le \frac{1}{a} B'_p.$$

and it is enough to estimate A'_p and B'_p.

Recall that for every $1 \le r \le q < \infty$ and every sequence $\{x_n\}_n \subset \mathbb{C}$, the inequality

$$\left(\sum_k |x_k|^q \right)^{\frac{1}{q}} \le \left(\sum_k |x_k|^r \right)^{\frac{1}{r}} \tag{27}$$

holds. From this inequality follows that $(\sum_k |x_k|)^r \ge \sum_k |x_k|^r$ whenever $r \ge 1$. Thus,

$$\sum_{k'} |\hat{h}(x - bk')|^{2p} \leq \sup_x \left(\sum_k |\hat{h}(x - bk)|^2 \right)^p$$

$$= \sup_x \left(\sum_k |\hat{h}(x - bk)|^2 \right)^p = (aB)^p,$$

and so $B'_p \leq (aB)^p$.

To estimate A'_p, we use Hölder's inequality and induction on p. Letting

$$N(h) =: \left(\sum_k |\hat{h}(x - bk)|^{\frac{4}{3}} \right)^{\frac{3}{4}},$$

we start with the inequality

$$\sum_k |\hat{h}(x - bk)|^2 \leq \left(\sum_k |\hat{h}(x - bk)|^4 \right)^{\frac{1}{4}} \left(\sum_k |\hat{h}(x - bk)|^{\frac{4}{3}} \right)^{\frac{3}{4}}$$

$$= \left(\sum_k |\hat{h}(x - bk)|^4 \right)^{\frac{1}{4}} N(h)$$

from which we obtain

$$\sum_k |\hat{h}(x - bk)|^4 \geq \left(\frac{\sum_k |\hat{h}(x - bk)|^2}{N(h)} \right)^4 \geq \left(\frac{aA}{N(h)} \right)^4.$$

and $(A_2^{prime})^{\frac{1}{2}} \geq \left(\frac{A_1^{prime}}{N(h)} \right)^2 = \left(\frac{aA}{N(h)} \right)^2$.

When $p \geq 2$, we can write

$$\left(\sum_k |\hat{h}(x - bk)|^{2p} \right)^{\frac{1}{p}} \leq \left(\sum_k |\hat{h}(x - bk)|^{2(p+1)} \right)^{\frac{1}{2(p+1)}} \left(\sum_k |\hat{h}(x - bk)|^q \right)^{\frac{1}{q}}$$

where q satisfies $\frac{1}{p} = \frac{1}{2(p-1)} + \frac{1}{q}$. Since $p \geq 2$, we have that $q \geq 2 > \frac{4}{3}$. By the inequality (27), $\left(\sum_k |\hat{h}(x - bk)|^q \right)^{\frac{1}{q}} \leq N(h)$; thus,

$$\left(\sum_k |\hat{h}(x - bk)|^{2p} \right)^{\frac{1}{p}} \leq \left(\sum_k |\hat{h}(x - bk)|^{2(p+1)} \right)^{\frac{1}{2(p+1)}} N(h)$$

from which follows that

$$\left(\sum_k |\hat{h}(x-bk)|^{2(p+1)}\right)^{\frac{1}{2(p+1)}} \geq \frac{\left(\sum_k |\hat{h}(x-bk)|^{2p}\right)^{\frac{1}{p}}}{N(h)} \geq \frac{(A_p^{prime})^{\frac{1}{p}}}{N(h)}$$

and $(A_{p+1}^{prime})^{\frac{1}{p+1}} \geq \left(\frac{(A_p^{prime})^{\frac{1}{p}}}{N(h)}\right)^2$. From this relation (25) follows. □

Remark Obtaining a closed-form estimate for the optimal frame bounds of $\mathcal{G}(h^{(p)}; pa, b/p)$ is generally not feasible, as these expressions depend heavily on the specific properties of the function h. The optimal frame bounds of $\mathcal{G}(\text{rect}^{(p)}; ap, b/p)$ were explicitly evaluated in [20].

The following corollary generalizes [8, Theorem 1.2].

Corollary 2 *Let h be as in Lemma 3; Let $\{\delta_{n,k}\} \subset (-\frac{1}{2}, \frac{1}{2})$ for which*

$$\lambda =: 4\|h\|_1^{p-1} \sup_k \sum_n \|h - h(. - ap\delta_{n,k})\|_2^2 < A_p, \tag{28}$$

where the A_p are defined by (25). *Then, the set $\tilde{B} = \{e^{2\pi ibnx/p} h^{(p)}(x - ap(k + \delta_{n,k}))\}_{n,k\in\mathbb{Z}}$ is a frame with bounds*

$$A_p^{prime} =: \left(1 - \sqrt{\frac{\lambda}{A_p}}\right) A_p, \quad B_p^{prime} =: \left(1 + \sqrt{\frac{\lambda}{A_p}}\right) B_p.$$

Proof We prove that

$$\|h^{(p)} - h^{(p)}(\cdot - ap\delta_{n,k})\|_2 \leq \|h\|_1^{p-1} \|h - h(. - ap\delta_{n,k})\|_2^2. \tag{29}$$

By (28) and by Theorem 1 (with $h^{(p)}$ replacing h and ap replacing a), the thesis of the corollary follows.

When $p = 1$ there is nothing to prove. When $p > 1$, we can let $f_{p,d}(t) = h^{(p)}(t) - h^{(p)}(t-d)$ and observe that

$$f_{p,d}(t) = \int_{-\infty}^{\infty} h(s) \left(h^{(p-1)}(t-s) - h^{(p-1)}(t-d-s)\right) ds = h * f_{p-1,d}(t).$$

By Young's inequality for convolution, $\|f_{p,d}\|_2 \leq \|h\|_1 \|f_{p-1,d}\|_2$, and, if we apply this inequality $p-1$ times, we obtain

$$\|f_{p,d}\|_2 \leq \|h\|_1^{p-1} \|f_{1,d}\|_2 = \|h\|_1^{p-1} \|h - h(\cdot - d)\|_2.$$

After replacing d with $ap\delta_{n,k}$, we obtain (29).

□

4 Other Results

In this section we state and prove stability results for Gabor frames $\mathcal{G}(a, b, h)$ for window functions h that do not necessarily have compact support.

4.1 h′ in $W(L^\infty, \ell^1)$

Recall that a function $g \in L^\infty(\mathbb{R})$ belongs to the *Wiener amalgam space* $W(L^\infty, \ell^1)$ if

$$\|g\|_{W(L^\infty,\ell^1)} =: \sum_m \|g(.+m)\|_{L^\infty(-\frac{1}{2},\frac{1}{2})} < \infty.$$

It is not difficult to prove that $\|g\|_{W(L^\infty,\ell^1)} = \|g(.-t)\|_{W(L^\infty,\ell^1)}$ for every $t \in \mathbb{R}$. Observe also that when g has compact support, $\|g\|_{W(L^\infty,\ell^1)} = \|g\|_\infty$.

In this section we prove the following

Theorem 3 *Let $\mathcal{G}(h; a, b)$ be a frame with bounds A, B; assume that $h : \mathbb{R} \to \mathbb{R}$ is differentiable a.e., and $h' \in W(L^\infty, \ell^1)$. If*

$$\lambda =: (ab)^{-\frac{1}{2}} \|h'\|_{W(L^\infty,\ell^1)} \left(\sum_k \sup_n |\delta_{n,k}|^2\right)^{\frac{1}{2}} < \sqrt{A},$$

the set $\tilde{\mathcal{B}} = \{e^{2\pi ibnx} h(x - a(k + \delta_{n,k}))\}_{n,k\in\mathbb{Z}}$ is a frame with bounds

$$A' = (1 - \frac{\lambda}{\sqrt{A}})A, \quad B' = (1 + \frac{\lambda}{\sqrt{A}})B.$$

By "a.e" we mean "with the possible exception of a set of measure zero".

If h satisfies the assumptions of Theorem 3 and has compact support, then h must be continuous. Indeed, the derivative of a discontinuous function is a distribution because a discontinuity creates a singularity in the derivative, typically represented by a Dirac delta function, and so h' cannot be in $L^\infty(\mathbb{R})$.

If h is continuous in $\mathbb{R}$ and differentiable a.e. on its support we can apply both Theorems 1 and 3.

Proof of Theorem 3 We let

$$f_n(x) = \sum_k a_{n,k} e^{2\pi iabn}(h(x-\mu_{n,k})-h(x-k)) = \sum_k a_{n,k} e^{2\pi iabn} \int_k^{\mu_{n,k}} h'(x-t)dt.$$

By Theorem 2, the thesis of Theorem 3 follows if we prove that $\|\sum_n f_n\|_2 \le \lambda$ for every finite sequence $\{a_{n,k}\} \subset \mathbb{C}$ such that $\sum_{n,k} |a_{n,k}|^2 = 1$.

By duality, $\|f\|_2 = \sup_{\substack{\psi\in C_0^\infty(\mathbb{R})\\ \|\psi\|_2=1}} \left|\int_{-\infty}^{\infty} f(x)\overline{\psi}(x)dx\right|$; with the triangle inequality and the fundamental theorem of calculus, we obtain

$$\begin{aligned}
\|\sum_n f_n\|_2 &\le \sum_m \|\sum_n f_n\|_{L^2(m-\frac{1}{2},m+\frac{1}{2})} \\
&= \sum_m \sup_{\substack{\psi\in C_0^\infty(-\frac{1}{2},+\frac{1}{2})\\ \|\psi\|_{L^2(-\frac{1}{2},+\frac{1}{2})}=1}} \left|\sum_n \int_{-\infty}^{\infty} f_n(x)\psi(m-x)dx\right| \qquad (30)\\
&= \sum_m \sup_{\substack{\psi\in C_0^\infty(\frac{1}{2},\frac{1}{2})\\ \|\psi\|_{L^2(-\frac{1}{2},\frac{1}{2})}=1}} \left|\sum_{n,k} a_{n,k}\int_k^{\mu_{n,k}}\int_{-\infty}^{\infty} e^{2\pi iabnx}\psi(m-x)h'(x-t)dx\,dt\right|
\end{aligned}$$

We let $d_j = \sup_n |\delta_{n,j}|$ and we observe that $(k,\ \mu_{n,k}) = (k,\ k+\delta_{n,k}) \subset (k,\ k+d_k)$ when $\delta_{n,k} \ge 0$, and $(k,\ k+\delta_{n,k}) \subset (k-d_k,\ k)$ when $\delta_{n,k} \le 0$. Assuming that all $\delta_{n,k}$ are ≥ 0 simplifies the notation, but will not change the substance of the proof.

Recalling the definition of short-term Fourier transform and using a change of variables in the second integral above, we gather

$$\begin{aligned}
&\|\sum_n f_n\|_{L^2(m-\frac{1}{2},m+\frac{1}{2})} \\
&= \sup_{\substack{\psi\in C_0^\infty(-\frac{1}{2},\frac{1}{2})\\ \|\psi\|_{L^2(-\frac{1}{2},+\frac{1}{2})}=1}} \left|\sum_{n,k} a_{n,k}\int_k^{\mu_{n,k}}\int_{-\infty}^{\infty} e^{2\pi iabn(x+t)}\psi(m-t-x)h'(x)dx\,dt\right| \\
&\le \sup_{\substack{\psi\in C_0^\infty(-\frac{1}{2},+\frac{1}{2})\\ \|\psi\|_{L^2(-\frac{1}{2},+\frac{1}{2})}=1}} \sum_k \int_k^{k+d_k} \left|\sum_n a_{n,k}\mathcal{F}^\psi(h)(abn,m-t)dt\right|.
\end{aligned}$$

Applying the Cauchy-Schwartz inequality to the sum in n inside the integral and Proposition 2 gives

$$\|\sum_n f_n\|_{L^2(m-\frac{1}{2},m+\frac{1}{2})}$$

$$\leq (ab)^{-\frac{1}{2}} \sup_{\substack{\psi \in C_0^\infty(-\frac{1}{2},\frac{1}{2}) \\ \|\psi\|_{L^2(-\frac{1}{2},\frac{1}{2})}=1}} \sum_k \left(\sum_n |a_{n,k}|^2\right)^{\frac{1}{2}} \int_k^{k+d_k}$$

$$\times \left(\sum_n |F^\psi(h)(abn, m-t)|^2\right)^{\frac{1}{2}} dt$$

$$= (ab)^{-\frac{1}{2}} \sup_{\substack{\psi \in C_0^\infty(-\frac{1}{2},\frac{1}{2}) \\ \|\psi\|_{L^2(-\frac{1}{2},\frac{1}{2})}=1}} \sum_k \left(\sum_n |a_{n,k}|^2\right)^{\frac{1}{2}} \int_k^{k+d_k}$$

$$\times \left(\int_{-\infty}^{\infty} |\psi(m-t-s)|^2 |h'(s)|^2 ds\right)^{\frac{1}{2}} dt.$$

After a change of variables in the second integral,

$$\|\sum_n f_n\|_{L^2(m-\frac{1}{2},m+\frac{1}{2})}$$

$$\leq (ab)^{-\frac{1}{2}} \sup_{\substack{\psi \in C_0^\infty(-\frac{1}{2},\frac{1}{2}) \\ \|\psi\|_{L^2(-\frac{1}{2},\frac{1}{2})}=1}} \sum_k \left(\sum_n |a_{n,k}|^2\right)^{\frac{1}{2}} \int_k^{k+d_k}$$

$$\times \left(\int_{-\infty}^{\infty} |\psi(s)|^2 |h'(m-t-s)|^2 ds\right)^{\frac{1}{2}} dt$$

$$\leq (ab)^{-\frac{1}{2}} \sup_{\substack{\psi \in C_0^\infty(-\frac{1}{2},\frac{1}{2}) \\ \|\psi\|_{L^2(-\frac{1}{2},\frac{1}{2})}=1}} \sum_k \left(\sum_n |a_{n,k}|^2\right)^{\frac{1}{2}} \int_k^{k+d_k} \sup_{y \in (m-\frac{1}{2},m+\frac{1}{2})} |h'(y-t)|$$

$$\times \left(\int_{-\infty}^{\infty} |\psi(s)|^2\right)^{\frac{1}{2}} dt$$

$$= (ab)^{-\frac{1}{2}} \sum_k \left(\sum_n |a_{n,k}|^2\right)^{\frac{1}{2}} \int_k^{k+d_k} \|h(.-t)\|_{L^\infty(m-\frac{1}{2},m+\frac{1}{2})} dt.$$

In view of (30), we gather

$$\sum_m \|\sum_n f_n\|_{L^2(m-\frac{1}{2},m+\frac{1}{2})}$$

$$\leq (ab)^{-\frac{1}{2}} \sum_k \left(\sum_n |a_{n,k}|^2 \right)^{\frac{1}{2}} \int_k^{k+d_k} \sum_m \|h'(. - t)\|_{L^\infty(m-\frac{1}{2},m+\frac{1}{2})} dt$$

$$= \|h'\|_{W(L^\infty,\ell^1)} (ab)^{-\frac{1}{2}} \sum_k d_k \left(\sum_n |a_{n,k}|^2 \right)^{\frac{1}{2}}.$$

By Cauchy-Schwartz

$$\sum_m \| \sum_n f_n \|_{L^2(m-\frac{1}{2}, m+\frac{1}{2})} \leq \|h'\|_{W(L^\infty,\ell^1)} (ab)^{-\frac{1}{2}} \left(\sum_k |d_k|^2 \right)^{\frac{1}{2}}.$$

By assumption, $\|h'\|_{W(L^\infty,\ell^1)} (ab)^{-\frac{1}{2}} \left(\sum_k |d_k|^2 \right)^{\frac{1}{2}} < \sqrt{A}$ and so we have proved that

$$\| \sum_n f_n \|_2 \leq \sum_m \| \sum_n f_n \|_{L^2(m-\frac{1}{2}, m+\frac{1}{2})} < \sqrt{A}.$$

By Theorem 2, the thesis of the theorem follows. □

4.2 *h* Band-limited

In this section we assume that $\hat{h}$ has compact support. As always, we assume that $\mathcal{G}(h; a, b)$ is a frame with bounds A, B. We prove the following

Theorem 4 *Let $h \in L^2(\mathbb{R})$ with Fourier transform supported in $(-\frac{M}{2}, \frac{M}{2})$, with $M \in \mathbb{N}$.*

Let $\{\delta_{n,k}\} \subset (-\frac{1}{4M}, \frac{1}{4M})$ be such that

$$\|\hat{h}\|_\infty \sqrt{\frac{M}{2\pi}} \left(\sum_n (1 - \cos(2\pi M D_n) + \sin(2\pi M D_n))^2 \right)^{\frac{1}{2}} = \mu < \sqrt{A}$$

where we have let $D_n = \sup_k |\delta_{n,k}|$. Then, the set $\tilde{\mathcal{B}} = \{e^{2\pi i bnx} h(x - a(k + \delta_{n,k}))\}_{n,k \in \mathbb{Z}}$ is a frame with bounds

$$A' = (1 - \frac{\mu}{\sqrt{A}})A, \quad B' = (1 + \frac{\mu}{\sqrt{A}})B.$$

Remark In view of (8), we have that $\frac{M}{2\pi}(1 - \cos(2\pi M D_n) + \sin(2\pi M D_n)) \geq \frac{M^2}{\pi} D_n$, and so it is enough to verify that $\frac{M}{\sqrt{\pi}} \|\hat{h}\|_\infty \left(\sum_n D_n^2\right)^{\frac{1}{2}} = \mu < \sqrt{A}$

Proof We have proved in section 3.1 that $\mathcal{G}(h; a, b)$ is a frame with bounds A and B if and only if the set

$$\mathcal{G}(h_a; 1, ab) = \{e^{2\pi i abnx} h_a(x - k)\}_{n,k \in \mathbb{Z}}$$

is a frame with bounds $a^{-1}A$ and $a^{-1}B$. Recall that $h_a(x) = h(ax)$, and $\mu_{n,k}$ may be used in place of $k + \delta_{n,k}$.

Let $\{a_{n,k}\} \subset \mathbb{C}$ with $\sum_{n,k} |a_{n,k}| = 1$, and let

$$\mathcal{I} =: \left\| \sum_{n,k} a_{n,k}\, e^{2\pi i abnx} \left(h_a\left(\cdot - \mu_{n,k}\right) - h_a\left(\cdot - k\right)\right) \right\|_2 .$$

By Theorem 2, the thesis of Theorem 4 follows if we prove that $\mathcal{I} \leq \lambda a^{-\frac{1}{2}} \sqrt{A}$, with $\lambda < 1$.

An application of Plancherel theorem gives

$$\mathcal{I} = \| \sum_{n,k} a_{n,k}\, \hat{h}_a(y - abn) \left(e^{2\pi i k(abn-y)} - e^{2\pi i \mu_{n,k}(abn-y)}\right) \|_2$$

$$= \| \sum_{n,k} a_{n,k}\, \hat{h}_a(y - abn) e^{2\pi i k(abn-y)} (1 - e^{2\pi i \delta_{n,k}(abn-y)}\big) \|_2 .$$

We apply the triangle inequality and the change of variables $y \to y + abn$ in the norm; letting $b_{n,k} = a_{n,k} e^{2\pi i nkab}$ and recalling that $\widehat{h}_a(yP) = \frac{1}{a}\hat{h}(\frac{y}{a})$, we can write the following chain of inequalities.

$$\mathcal{I} \leq \sum_n \|\widehat{h}_a(y - abn) \sum_k a_{n,k}\, e^{-2\pi i k y} \left(1 - e^{2\pi i \delta_{n,k}(abn-y)}\right) \|_2$$

$$= \sum_n \|\widehat{h}_a \sum_k b_{n,k}\, e^{-2\pi i k y} \left(1 - e^{2\pi i \delta_{n,k} y}\right) \|_2 .$$

$$= a^{-1} \sum_n \|\hat{h}(\frac{y}{a}) \sum_k b_{n,k}\, e^{-2\pi i k y} \left(1 - e^{2\pi i \delta_{n,k} y}\right) \|_{L^2(-\frac{aM}{2}, \frac{aM}{2})}$$

$$\leq a^{-1} \|\hat{h}\|_\infty \sum_n \| \sum_k b_{n,k}\, e^{-2\pi i k y} \left(1 - e^{2\pi i \delta_{n,k} y}\right) \|_{L^2(-\frac{aM}{2}, \frac{aM}{2})} .$$

By Lemma (1),

$$\mathcal{I} \leq a^{-1}\|\hat{h}\|_{\infty}\sqrt{\frac{aM}{2\pi}}\sum_{n}(1-\cos(\pi D_n aM)+\sin(\pi D_n aM))\left(\sum_{k}|b_{n,k}|^2\right)^{\frac{1}{2}}$$

After applying the Cauchy-Schwartz inequality, and observing that $|b_{n,k}| = |a_{n,k}|$, we obtain

$$\mathcal{I} \leq \|\hat{h}\|_{\infty}\sqrt{\frac{M}{2a\pi}}\left(\sum_{n}(1-\cos(2\pi D_n)+\sin(2\pi D_n))^2\right)^{\frac{1}{2}}\left(\sum_{n,k}|a_{n,k}|^2\right)^{\frac{1}{2}}.$$

By assumption, $\sqrt{\frac{M}{2\pi}}\left(\sum_n(1-\cos(2\pi D_n)+\sin(2\pi D_n))^2\right)^{\frac{1}{2}} < \sqrt{A}$ and so we have proved that $\mathcal{I} < a^{-\frac{1}{2}}\sqrt{A}$, as required □

Remark Recalling that a Gabor system $\mathcal{G} = \{e^{2\pi i b_n t}h(t-a_k)\}_{n,k\in\mathbb{Z}}$ is a frame if and only if $\hat{\mathcal{G}} = \{e^{2\pi i a_k s}\hat{h}(s-b_n)\}_{n,k\in\mathbb{Z}}$ is a frame, we can use Theorems 1, 4 and 3 to prove stability results for frames $\{e^{2\pi i b(n+\omega_{k,n})t}g(t-ak)\}_{n,k\in\mathbb{Z}}$ when g or $\hat{g}$ have compact support, or when $\hat{g}'$ is in a Wiener amalgam space. We leave these generalizations to the interested reader.

References

1. R. Balan, *Stability theorems for Fourier frames and wavelet Riesz bases*, Journal of Fourier Analysis and Applications 3 (1997) 499–504.
2. P. Balazs, M. Dörfler, F. Jaillet, N. Holighaus, G. Velasco, *Theory, implementation and applications of nonstationary Gabor frames*. J. Comput. Appl. Math. 236 (2011), no.6, 1481–1496.
3. O. Christensen *An Introduction to Frames and Riesz Bases,* Birkäuser (2016).
4. O.Christensen, *A Paley-Wiener Theorem for Frames*. Proc. Amer. Math. Soc. 123 (1995), 2199–2202.
5. O.Christensen, *Moment problems and stability results for frames with applications to irregular sampling and Gabor frames*. Appl. Comput. Harmon. Anal. 3 (1996), no.1, 82–86.
6. I. Daubechies, A. Grossmann, and Y. Meyer, *Painless nonorthogonal expansions* J. Math. Phys. 27 (1986), no. 5, 1271–1283.
7. C. de Boor, *A practical guide to splines* (Revised ed.). Springer-Verlag (2001).
8. L. De Carli, P. Vellucci, *Stability theorems for the n-order hold model*, Linear Algebra and its Applications 536 (2018), 186–200.
9. M. Dörfler, E. Matusiak, *Nonstationary Gabor frames—existence and construction.* Int. J. Wavelets Multiresolut. Inf. Process. 12 (2014), no.3, 1450032.
10. H. Feichtinger, *Varying the time-frequency lattice of Gabor frames*, Transactions of the American Mathematical Society 356 (2004), no. 5, 2001–2023.
11. H. Feichtinger, K. Gröchenig, *Gabor frames and time-frequency analysis of distributions.* Journal of Functional Analysis, 146 (1992), no. 2, 464–495.

12. K. Gröchenig. *Foundations of Time-Frequency Analysis*. Appl. Numer. Harmon. Anal. Birkhäuser Boston, 2001
13. K. Hamm, C. Heil, D. Walnut, *Continuous and discrete wavelet transforms*, SIAM Rev. 31 (1989), 403–436.
14. C. Heil, *A basis theory primer*, Appl. Num. Harm. Analysis, Birkhäuser 2011.
15. F. Jaillet, P. Balazs, M. Dörfler, N. Engelputzeder, *Nonstationary Gabor frames*, in: Proceedings of the 8th International Conference on Sampling Theory and Applications, SampTA (2009), 227–230.
16. M.I. Kadec, *The exact value of the Paley-Wiener constant*, Soviet Math. Dokl., 5 (1964), 559–561.
17. I. Krishtal, B. Miller, *Kadec-type theorems for sampled group orbits*, https://arxiv.org/abs/2407.15827 (2024)
18. J. Lakey, Y. Wang, *On perturbations of irregular Gabor frames*, Journal of Computational and Applied Mathematics 155 (2003), 111– 129.
19. Q. Lian, L. Song, *Vector-valued nonstationary Gabor frames*. J. Math. Anal. Appl. 469 (2019), no.1, 358–377.
20. E.V. Mischenko, *Determination of Riesz bounds for the spline basis with the help of trigonometric polynomials*. Sib. Math. J. 51 (2010) no. 4, 660–666.
21. H. Prautzsch, W. Boehm, M. Paluszny, *Bezier and B-spline techniques*, Springer Science & Business Media, 2002.
22. R. Paley and N. Wiener, *Fourier transforms in the complex domain*, American Mathematical Soc., 1934.
23. D. Rudoy, B. Prabahan, P. Wolfe, *Superposition frames for adaptive time–frequency analysis and fast reconstruction*, IEEE Trans. Signal Process. 58 (2010), 2581–2596.
24. W. Sun, X. Zhou, *On the stability of Gabor frames*, Advances in Applied Mathematics 26 (2001), no. 3, 181–191.
25. W. Sun, X. Zhou, *Irregular Gabor frames and their stability*, Proceedings of the American Mathematical Society 131 (2003), no. 9, 2883–2893.
26. R. M. Young, *An introduction to nonharmonic Fourier series*, Academic Press, (2001).

A Nonnegative Even Kernel Arising from Extremal Sine Polynomials

Dmitriy Dmitrishin, Alexander M. Stokolos, and Walter Trebels

Dedicated to Ed Saff on the occasion of his 80th birthday.

1 Introduction

In 1900 L. Fejér [6] introduced the even, continuous, nonnegative trigonometric polynomials

$$F_N(t) = \frac{1}{N+1}\left(\frac{\sin\frac{N+1}{2}t}{\sin\frac{t}{2}}\right)^2 = 1 + 2\sum_{k=1}^{N}\left(1 - \frac{k}{N+1}\right)\cos kt, \qquad (1)$$

the famous Fejér kernel. With this, he obtained two fundamental results.

I. Given the Fourier series of a continuous, 2π periodic function $f \in C[0, 2\pi]$

$$f(t) \sim \sum_{k=-\infty}^{\infty} \hat{f}_k e^{ikt}, \qquad \hat{f}_k = \frac{1}{2\pi}\int_{-\pi}^{\pi} f(t)e^{-ikt}dt, \quad k \in \mathbb{Z}.$$

D. Dmitrishin (✉)
Odessa National Polytechnic University, Odesa, Ukraine
e-mail: dmitrishin@op.edu.ua

A. M. Stokolos
Georgia Southern University, Statesboro, GA, USA
e-mail: astokolos@georgiasouthern.edu

W. Trebels
Department of Mathematics, AG Algebra, Technical University of Darmstadt, Darmstadt, Germany
e-mail: trebels@mathematik.tu-darmstadt.de

D. Bilyk et al. (eds.), *Recent Advances in Approximation and Potential Theory*, Applied and Numerical Harmonic Analysis,
https://doi.org/10.1007/978-3-031-95551-8_9

Then the first arithmetic means of the Fourier series converge to $f(t)$,

$$\lim_{N\to\infty} \max_t \left| f(t) - \sum_{|k|\leq N} \left(1 - \frac{|k|}{N+1}\right) \hat{f}_k e^{ikt} \right| = 0.$$

The best possible approximation order (saturation order) by the Fejér means is $O(N^{-1})$ as the example $e^{it} \in C^\infty[0, 2\pi]$ shows.

II. In 1913 Fejér [7] realized that $K_N(t)$ is the extremal trigonometric polynomial that attains the maximum over all even, nonnegative normalized trigonometric polynomials $C_N(t)$, i.e.

$$\sup_{C_N(t)} \max_t \left\{ C_N(t) = 1 + 2\sum_{k=1}^{N} a_k \cos kt : C_N(t) \geq 0 \right\} = K_N(0).$$

Further, in [8] he considered the problem

$$\max\{a_1 : 1 + 2\sum_{k=1}^{N} a_k \cos kt \geq 0\} \tag{2}$$

and found the unique extremizer

$$\mathcal{K}_N(t) = \frac{1}{N+2} \left(\frac{\sin \frac{\pi}{N+2} \cos(N+2)\frac{t}{2}}{\cos t - \cos \frac{\pi}{N+2}} \right)^2. \tag{3}$$

In [1, p. 438,9] the kernel is named Fejér-Korovkin due to their deep results involving this kernel. In 1983 E.L. Stark [10] gave a careful account on this kernel as a basic tool in the constructive theory of functions. For the first time its Fourier coefficients were computed in 1928 by Egerváry and Szász [5] and can be found in [1, p.438]. $\mathcal{K}_N$ is also an approximate identity since $a_1 = \cos \frac{\pi}{N+2} \to 1$ for $N \to \infty$ (cf. [1, pp. 59, 438]).

The problem for all a_k, $2 \leq k \leq N$, corresponding to (2), is solved in [2, p.7].

As a counterpart of these results, in 1950 Rogosinski and Szegö [9] dealt with the analogous problem for the conjugate normalized sine polynomials, nonnegative on $[0, \pi]$,

$$S(t) = \sin t + \sum_{k=2}^{N} a_k \sin kt \geq 0, \qquad a_N \neq 0, \quad t \in [0, \pi].$$

For a_2 and a_3 they obtained sharp upper and lower bounds, but no extremizers, i.e., extremal sine polynomials attaining these bounds. In [4] constructive and compact representations for such extremizers are given for a_2. This allows to derive nonnegative cosine kernels which, after appropriate remodelling, turn out to be approximate identities. We start with some general considerations.

Thus let us assume that S_N^{extr} is a sine extremizer. Then, on $[0, \pi]$,

$$0 \le S_N^{extr}(t) = (\sin t)\cdot(\widetilde{C}_N(t)), \qquad \widetilde{C}_N(t) = \gamma_{1,N} + 2\sum_{j=2}^{N} \gamma_{j,N} \cos(j-1)t. \tag{4}$$

In order to arrive at an approximate identity we divide S_N^{extr} by $\sin t$ and then determine $\gamma_{1,N}$ and $\gamma_{2,N}$. This leads to the even, nonnegative cosine polynomial

$$C_{N-1}(t) = 1 + 2\sum_{j=1}^{N-1} b_{j,N} \cos jt, \qquad b_{j,N} = \gamma_{j+1,N}/\gamma_{1,N}. \tag{5}$$

Now C_{N-1} is an approximate identity if $\lim_{N\to\infty} \gamma_{2,N}/\gamma_{1,N} = 1$ (cf. [1, p. 59]).

To compute $\gamma_{1,N}$ and $\gamma_{2,N}$ we have briefly to describe the setting in which the extremizers are derived. To this end consider typically real polynomials on the unit disc $\mathbb{D} = \{z \in \mathbb{C} : |z| \le 1\}$ with

$$P(z) = z + \sum_{k=2}^{N} a_k z^k, \quad a_k \in \mathbb{R}, \quad \big(\Im m\{P(z)\}\big)\cdot\big(\Im m\{z\}\big) > 0, \ z \in \mathbb{D}\setminus\mathbb{R}.$$

As soon as one has found an extremizer P^{extr} in this framework then $S^{\text{extr}}(t) := \Im m\{P^{\text{extr}}(e^{it})\}$ is an extremizer for the corresponding sine polynomials $S(t)$. Thus we arrive at

$$\widetilde{C}(t) := \Im m\{P^{extr}(e^{it})\}/\sin t\,, \tag{6}$$

where

$$\gamma_{k,N} = \sum_{\substack{k\le j\le N \\ j\ odd}} a_j \quad \text{if } k \text{ is odd}, \qquad \gamma_{k,N} = \sum_{\substack{k\le j\le N \\ j\ even}} a_j \quad \text{if } k \text{ is even}.$$

The latter formulas are written in a sloppy, intuitive way to read

$$\begin{aligned}
\gamma_{1,N} &= a_1 + a_3 + a_5 + \dots, && \text{(observe } a_1 = 1\text{)},\\
\gamma_{2,N} &= a_2 + a_4 + a_6 + \dots, &&\\
\gamma_{3,N} &= a_3 + a_5 + a_7 + \dots, && \text{(observe } \gamma_{3,N} = \gamma_{1,N} - 1\text{)},\\
\gamma_{4,N} &= a_4 + a_6 + a_8 + \dots, && \text{(observe } \gamma_{4,N} = \gamma_{2,N} - a_2\text{)}.
\end{aligned}$$

For later use we note that

$$\frac{b_{2,N}-1}{b_{1,N}-1}=\frac{\gamma_{3,N}/\gamma_{1,N}-1}{\gamma_{2,N}/\gamma_{1,N}-1}=\frac{1}{\gamma_{1,N}-\gamma_{2,N}}.$$

The explicit values of the γ's follow from $P_N^{extr}(1)$ and $P_N^{extr}(-1)$, in accordance with N odd and N even, as follows

$$\gamma_{1,N}=\frac{P_N^{extr}(1)-P_N^{extr}(-1)}{2},\quad \gamma_{2,N}=\frac{P_N^{extr}(1)+P_N^{extr}(-1)}{2}. \tag{7}$$

For the sake of simplicity we restrict ourselves to the extremal sine polynomials related to the coefficient a_2. Further observe that it is sufficient to consider only the $\max\{a_2\}$ case, since $P_2^{\min}(z)=-P_2^{\max}(-z)$—see [4, Thms 4.3,5.2].

2 Explicit Representations of C_N

In [4, Thrm 4.3, Cor 4.3.1.] the following result is proved

Theorem 1 *If N is odd, then*

$$P_{max}^{odd}(1)=\frac{\cos(\frac{2\pi}{N+3})+2}{2-2\cos(\frac{2\pi}{N+3})},\quad \textit{and}\quad P_{max}^{odd}(-1)=\frac{\cos(\frac{2\pi}{N+3})-2}{2+2\cos(\frac{2\pi}{N+3})}, \tag{8}$$

the associate non-negative extremal trigonometric polynomial has the form

$$\Im m\left(P_{\max}^{odd}(e^{it})\right)=\frac{\sin^2\frac{2\pi}{N+3}}{N+3}\cdot\frac{1}{\sin t}\cdot\frac{\sin^2\frac{N+3}{2}t}{\left(\cos t-\cos\frac{2\pi}{N+3}\right)^2}. \tag{9}$$

By (7) and (8) we obtain

$$\gamma_{1,N}=\frac{3-\sin^2\left(\frac{2\pi}{N+3}\right)}{2\sin^2\left(\frac{2\pi}{N+3}\right)},\quad \gamma_{2,N}=\frac{3}{2}\frac{\cos\left(\frac{2\pi}{N+3}\right)}{\sin^2\left(\frac{2\pi}{N+3}\right)},\quad b_{1,N}=\frac{3\cos\left(\frac{2\pi}{N+3}\right)}{3-\sin^2\left(\frac{2\pi}{N+3}\right)}, \tag{10}$$

hence $b_{1,N}\to 1$ for $N\to\infty$. Thus for the subsequence of odd $N=2k-1,\ k\in\mathbb{N}$ the associated kernel (5) is an approximate identity which we denote by K_{N-1}. Its compact representation reads

$$\mathsf{K}_{N-1}^{odd}(t)=\frac{1}{N+3}\cdot\frac{2\sin^4\left(\frac{2\pi}{N+3}\right)}{3-\sin^2\left(\frac{2\pi}{N+3}\right)}\cdot\frac{1}{\sin^2 t}\cdot\left(\frac{\sin\frac{N+3}{2}t}{\cos t-\cos\frac{2\pi}{N+3}}\right)^2, \tag{11}$$

being obviously even and nonnegative.

If N is even, the first derivatives U'_N of the Chebyshev polynomials of the second kind U_N are required. We recall

$$U_N(x) = \frac{\sin(j+1)t}{\sin t}, \qquad x = \cos t$$

and now make use of [4, Thrm 5.2, Cor 5.2] which leads to

Theorem 2 *If N is even and η_N the maximal root of*

$$U'_{\frac{N}{2}+1}(x) - U'_{\frac{N}{2}}(x) = 0,$$

then

$$P_{max}^{even}(1) = \frac{\eta_N + 2}{2 - 2\eta_N}, \qquad and \qquad P_{max}^{even}(-1) = \frac{\eta_N - 4}{6\eta_N + 6}. \tag{12}$$

The associate non-negative trigonometric polynomial has the form

$$\Im m\left(P_{\max}^{even}(e^{it})\right) = Q_N \frac{1}{1+\cos t}\frac{1}{\sin t}\frac{\left(\frac{N+4}{2}\sin\frac{N+2}{2}t + \frac{N+2}{2}\sin\frac{N+4}{2}t\right)^2}{(\cos t - \eta_N)^2}, \tag{13}$$

where

$$Q_N = \frac{2(1-\eta_N^2)}{(N+2)(N+3)(N+4)}.$$

For even $N = 2k,\ k \in \mathbb{N}$, (7) and (12) yield

$$\gamma_{1,N} = \frac{2\eta_N^2 + 2\eta_N + 5}{6 - 6\eta_N^2}, \quad \gamma_{2,N} = \frac{\eta_N^2 + 7\eta_N + 1}{6 - 6\eta_N^2}, \quad b_{1,N} = \frac{\eta_N^2 + 7\eta_N + 1}{2\eta_N^2 + 2\eta_N + 5}. \tag{14}$$

Since $\eta_N \to 1$ for $N \to \infty$ it follows that also in the case of even N there holds $b_{1,N} \to 1$ for $N \to \infty$, i.e. the subsequence K_{N-1}^{even},

$$\mathsf{K}_{N-1}^{even}(t) = \left(\prod_{j=2}^{4}\frac{1}{N+j}\right)\frac{12(1-\eta_N^2)^2}{2\eta_N^2 + 2\eta_N + 5}\frac{1}{1+\cos t}$$
$$\left(\frac{\frac{N+4}{2}\sin\frac{N+2}{2}t + \frac{N+2}{2}\sin\frac{N+4}{2}t}{(\cos t - \eta_N)\sin t}\right)^2 \tag{15}$$

generates an even, nonnegative approximate identity. Thus, by (11) and (15), this is also true for

$$\mathsf{K}_{N-1}(t) = \begin{cases} \mathsf{K}^{odd}_{N-1}(t), \ N \text{ odd} \\ \\ \mathsf{K}^{even}_{N-1}(t), \ N \text{ even}, \end{cases} \qquad N \in \mathbb{N}. \tag{16}$$

Its peak at the origin behaves like

$$\mathsf{K}_{N-1}(0+) = \frac{2}{3}(N+3) + o(N+3).$$

For odd N this is trivial, for even N observe

$$\mathsf{K}^{even}_{N-1}(0+) = c_\eta \frac{(N+2)(N+4)}{N+3}, \quad c_\eta := \frac{3}{2} \cdot \frac{(1+\eta_N)^2}{2\eta_N^2 + 2\eta_N + 5} \to \frac{2}{3} \text{ for } N \to \infty.$$

The peaks of $F_N(t)$, $\mathcal{K}_N(t)$ are

$$F_N(0) = N+1, \quad \mathcal{K}_N(0) = \frac{8}{\pi^2}(N+2).$$

We illustrate the results obtained so far in Fig. 1.

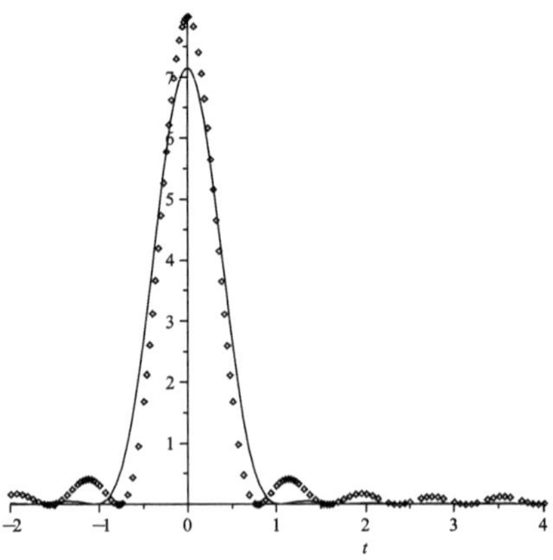

a)

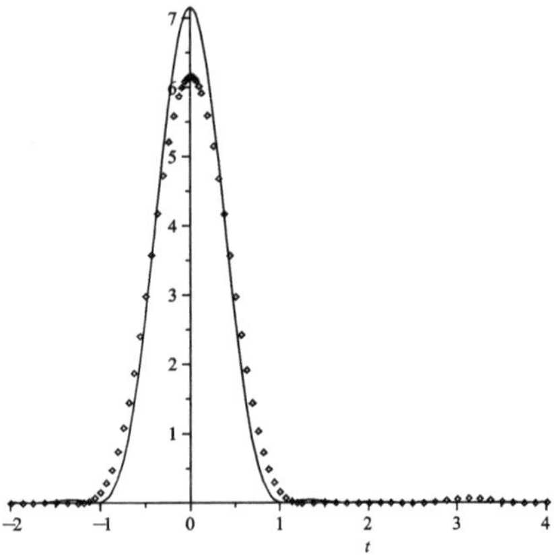

b)

Fig. 1 Kernels: (**a**) Fejèr $F_7(t)$ (point), and Egerváry-Szász $\mathcal{K}_7(t)$ (solid); (**b**) $\mathsf{K}_7(t)$ (point), and $\mathcal{K}_7(t)$ Egerváry-Szász (solid)

3 Approximation Behavior of C_N

The representations of $\mathcal{K}_N(t)$ and $K_N(t)$ evoke the impression that both approximate identities are somehow related; also both are even, nonnegative so that the classical theory can be applied in the present case. Let us briefly recall some notation in connection with even, nonnegative kernels χ_N,

$$\chi_N(t) = 1 + 2\sum_{k=1}^{N} \hat{\chi_N}(k)\cos kt, \quad \hat{\chi_N}(k) = \frac{1}{\pi}\int_0^{\pi} \chi_N(t)\cos kt\, dt, \qquad k \in \mathbb{N};$$

the associated approximation process (on $C_{2\pi}$) is defined by

$$\chi_N * f(x) = \frac{1}{2\pi}\int_{-\pi}^{\pi} f(x-t)\chi_N(t)\, dt, \qquad f \in C_{2\pi};$$

the modulus of smoothness of the second order by

$$\omega_2(f,\delta) = \sup_{|h|\le\delta} \|\Delta_h^2 f\|_{C_{2\pi}}, \qquad \Delta_h^2 f(x) = f(x+h) - 2f(x) + f(x+h).$$

Then [1, Thm 1.5.8] states that

$$\|\chi_N * f - f\| \lesssim \omega_2\left(f, \sqrt{1-\hat{\chi_N}(1)}\right), \qquad N \to \infty.$$

Its application serves the Egerváry-Szász / Fejér-Korovkin process—see [1, Cor. 1.6.1]

$$\|\mathcal{K}_N * f - f\| \lesssim \omega_2\left(f, N^{-1}\right).$$

Corollary 12.1.7 of [1] deals with the saturation problem. The best possible approximation order is O(N^{-2}), the saturation class, i.e. the set of those $f \in C_{2\pi}$ attaining the approximation order $O(N^{-2})$, is contained in

$$V_{2\pi} := \{f \in C_{2\pi} : \exists g \in L_{2\pi}^{\infty},\ ck^2\hat{f}_k = \hat{g}_k,\ \forall k \in \mathbb{Z}\}$$

with $c = -\frac{\pi^2}{2}$.

Concerning the new kernel K_N let us first consider the case of odd N given by (11). Then $\hat{\chi_N}(1) = b_{1,N}$ is given in (10) and

$$1 - b_{1,N} \approx \frac{2}{3}\pi^2(N+3)^{-2},$$

hence

$$\|\mathsf{K}_N * f - f\| \lesssim \omega_2\left(f, N^{-1}\right), \qquad N \to \infty, \quad N \text{ odd.} \tag{17}$$

Concerning saturation, by the preceding it is clear that condition (i) of [1, Prop. 12.1.6] is satisfied. There remains to check [1, (12.1.7)] to conclude that the saturation order is $O(N^{-2})$ and that the saturation class is contained in $V_{2\pi}$ with $c = -\frac{2}{3}\pi^2$. By (8) and (10) it immediately follows that

$$\lim_{N\to\infty} \frac{b_{2,N} - 1}{b_{1,N} - 1} = \lim_{N\to\infty} \frac{-1}{\gamma_{2,N} - \gamma_{1,N}} = 4.$$

Thus, [1, (12.1.7)] is verified for odd N.

For even N consider

$$\frac{b_{2,N} - 1}{b_{1,N} - 1} = \frac{1}{\gamma_{1,N} - \gamma_{2,N}} = \frac{6 - 6\eta_N^2}{\eta^2 - 5\eta + 4}.$$

Since $\eta_N \to 1-$ for $N \to \infty$, L'Hopital's rule can be applied—giving, as desired,

$$\lim_{t\to 1-} \frac{6 - 6t^2}{t^2 - 5t + 4} = 4.$$

There remains to discuss the behavior of $1 - b_{1,N}$, N even. The Corollary 3.2.1.from [4] implies that $\eta_N = 1 - 2\nu_N^2$, where ν_N is the smallest positive root of the equation $U'_{N+2}(x) = 0$. Now use (see [3, Lemma 3])

Lemma 1 *Suppose N is even and ν is the minimal positive root of $U'_{N+1}(x)$. Then, $\nu \in (\cos\tau_1, \cos\tau_2)$, where $\tau_1 = \frac{(N+1)\pi}{2(N+2)}$ and $\tau_2 = \frac{N\pi}{2(N+1)}$.*

Hence

$$1 - b_{1,N} \sim C(N+3)^{-2},$$

with some constant $C > 0$. Thus, also the Jackson-type estimate (17) is proven for K_N as given by (16).

Acknowledgments Alexander Stokolos is partially supported by the AMS-Simons Research Enhancement Grant for PUI Faculty.

References

1. P. Butzer and R. Nessel. Fourier Analysis and Approximation. Birkhäuser Basel, 1971.
2. D. K. Dimitrov. Extremal positive trigonometric polynomials. In Approximation theory, pages 136–157. Darba, Sofia, 2002. ISBN 954-90126-5-4

3. D. Dmitrishin, A. Smorodin, and A. Stokolos. On C. Michel's hypothesis about the modulus of typically real polynomials. *J. Approx. Theory*, 289, 2023.
4. D. Dmitrishin, A. Stokolos, and W. Trebels. Extremizers for the Rogosinski - Szegö estimate of the second coefficient in nonnegative sine polynomials, ArXiv 2405.02932v2, 2025. Constr. Appr. (accepted)
5. E. Egerváry and O. Szász. Einige Extremalprobleme im Bereiche der trigonometrischen Polynome. *Math. Z.*, 27(1):641–652, 1928.
6. L. Fejér. Sur les fonctions bornées et intégrables. *C. R. Acad. Sci. Paris*, 131:984–987, 1900.
7. L. Fejér. Sur les polynomes trigonométriques. *C. R. Acad. Sci. Paris*, 157:571–574, 1913.
8. L. Fejér. Über trigonometrische Polynome. *J. für die reine und angew. Math.*, 146:53–82, 1915.
9. W. W. Rogosinski and G. Szegö. Extremum problems for non-negative sine polynomials. *Acta Sci. Math. (Szeged)*, 12:112–124, 1950.
10. E. Stark. The kernel of Fejér-Korovkin: a basic tool in the constructive theory of functions. *Colloq. Math. Soc. Janos Bolya*, 35:1095–1123, 1983.

The Number of Real Zeros of Polynomials with Constrained Coefficients

Tamás Erdélyi

Dedicated to Edward Saff on the occasion of his 80th birthday.

1 Introduction

We study the location of the zeros of polynomials of the form

$$P(z) = \sum_{j=0}^{n} a_j z^j\,, \quad |a_0| = 1\,, \quad |a_j| \le M\,, \quad a_j \in \mathbb{C}\,, \quad M \ge 1\,. \tag{1}$$

For some of closely related results see the List of References and/or the Introduction in [5].

2 Notation

For a complex-valued bounded function defined on an interval I we define

$$\|f\|_I := \sup_{x \in I} |f(x)|\,.$$

For $z_0 \in \mathbb{C}$ and $r > 0$ let $D(z_0, r)$ denote the open disk of the complex plane centered at z_0 of radius r. The closed disk centered at z_0 of radius r will be denoted by $\overline{D}(z_0, r)$. For $M \ge 1$ let $\mathcal{S}_M$ denote the collection of all analytic functions f defined in the open unit disk $D(0, 1)$ satisfying the growth condition

T. Erdélyi (✉)
Department of Mathematics, Texas A&M University, College Station, TX, USA
e-mail: terdelyi@tamu.edu

D. Bilyk et al. (eds.), *Recent Advances in Approximation and Potential Theory*,
Applied and Numerical Harmonic Analysis,
https://doi.org/10.1007/978-3-031-95551-8_10

$$|f(z)| \le \frac{M}{1-|z|}, \qquad z \in D(0,1). \tag{2}$$

In this paper zeros are counted with multiplicities, and c_j denotes a suitable positive absolute constant.

3 Results

Our first theorem gives an upper bound for the number of zeros of a polynomial P of the form (1) in the interval $[-1, 1]$. This has been proved with the factor $(1+\log M)$ rather than $(1 + \log M)^{1/2}$ in the Appendix of a recent paper [11] by Jacob and Nazarov, while the case $M = 1$ was proved in [5, Theorem 4.1]. We offer a proof in the case of $M \ge 1$ by using Jensen's formula (Lemma 1) and improving the dependence on M.

Theorem 1 *There is an absolute constant $c > 0$ such that every polynomial P of the form* (1) *has at most $c(n(1 + \log M))^{1/2}$ zeros in* $[-1, 1]$.

Our second theorem gives an upper bound for the number of zeros of a polynomial P of the form (1) in the interval $[-1+a, 1-a]$, $a \in (0, 1]$. For $M = 1$ the theorem below has been proved in [5, Theorem 4.2].

Theorem 2 *Let $a \in (0, 1]$. There is an absolute constant $c > 0$ such that every polynomial P of the form* (1) *has at most $(c/a)(1+\log M)$ zeros in* $[-1+a, 1-a]$.

The sharpness of Theorem 1 may be compared with [5, Theorem 2.3] stating that there exist polynomials P of the form

$$P(z) = \sum_{j=0}^{n} a_j z^j, \quad a_0 \ne 0, \quad |a_j| \le 1, \quad a_j \in \mathbb{R}, \tag{3}$$

such that P has a zero at 1 with multiplicity at least

$$\min\left\{\frac{1}{6}\,(n\,(1-\log|a_0|))^{1/2} - 1\,,\, n\right\}.$$

Multiplying by $M := 1/|a_0|$ in (3) shows that there exist polynomials P of the form (1) such that P has a zero at 1 with multiplicity at least

$$\min\left\{\frac{1}{6}\,(n(1+\log M))^{1/2} - 1\,,\, n\right\}.$$

It is also shown in [5, Theorem 2.8] that if $|a_0| \ge \exp(-L^2)$ and $|a_j| \le 1$ for each $j = L^2 + 1, L^2 + 2, \ldots, n$, then the polynomial

$$P(x) = \sum_{j=0}^{n} a_j x^j\,, \qquad a_j \in \mathbb{C}\,,$$

has at most $\frac{44}{7}(L+1)n^{1/2} + 5$ zeros at 1.

Jacob and Nazarov [11] proved a general theorem recently including that there is an absolute constant $c > 0$ such that there are polynomials of degree n with coefficients in the set $\{-1, 1\}$ having at least $cn^{1/2}$ *distinct* zeros in the interval $[-1, 1]$; and there are polynomials of degree at most n with coefficients in the set $\{0, 1\}$ having at least $cn^{1/2}$ *distinct* zeros in the interval $[-1, 1]$.

The proof of our next theorem corrects a somewhat incorrect proof of an earlier result in [4, Theorem 2.1].

Theorem 3 *There is a constant $\eta > 0$ such that every polynomial of the form* (1) *with $M = 1$ has at most $\eta n^{1/2}$ zeros inside any polygon with vertices on the unit circle, where the multiplicative constant $\eta > 0$ depends only on the polygon.*

Theorems 1–3 above may be compared with results in [1–16].

4 Lemmas

One of our main tools is Jensen's formula stated as Lemma 1 below. See [3] for instance.

Lemma 1 *Suppose h is a nonnegative integer, $z_0 \in \mathbb{C}$, and*

$$f(z) = \sum_{j=h}^{m} a_j (z - z_0)^j\,, \quad a_j \in \mathbb{C}, \quad a_h \neq 0\,,$$

is analytic in an open disk centered at z_0 of radius greater than $R > 0$, and suppose that the zeros of f in $\overline{D}(z_0, R) \setminus \{z_0\}$ are $z_1, z_2, \ldots, z_m$, where each zero is listed as many times as its multiplicity. Then

$$\log|a_h| + h \log R + \sum_{j=1}^{m} \log\left(\frac{R}{|z_j|}\right) = \frac{1}{2\pi} \int_0^{2\pi} \log|f(Re^{i\theta})|\, d\theta\,.$$

Our second main tool is [5, Theorem 5.1] stated below.

Lemma 2 *There are absolute constants $c_1 > 0$ and $c_2 > 0$ such that*

$$|f(0)|^{c_1/b} \le \exp\left(\frac{c_2}{b}\right) \|f\|_{[1-b,1)}$$

for every $f \in \mathcal{S}_1$ and $b \in (0, 1]$.

Lemma 2 implies the following by a simple linear scaling.

Lemma 3 *Let* $I_a := [1 - 5a/4, 1 - a]$, $a \in (0, 4/5]$. *There is an absolute constant* $c_3 > 0$ *such that*

$$(eM)^{-c_3/a} \le \|P\|_{I_a}$$

for every polynomial P *of the form* (1).

Proof of Lemma 3 A simple estimation shows that every polynomial P of the form (1) is contained in $\mathcal{S}_M$. It is also straightforward that $f \in \mathcal{S}_M$ implies that g_r defined by $g_r(z) := f(rz)$ with $0 < r \le 1$ satisfies $g_r \in \mathcal{S}_M$ and $g_r(0) = f(0)$. We apply this observation with $r := 1 - a$ and $b := a/4$ which implies that the interval $I_a = [1 - 5a/4, 1 - a]$ contains the interval $[(1 - b)(1 - a), 1 - a]$. Observe also that $f \in \mathcal{S}_M$ implies that $h_M := f/M \in \mathcal{S}_1$ and $h_M(0) = f(0)/M$ to conclude that the lemma follows from Lemma 2. □

To prove Theorem 1 we need the lemma below.

Lemma 4 *Assume that in addition to* $M \ge 1$ *we have* $1 + \log M \le n$. *Let* $J_{n,M} := [1 - n^{-1/2}(1 + \log M)^{1/2}, 1]$. *There is an absolute constant* $c_4 > 0$ *such that every polynomial* P *of the form* (1) *has at most* $c_4(n(1 + \log M))^{1/2}$ *zeros in the interval* $J_{n,M}$.

Proof of Lemma 4 Let P be a polynomial of the form (1). It follows from Lemma 3 with $a := (4/5)n^{-1/2}(1 + \log M)^{1/2}$ that there is a point $z_0 \in J_{n,M}$ such that

$$\begin{aligned}\log |P(z_0)| &\ge -(5/4)c_3 n^{1/2}(1 + \log M)^{-1/2}\,(1 + \log M) \\ &\ge -2c_3 n^{1/2}(1 + \log M)^{1/2}\,. \end{aligned} \tag{4}$$

We will apply Jensen's formula (Lemma 1) with the disk

$$\overline{D}(z_0, 2n^{-1/2}(1 + \log M)^{1/2}).$$

Using the bounds for the coefficients of P of the form (1), we get

$$|P(z)| \le (n+1)M\big(1+2n^{-1/2}(1+\log M)^{1/2}\big)^n \le (n+1)M\exp(2n^{1/2}(1+\log M)^{1/2})$$

for every $z \in \overline{D}(z_0, 2n^{-1/2}(1 + \log M)^{1/2})$. Hence

$$\log |P(z)| \le \log(n + 1) + \log M + 2n^{1/2}(1 + \log M)^{1/2} \tag{5}$$

for every $z \in \overline{D}(z_0, 2n^{-1/2}(1 + \log M)^{1/2})$. Denote the zeros of P in

$$\overline{D}(z_0, n^{-1/2}(1 + \log M)^{1/2})$$

by $z_1, z_2, \dots, z_m$, where each zero is listed as many times as its multiplicity. Estimating by Jensen's formula (Lemma 1) with the disk $\overline{D}(z_0, 2n^{-1/2}(1+\log M)^{1/2})$ and recalling (4) and (5), we obtain

$$\begin{aligned}
&-2c_3 n^{1/2}(1+\log M)^{1/2} + 0 + m\log 2 \\
&\qquad \le -2c_3 n^{1/2}(1+\log M)^{1/2} + \sum_{j=1}^{m} \log\left(\frac{2n^{-1/2}}{|z_j|}\right) \\
&\qquad \le \frac{1}{2\pi} 2\pi(\log(n+1) + \log M + 2n^{1/2}(1+\log M)^{1/2}) .
\end{aligned}$$

Therefore

$$\begin{aligned}
m &\le \frac{1}{\log 2}\big(\log(n+1) + \log M + 2n^{1/2}(1+\log M)^{1/2} + 2c_3 n^{1/2}(1+\log M)^{1/2}\big) \\
&\le c_4 n^{1/2}(1+\log M)^{1/2} ,
\end{aligned}$$

where the inequalities

$$\log M = (\log M)^{1/2}(\log M)^{1/2} \le n^{1/2}(\log M)^{1/2}$$

and $\log(n+1) \le n^{1/2}$ have also been used. Finally observe that the interval $J_{n,M}$ is contained in the disk $\overline{D}(z_0, n^{-1/2}(1+\log M)^{1/2})$, and the lemma follows. □

The proof of our next lemma is quite similar to that of Lemma 4.

Lemma 5 *Let* $a \in (0, 4/5]$ *and* $I_a := [1 - 5a/4, 1 - a]$. *There is an absolute constant* $c_5 > 0$ *such that every polynomial* P *of the form* (1) *has at most*

$$(c_5/a)(1+\log M)$$

zeros in the interval I_a.

Proof of Lemma 5 Let P be a polynomial of the form (1). By Lemma 3 there is a point $z_0 \in I_a$ such that

$$\log|P(z_0)| \ge -(c_3/a)(1+\log M) . \tag{6}$$

We will apply Jensen's formula (Lemma 1) with the disk $\overline{D}(z_0, a/2)$. Observe that

$$|P(z)| \le \frac{2M}{a} , \qquad z \in \overline{D}(z_0, a/2) ,$$

hence

$$\log|P(z)| \le \log(2M) + \log(1/a)\,, \qquad z \in \overline{D}(z_0, a/2)\,. \tag{7}$$

Denote the zeros of P in $\overline{D}(z_0, a/4)$ by $z_1, z_2, \dots, z_m$, where each zero is listed as many times as its multiplicity. Estimating by Jensen's formula (Lemma 1) with the disk $\overline{D}(z_0, a/2)$ and recalling (6) and (7), we obtain

$$\begin{aligned} -(c_3/a)(1+\log M) + 0 + m\log 2 &\le -(c_3/a)(1+\log M) + \sum_{j=1}^{m} \log\left(\frac{a/2}{|z_j|}\right) \\ &\le \frac{1}{2\pi}\, 2\pi(\log(2M) + \log(1/a))\,, \end{aligned}$$

hence

$$m \le \frac{1}{\log 2}\big(\log(2M) + \log(1/a) + (c_3/a)(1+\log M)\big) \le (c_5/a)(1+\log M)\,.$$

Finally observe that the interval $I_a = [1 - 5a/4, 1-a]$ is contained in the disk $\overline{D}(z_0, a/4)$, and the lemma follows. □

Lemma 6 *Every polynomial P of the form* (1) *has at most* $\dfrac{\log(2M)}{\log 2}$ *zeros in the disk* $\overline{D}(0, 1/4)$.

Proof of Lemma 6 Let P be a polynomial of the form (1). Let $z_1, z_2, \dots, z_m$ be the zeros of P in the disk $\overline{D}(0, 1/4)$, where each zero is listed as many times as its multiplicity. Using Jensen's formula (Lemma 1) with the disk $\overline{D}(0, 1/2)$ we have

$$0 + 0 + m\log 2 \le \sum_{j=1}^{m} \log\left(\frac{1/2}{|z_j|}\right) \le \frac{1}{2\pi}\, 2\pi \log(2M)\,,$$

and the lemma follows. □

To prove Theorem 3 we need Lemmas 3.2, 3.3, and 3.5 from [4], which are formulated below as Lemmas 7, 8, and 9. For the sake of completeness we present the proof of Lemmas 7, 8, and 9.

Lemma 7 *Suppose $f \in \mathcal{S}_1$, $\overline{D}(z_0, r) \subset \overline{D}(0,1)$ with $r \in [3/4, 1)$, and the boundary circles have a common point. There is a constant $c(r) > 0$ depending only on $r \in [3/4, 1)$ such that*

$$\frac{1}{2\pi}\int_0^{2\pi} \log|f(z_0 + re^{i\theta})|\, d\theta \le c(r)\,.$$

Proof of Lemma 7 This follows simply from Lemmas 10 and 11 and the growth condition (2) with $M = 1$. □

The lemma below is a consequence of Jensen's formula (Lemma 1) and Lemma 2.

Lemma 8 *There is an absolute constant* $c_6 > 0$ *such that every polynomial* P *of the form* (1) *with* $M = 1$ *has at most* $c_6(n^{1/2} + nr)$ *zeros in any disk* $\overline{D}(u, r)$ *with* $u \in \mathbb{C}$, $|u| = 1$, *and* $r > 0$.

Proof of Lemma 8 Without loss of generality we may assume that $u = 1$. We may also assume that and $n^{-1/2} \le r \le 1$ as the case $0 < r < n^{-1/2}$ follows from the case $r = n^{-1/2}$, and the case $r > 1$ is obvious. Let P be a polynomial of the form (1) with $M = 1$. Observe that such a polynomial satisfies the growth condition (2). Choose a point $z_0 \in [1 - r, 1]$ such that

$$|P(z_0)| \ge \exp\left(\frac{-c_2}{r}\right). \tag{8}$$

There is such a point z_0 by Lemma 2. Using the bounds for the coefficients of P of the form (1), we have

$$\log|P(z)| \le \log((n+1)(1+4r)^n) \le \log(n+1) + 4nr\,, \qquad |z| \le 1 + 4r\,. \tag{9}$$

Let m denote the number of zeros of P in the disk $\overline{D}(z_0, 2r)$. Applying Jensen's formula (Lemma 1) with the disk $\overline{D}(z_0, 4r)$ and recalling (8) and (9), we obtain

$$\frac{-c_2}{r} + m\log 2 \le \log|P(z_0)| + m\log 2 \le \frac{1}{2\pi} 2\pi(\log(n+1) + 4nr)\,.$$

This, together with $n^{-1/2} \le r \le 1$, implies

$$m \le \frac{1}{\log 2}\left(\frac{c_2}{r} + \log(n+1) + 4nr\right) \le c_6(n^{1/2} + nr)\,.$$

Now observe that $\overline{D}(1, r)$ is contained in $\overline{D}(z_0, 2r)$, and the lemma follows. □

The lemma below is a consequence of Jensen's formula (Lemma 1) and Lemma 7.

Lemma 9 *Suppose that* P *is a polynomial of the form* (1) *with* $M = 1$. *Let* $\overline{D}(z_0, r) \subset \overline{D}(0, 1)$, $r \in [3/4, 1)$, $0 < \delta < r$, *and the boundary circles have a common point. There is a constant* $c(r) > 0$ *depending only on* $r \in [3/4, 1)$ *such that* P *has at most* $c(r)\delta^{-1}$ *zeros in the disk* $\overline{D}(z_0, r - \delta)$.

Proof of Lemma 9 Let P be a polynomial of the form (1) with $M = 1$. Observe that $r \in [3/4, 1)$ and $\overline{D}(z_0, r) \subset \overline{D}(0, 1)$ imply $|z_0| \le 1/4$, hence

$$|P(z_0)| \ge 1 - \sum_{j=1}^{\infty}\left(\frac{1}{4}\right)^j = 1 - 1/3 = 2/3. \tag{10}$$

Let m denote the number of zeros of P in the disk $\overline{D}(z_0, r-\delta)$. Applying Jensen's formula (Lemma 1) with the disk $\overline{D}(z_0, r)$, (10), and Lemma 7, we obtain

$$\log\left(\frac{2}{3}\right) + m\log\left(\frac{r}{r-\delta}\right) \le \log|P(z_0)| + m\log\left(\frac{r}{r-\delta}\right)$$
$$\le \frac{1}{2\pi}\int_0^{2\pi} \log|P(z_0+re^{i\theta})|\,d\theta \le c(r)\,.$$

As $3/4 \le r < 1$ and $0 < \delta < r$, we have

$$\log\left(\frac{r}{r-\delta}\right) = \log\left(\frac{1}{1-(\delta/r)}\right) \ge \frac{\delta}{r} \ge \delta\,,$$

and, together with the previous inequality, this implies that $m \le (c(r)+1)\delta^{-1}$. □

Lemmas 10 and 11 below are technical lemmas needed in the proof of Lemma 7. We study the closed disk centered at $(0,1)$ of radius 1, and the closed disk centered at $(0,r)$ of radius $r \in (0,1)$ in the Cartesian plane, so the larger disk contains the smaller disk, and the origin $(0,0)$ is the common point of the boundary circles.

Lemma 10 *Let $r \in (0,1)$. Let $d(x)$ be the distance between the points $(0,1)$ and $(x, r-(r^2-x^2)^{1/2})$, where $x \in [-r,r]$. Then $1-d(x) \ge c(r)x^2$ where $c(r) = \dfrac{1-r}{2r} > 0$ is a suitable constant depending only on $r \in (0,1)$.*

Proof of Lemma 10 Let $x \in [-r,r]$ and $r \in (0,1)$. We want to see that

$$1-d(x) = 1-\left((x-0)^2 + ((1-r)+(r^2-x^2)^{1/2})^2\right)^{1/2} \ge cx^2\,,$$

so it is sufficient to show that

$$1-cx^2 \ge \left(x^2+(1-r)^2+r^2-x^2+2(1-r)(r^2-x^2)^{1/2}\right)^{1/2},$$

so it is sufficient to show that

$$1-2cx^2 \ge (1-r)^2+r^2+2(1-r)(r^2-x^2)^{1/2}\,,$$

so it is sufficient to show that

$$cx^2 \le (1-r)(r-(r^2-x^2)^{1/2}) = (1-r)\frac{x^2}{r+(r^2-x^2)^{1/2}}\,,$$

and this is true if $c = c(r) = (1-r)/(2r)$. □

Lemma 11 *Let $r \in (0,1)$. Let $d(x)$ be the distance between the points $(0,1)$ and $(x, r+(r^2-x^2)^{1/2})$, where $x \in [-r,r]$. Then $1-d(x) \geq c(r)$, where $c(r) = r(1-r) > 0$ is a suitable constant depending only on $r \in (0,1)$.*

Proof of Lemma 11 Let $x \in [-r,r]$ and $r \in (0,1)$. We have

$$\begin{aligned}1-d(x) =&1-\left((x-0)^2+((1-r)-(r^2-x^2)^{1/2})^2\right)^{1/2}\\ \geq&1-\left(1-2r+2r^2-2(1-r)(r^2-x^2)^{1/2}\right)^{1/2}\\ \geq&1-(1-2r+2r^2)^{1/2}=\frac{1-1+2r-2r^2}{1+(1-2r+2r^2)^{1/2}}\\ \geq&\frac{2r(1-r)}{1+1}=r(1-r)\,.\end{aligned}$$

□

5 Proof of the Theorems

Proof of Theorem 2 Let P be a polynomial of the form (1). Without loss of generality we may assume that $a \in (0,1/4]$ otherwise the theorem follows from the case $a = 1/4$. By repeated applications of Lemma 5 we count the zeros of P in the intervals

$$[1-(5/4)a, 1-a]\,,\ [1-(5/4)^2a, 1-(5/4)a]\,,\ \ldots\,,\ [1-(5/4)^k a, 1-(5/4)^{k-1}a]\,,$$

where k is the smallest positive integer for which $(5/4)^k a \geq 3/4$. We obtain that P has at most

$$\sum_{j=0}^{k-1}\frac{c_5(4/5)^j}{a}(1+\log M) \leq \frac{5c_5}{a}(1+\log M)$$

zeros in the interval $[1/4,1]$. Combining this with Lemma 6 we obtain that P has at most

$$\frac{5c_5}{a}(1+\log M)+\frac{\log(2M)}{\log 2} \leq (c_7/a)(1+\log M)$$

zeros in the interval $[0, 1-a]$. Replacing P by Q defined by $Q(z) = P(-z)$ we get that P has at most $(c_7/a)(1+\log M)$ zeros in $[-1+a, 0]$ and the theorem follows.

□

Proof of Theorem 1 Assume that in addition to $M \geq 1$ we have $\log M \leq n-1$, otherwise the theorem is trivial. The theorem follows from Lemma 4 and Theorem 2, as with the choice of $a := n^{-1/2}(1+\log M)^{1/2}$ we have

$$(c/a)(1+\log M) = cn^{1/2}(1+\log M)^{-1/2}(1+\log M) = cn^{1/2}(1+\log M)^{1/2}.$$

□

Proof of Theorem 3 It is sufficient to prove the theorem for the number of zeros in a triangle with vertices 0, w, and w^{-1}, where $w = e^{i\alpha}$ with $\alpha \in (0, \pi/3]$. Choose a $\gamma = \gamma(\alpha) \in (0, 1/4]$ so that

$$1-\gamma \geq \left|\gamma w - \left(\cos\alpha + \frac{1}{2}(1-\cos\alpha)\right)\right|.$$

Let

$$S_1 := \overline{D}(w, \beta n^{-1/2}), \qquad S_2 := \overline{D}(w^{-1}, \beta n^{-1/2}),$$

$$S_3 := \overline{D}(\gamma w, 1-\gamma-n^{-1/2}), \qquad S_4 := \overline{D}(\gamma w^{-1}, 1-\gamma-n^{-1/2}).$$

Note that if $\beta = \beta(\alpha)$ and $n = n(\alpha)$ are sufficiently large then the triangle with vertices 0, w, and w^{-1} is covered by the union of S_1, S_2, S_3, and S_4. Hence the theorem follows from Lemmas 8 and 9. □

Acknowledgments The author thanks Fedor Nazarov for the discussions on the topic and his calling my attention to his paper [11].

References

1. P. Borwein Computational Excursions in Analysis and Number Theory *Spring-Verlag*, New York, 2002.
2. A. Bloch and G. Pólya On the roots of certain algebraic equations *Proc. London Math. Soc.*, 33: 102–114, 1932.
3. P. Borwein and T. Erdélyi Polynomials and Polynomial Inequalities *Springer-Verlag*, New York, 1995.
4. P. Borwein and T. Erdélyi On the zeros of polynomials with restricted coefficients *Illinois J. Math.*, 41: 667–675, 1997.
5. P. Borwein, T. Erdélyi, and G. Kós Littlewood-type problems on [0, 1] *Proc. London Math. Soc.*, 79: 22–46, 1999.
6. D.W. Boyd On a problem of Byrne's concerning polynomials with restricted coefficients *Math. Comp.*, 66: 1697–1703, 1997.
7. T. Erdélyi Extensions of the Bloch-Pólya theorem on the number of real zeros of polynomials *Journal de théorie des nombres de Bordeaux*, 20(2): 281–287, 2008.
8. T. Erdélyi, An improvement of the Erdős-Turán theorem on the zero distribution of polynomials *C. R. Acad. Sci. Paris Sér. I Math.*, 346(5–6): 267–270, 2008.
9. T. Erdélyi Extensions of the Bloch-Pólya theorem on the number of real zeros of polynomials (II) *J. Approx. Theory*, 306: 106122, 2025.

10. P. Erdős and P. Turán On the distribution of roots of polynomials *Ann. Math.*, 57: 105–119, 1950.
11. M. Jacob and F. Nazarov The Newman algorithm for constructing polynomials with restricted coefficients and many real roots https://arxiv.org/pdf/2404.07971. An early version is available at: https://inmabb.criba.edu.ar/revuma/pdf/earlyview/4883.pdf
12. M. Kac On the average number of real roots of a random algebraic equation (II) *Proc. London Math. Soc.*, 50: 390–408, 1948.
13. J.E. Littlewood and A.C. Offord On the number of real roots of a random algebraic equation. II *Proc. Cam. Phil. Soc.*, 35: 133–148, 1939.
14. E. Schmidt Über algebraische Gleichungen vom Pólya-Bloch-Typos *Sitz. Preuss. Akad. Wiss., Phys.-Math. Kl.*, 321, 1932.
15. I. Schur Untersuchungen über algebraische Gleichungen. *Sitz. Preuss. Akad. Wiss., Phys.-Math. Kl.*, 403–428, 1933.
16. G. Szegő Bemerkungen zu einem Satz von E. Schmidt über algebraische Gleichungen. *Sitz. Preuss. Akad. Wiss., Phys.-Math. Kl.*, 86–98, 1934.

Approximate Orthogonality, Bourgain's Pinned Distance Theorem, and Exponential Frames

Alex Iosevich and Azita Mayeli

1 Introduction

The primary goal of this paper is to continue the investigation of the structure of discrete sets A such that $\{e^{2\pi i x\cdot a}\}_{a\in A}$ is a frame for $L^2(K)$, where K is a bounded symmetric convex body with a smooth boundary and non-vanishing curvature. Recall that such a collection of exponentials is a frame if there exist $0 < C_L \leq C_U < \infty$ so that for every $f \in L^2(K)$,

$$C_L|K| \cdot ||f||^2_{L^2(K)} \leq \sum_{a\in A} \left|\widehat{f}(a)\right|^2 \leq C_U|K| \cdot ||f||^2_{L^2(K)}. \tag{1}$$

Let us recall some definitions.

Definition 1 Let S be a separated subset of $\mathbb{R}^d$, in the sense that there exists $\delta > 0$ such that $|s - s'| \geq \delta > 0$ for all $s, s' \in S$, $s \neq s'$. The upper density of S, is given by

$$\overline{dens}(S) = \limsup_{R\to\infty} \frac{\#\{S \cap [-R, R]^d\}}{2^d R^d}.$$

A. Iosevich (✉)
Department of Mathematics, University of Rochester, Rochester, NY, USA
e-mail: iosevich@math.rochester.edu

A. Mayeli
City University of New York, New York, NY, USA
e-mail: amayeli@gc.cuny.edu

D. Bilyk et al. (eds.), *Recent Advances in Approximation and Potential Theory*, Applied and Numerical Harmonic Analysis,
https://doi.org/10.1007/978-3-031-95551-8_11

Similarly, the lower density of S is given by

$$\underline{dens}(S) = \liminf_{R\to\infty} \frac{\#\{S \cap [-R, R]^d\}}{2^d R^d}.$$

If $\overline{dens}(S) = \underline{dens}(S)$, then we say that the Beurling density $dens(S)$ exists and is equal to this quantity.

It is not difficult to construct a frame for $L^2(K)$ by enclosing K in a rectangular box and restricting the orthogonal exponential basis for this box to K. In this case, the frame is tight, i.e., in the definition of the frame in (1) we have $C_L = C_U$. One unfortunate feature of such a frame is that its density is much higher than the volume of K.

Also recall that $\mathcal{E}(A) = \{e^{2\pi i x\cdot a}\}_{a\in A}$ is a Riesz basis for $L^2(K)$ if it is equivalent (via an invertible and bounded map) to an orthonormal basis (not necessarily exponential) for $L^2(K)$. It is an old open question of whether $L^2(B_d)$, $d \geq 2$, possesses a Riesz basis of exponentials, where B_d denotes the unit ball.

It is well-known that if $\mathcal{E}(A)$ is a Riesz basis for $L^2(K)$, then $dens(A) = |K|$. In hopes of shedding some light on the question of a possible Riesz basis for $L^2(K)$, where K is bounded, symmetric, convex and has a smooth boundary with non-vanishing curvature, we prove a result that puts a significant restriction on the structure of a set A that generates a frame for $L^2(K)$.

A well-known stationary phase formula due to Herz [7] (see also [5] and Lemma 1 below for a precise formulation) says that

$$\widehat{\chi}_K(\xi) = \kappa^{-\frac{1}{2}}\left(\frac{\xi}{|\xi|}\right)\sin\left(2\pi\left(\rho^*(\xi) - \frac{d-1}{8}\right)\right)|\xi|^{-\frac{d+1}{2}} \tag{2}$$

plus an error term, where ρ^* is the Minkowski functional on K^*, the dual body of K. We are going to show in Theorem 3 below that if $\mathcal{E}(A)$ is a frame, $|\widehat{\chi}_K(a-a')|$, with $a \neq a'$, $a, a' \in A$, cannot be much smaller than $|a-a'|^{-\frac{d+1}{2}}$ very often. In other words, the difference set $A - A$ must stay away from the zeroes of $\widehat{\chi}_K$.

It was shown in [12] (see also the case of the ball in [11]) that with K as above, $L^2(K)$ does not possess an orthogonal basis of exponentials. Orthogonality implies that $(A - A)\backslash\{\mathbf{0}\}$ is strictly contained in the zero set of $\widehat{\chi}_K$. What we prove in this paper is that $A - A$ cannot be close to the zeroes of $\widehat{\chi}_K$ very often, even on average, For $\mathcal{E}(A)$ to be a frame for $L^2(K)$. More precisely, the main results of the current paper are the following.

1.1 Our Contributions

Theorem 1 *Let A be a countable and discrete subset of $\mathbb{R}^d$, $d \geq 2$, and let $p \in [1, \infty]$. Let K denote a bounded symmetric convex set with a smooth boundary* *and everywhere non-vanishing Gaussian curvature. Let* $\phi \geq 0$ *be a continuous monotonically non-increasing function on* $[0, \infty)$ *such that*

$$\left(\frac{1}{2^j}\int_{2^j}^{2^{j+1}} \phi^p(t)dt\right)^{\frac{1}{p}} \leq c_j\left(2^{-j\frac{d+1}{2}}\right) \quad \forall j \in \mathbb{Z}, \tag{3}$$

for some constants $c_j \geq 0$.

Further, suppose that

$$|\widehat{\chi}_K(a - a')| \leq \phi(\rho^*(a - a')) \tag{4}$$

for all $a \neq a'$, $a, a' \in A$.

Then if

$$\limsup_{j\to\infty} c_j = 0,$$

the upper density of A *is equal to* 0*, and, consequently,* $\mathcal{E}(A)$ ***is not*** *a frame for* $L^2(K)$.

Remark 1 To understand the meaning of the assumptions of Theorem 1, recall that under the assumption on K, $|\widehat{\chi}_K(\xi)| \leq C(1 + |\xi|)^{-\frac{d+1}{2}}$ by the classical result due to Herz [7]. This means that the assumptions of Theorem 1 would automatically hold if we did not require the coefficients c_j to tend to 0. The assumption of Theorem 1 thus say that if $|\widehat{\chi}_K(a - a')$ is $o(|a - a'|^{-\frac{d+1}{2}}$ even in the L^p-average sense, then the set $\{e^{2\pi i x\cdot a}\}_{a\in A}$ cannot be a frame for $L^2(K)$. This is a significant weakening of the assumption from the previous results of this type. See [9] and the references contained therein.

Remark 2 Note that if $\mathcal{E}(A)$ is an orthogonal system, then $\widehat{\chi}_K(a - a') = 0$ if $a \neq a'$. Therefore, Theorem 1 is a generalization of the fact that $L^2(K)$ does not possess any orthogonal basis of exponentials if K is a symmetric convex body with a smooth boundary and non-vanishing Gaussian curvature, established in [12].

Remark 3 An immediate consequence of Theorem 1 is that if $\mathcal{E}(A)$ is a frame for $L^2(K)$, with K as above, then for any subset $A' \subset A$ on which (3) and (4) hold, with $\limsup_{j\to\infty} c_j = 0$, A' has zero positive density. In other words, these properties can hold only on a subset of zero density. It would be interesting to determine if a stronger statement is true, namely that if $\mathcal{E}(A)$ is a frame for $L^2(K)$, and the above hypotheses hold on $A' \subset A$, then for some $0 < \alpha < d$,

$$\limsup_{R\to\infty} \frac{\#\{A' \cap B_R\}}{R^{d-\alpha}} = 0$$

where B_R denote the ball of radius R. For example, if A were contained in a line, we could take any $\alpha < d-1$. We partially address this question in Theorem 2.

Remark 4 It is proved in ([15]; see also [8]) that if Ω is a bounded domain of positive Lebesgue measure in $\mathbb{R}^d$, $d \geq 2$, and $\{e^{2\pi i x\cdot a}\}_{a\in A}$ is a frame for $L^2(\Omega)$ with upper and lower frame constants C_U, C_L, respectively, then every d-dimensional cube of side-length ℓ with

$$\ell > \left(\frac{C_U|\partial\Omega|_\alpha}{C_L|\Omega|}\right)^{\frac{1}{d-\alpha}}$$

contains at least one point of A, where $0 < \alpha < d$ and $|\partial\Omega|_\alpha$ denotes the α-dimensional Minkowski content of $\partial\Omega$, the boundary of Ω. In particular, if Ω is convex, as it is assumed in Theorem 1, every cube of side-length ℓ with

$$\ell > \frac{C_U|\partial\Omega|}{C_L|\Omega|}$$

contains at least one point of A, where $|\partial\Omega|$ denotes the surface measure of $\partial\Omega$. This can be viewed as a quantitative version of the fact (due to Beurling [1]), which states that if Ω has positive Lebesgue measure and $\mathcal{E}(A)$ is frame for $L^2(\Omega)$, then A has positive upper density.

Remark 5 The case $p = \infty$, $K = B_d$, the unit ball in $\mathbb{R}^d$, is essentially contained in the prequel to this paper [9]. More precisely, the authors showed that if $\phi : [0,\infty) \to [0,\infty)$ is any bounded and positive measurable function such that

$$\lim_{t\to\infty}(1+t)^{\frac{d+1}{2}}\phi(t) = 0, \quad \text{and} \quad |\widehat{\chi}_{B_d}(a-a')| \leq \phi(|a-a'|)$$
$$\forall\, a, a' \in A,\ a \neq a', \tag{5}$$

then A has zero upper density. Making a change of variables $s = \frac{t}{2^j}$ in (3), it turns the inequality (3) into

$$\left(\int_1^2 \phi^p(2^j s)ds\right)^{\frac{1}{p}} \leq c_j 2^{-j\frac{d+1}{2}}.$$

The L^∞ version of this statement is

$$\sup_{s\in[1,2]} 2^{j\frac{d+1}{2}}\phi(2^j s) \leq c_j,$$

so in the case $p = \infty$, (3) is a slightly weaker assumption than the (5), except that the monotonicity assumption is not needed in the L^∞ setting. However, as the reader is about to see, we can strengthen the L^∞ result considerably by using different methods.

Our second result is a significant improvement of a previous theorem by the authors of this paper described in Remark 5 above.

Theorem 2 *Suppose that if $\phi : [0, \infty) \to [0, \infty)$ is any bounded positive measurable function such that (5) holds for all $a, a' \in A$, $a \neq a'$. Then A is contained in a line. Moreover, if $d \neq 1 \mod 4$, then A is finite.*

Remark 6 As the reader will see, the proof of Theorem 2 is based on the techniques developed in [13]. The re-examantion of these techniques in the context of Theorem 2 allowed the authors of this paper to realize that the approximate version of the Erdős Integer Distance Principle proved in [13] holds under considerably weaker hypotheses.

2 Outline of the Proofs and the Structure of the Paper

In this section we outline the proof of Theorems 1 and 2.

2.1 Outline of the Proof of Theorem 1

By the method of stationary phase (see Lemma 1 below), with K as above,

$$|\widehat{\chi}_K(a - a')| \leq C(1 + |a - a'|)^{-\frac{d+1}{2}}$$

for some constant $C > 0$. It follows that (6) always holds with $c = C$. The point is that if it holds with a small enough constant, we can invoke the asymptotic formula (9) (see below) and argue that

$$\sin\left(2\pi\left(\rho^*(a - a') + \frac{d-1}{8}\right)\right)$$

is small quite often for $a \neq a'$, $a, a' \in A$. This will be accomplished by reducing the proof of Theorem 1 to the following statement.

Theorem 3 *Let A be a discrete and countable subset of $\mathbb{R}^d$, $d \geq 2$. Let K denote a bounded symmetric convex set with a smooth boundary and everywhere non-vanishing Gaussian curvature. Let $\mathcal{A}_j$ denote annuli of inner radius 2^j, and outer radius 2^{j+1}, centered at $a' \in A$. Suppose that*

$$\left(\frac{1}{2^{dj}} \sum_{a \in A \cap \mathcal{A}_j} |\widehat{\chi}_K(a - a')|^p\right)^{\frac{1}{p}} \leq c_j 2^{-j\frac{d+1}{2}} \tag{6}$$

with c_j independent of a'. Then if

$$\limsup_{j\to\infty} c_j = 0, \tag{7}$$

then the upper density of A is equal to 0, hence $\mathcal{E}(A)$ ***is not*** *a frame for $L^2(K)$.*

By pigeonholing, we will be able to conclude that there exists a point x_0 living in a small neighborhood of an $a_0 \in A$, such that pinned ρ^*-distances from x_0 to the small neighborhood of a positive proportion of A is clustered around $\frac{k}{2} + \frac{d-1}{8}$, where k ranges over the integers. This violates a variant of the result due to Bourgain which says that in sets of positive upper Lebesgue density, a pinned distance set with respect to almost every point contains every sufficiently large positive real number. In order to invoke Bourgain's result we will further have to refine the previously pigeon-holed subset of A.

2.2 Outline of the Proof of Theorem 2

The authors of [13] proved that if $\mathcal{E}(A)$ is an infinite orthogonal set in $L^2(K)$, where $K \subset \mathbb{R}^d$ is a symmetric convex body with a smooth boundary and everywhere non-vanishing curvature, then A is contained in a line, thus, $\mathcal{E}(A)$ is incomplete in $L^2(K)$. Moreover, they showed that if $d \neq 1 \mod 4$, then A must be finite. We are going to see that the proof of this result still goes through under the weaker assumption given by (5) above.

2.3 Structure of This Paper

This paper is organized as follows. In the Section 3 below, we prove Theorem 3. At the end of that section, we comment on the modification of the pinned distance set result due to Bourgain that is needed for our proof. In Sect. 4 we deduce Theorem 1 from Theorem 3.

3 Proof of Theorem 3

Suppose that (6) and (7) hold and A has finite upper density $\alpha_0 > 0$. By assumption, for a fixed $a' \in A$

$$\sum_{a\in A\cap\mathcal{A}_j} |\widehat{\chi}_K(a-a')|^p \leq c_j^p 2^{j(d-p\frac{d+1}{2})}. \tag{8}$$

We shall need the following well-known stationary phase result.

Lemma 1 ([7]) *Let K be a bounded symmetric convex body with a smooth boundary and everywhere non-vanishing Gaussian curvature. Let*

$$\rho^*(\xi) = \sup_{x \in K} x \cdot \xi.$$

Given $\omega \in S^{d-1}$, let $\kappa(\omega)$ denote the Gaussian curvature of ∂K at the (unique) point where the unit normal is ω. Then

$$\widehat{\chi}_K(\xi) = \kappa^{-\frac{1}{2}}\left(\frac{\xi}{|\xi|}\right) \sin\left(2\pi\left(\rho^*(\xi) - \frac{d-1}{8}\right)\right) |\xi|^{-\frac{d+1}{2}} + \mathcal{D}_K(\xi), \tag{9}$$

where

$$|\mathcal{D}_K(\xi)| \leq C_K |\xi|^{-\frac{d+3}{2}}.$$

Plugging this into (8) yields

$$\begin{aligned} \sum_{a \in A \cap \mathcal{A}_j} \kappa^{-\frac{p}{2}}\left(\frac{a-a'}{|a-a'|}\right) \left|\sin\left(2\pi\left(\rho^*(a-a') - \frac{d-1}{8}\right)\right)\right|^p |a-a'|^{-p\frac{d+1}{2}} \\ \leq c_j^p C_1 2^{j(d-p\frac{d+1}{2})} \end{aligned} \tag{10}$$

for some fixed $C_1 > 0$, and it follows that

$$\sum_{a \in A \cap \mathcal{A}_j} \left|\sin\left(2\pi\left(\rho^*(a-a') - \frac{d-1}{8}\right)\right)\right|^p \leq C_2 c_j^p 2^{dj},$$

for some fixed $C_2 > 0$ that depends on the maximum value of the curvature function on ∂K. Note that the error term $\mathcal{D}_K$ from (9), which is $O(2^{-j\frac{d+3}{2}})$ when $a \in \mathcal{A}_j$, can be absorbed into the constant C_1 on the right hand side of (10) if j is sufficiently large.

We write the left hand side in the form $I + II$, where I is the sum over

$$\left\{a \in A \cap \mathcal{A}_j : \left|\sin\left(2\pi\left(\rho^*(a-a') - \frac{d-1}{8}\right)\right)\right|^p \leq \delta\right\}$$

and II is the sum over the complement of this set, with $0 < \delta << 1$ to be determined later. We have

$$II > \delta \cdot \# \left\{ a \in A \cap \mathcal{A}_j : \left| \sin \left(2\pi \left(\rho^*(a - a') - \frac{d-1}{8} \right) \right) \right|^p > \delta \right\},$$

hence

$$\# \left\{ a \in A \cap \mathcal{A}_j : \left| \sin \left(2\pi \left(\rho^*(a - a') - \frac{d-1}{8} \right) \right) \right|^p > \delta \right\} < C_3 c_j^p \delta^{-1} 2^{dj}, \tag{11}$$

for some fixed $C_3 > 0$.

Suppose that $j \geq j_0$, to be determined later. If $\sup_{j \geq j_0} c_j^p$ is sufficiently small, choosing $\delta = \frac{1}{1000}$ we have that the right hand side of (11) is $< (1-r)\#\{A \cap \mathcal{A}_j\}$, where $r \in (0, 1)$, close to 1, to be chosen later. This can be done since A has positive upper density and

$$\limsup_{j \to \infty} \frac{\#\{A \cap \mathcal{A}_j\}}{|\mathcal{A}_1| 2^{dj}} = \alpha_0 > 0.$$

Note that the upper density is not affected by taking only the annuli $\mathcal{A}_j$ with $j \geq j_0$ for any finite j_0. More precisely, the upper density of A with $\cup_{j \leq j_0} \{\mathcal{A}_j \cap A\}$ removed is the same as the upper density of A,

It follows that if we throw these points of A which belong to the set in (11) out, we will have an r fraction of the points of $A \cap \mathcal{A}_j$ left. Doing this for every $j \geq j_0$, with j_0 to be determined, and taking the union, we obtain a set $A_{good, j_0}(a')$ of positive upper density, such that

$$\left| \sin \left(2\pi \left(\rho^*(a - a') - \frac{d-1}{8} \right) \right) \right|^p \leq \delta \text{ for all } a \in A_{good, j_0}(a'). \tag{12}$$

It follows from (12) that if E denote the δ-neighborhood of $A_{good, j_0}(a')$, with δ as above, then there does not exist L_0 such that for every $L > L_0$, $\rho^*(x - a') = L$. If a' is replaced by any point in the ball of radius δ centered at a', the conclusion remains the same. We shall use a generalization of the following result due to Bourgain.

Theorem 4 *Let E be a set of positive upper Lebesgue density in $\mathbb{R}^2$. Then for almost every $x_0 \in E$, there exists $L_0 = L_0(x_0) > 0$ such that for all $L > L_0$, there exists $x \in E$ with $|x - x_0| = L$.*

As we shall sketch below in Sect. 5, Bourgain result can be generalized as follows. Let E be a set of positive upper Lebesgue density in $\mathbb{R}^d$, $d \geq 2$. Let K and ρ^* be as above. Then for almost every $x_0 \in E$, there exists $L_0 > 0$ such that for all $L > L_0$, there exists $x \in E$ with $\rho^*(x - x_0) = L$.

This result was established by Bourgain in [2] (see a related result in [4]). However, as we shall indicate below, the generalization follows from Bourgain's method with only minor modifications. The proof only relies on the L^p estimate for the circular maximal function, which Bourgain proved in 1986 [3]. However, the same estimate holds if the circle is replaced with any smooth curve with non-vanishing curvature. Since

$$\{x \in \mathbb{R}^d : \rho^*(x) = 1\}$$

is a smooth surface, with ρ^* as above, the two-dimensional argument follows. In higher dimensions, the argument is much easier as the needed maximal estimate was proved by Stein [19] in the case of the sphere, and by Greenleaf [6] for general compact surfaces with non-vanishing Gaussian curvature. We shall sketch the proof of this in Sect. 5 below.

In order to use Bourgain's result, we need to make sure that if E is as in the previous paragraph, then one of the points in the $\frac{1}{1000}$ neighborhood of a' can serve as a pin in Theorem 4. The proof of Bourgain's theorem yields a slightly stronger results than the conclusion of Theorem 4, namely that there exists $x_0 \in E$ such that for every subset of E of relative density r sufficiently close to 1, there exists $L_0 > 0$ such that for all $L > L_0$, there exists $x \in E$ with $\rho^*(x - x_0) = L$. In view of this, we complete the proof by finding such an x_0 in the δ-neighborhood of A and choosing a' to be the point in A in the δ-neighborhood of which x_0 resides.

4 Proof of Theorem 1

Let $\mathcal{A}_j$ be as above. By the assumption of Theorem 1, we have

$$\sum_{a \in A \cap \mathcal{A}_j} |\widehat{\chi}_K(a - a')|^p \leq \sum_{a \in A \cap \mathcal{A}_j} \phi^p(\rho^*(a - a')).$$

Since ϕ is monotonic, the right expression above is bounded by

$$C \int_{c_1 2^j \leq \rho^*(u-a') \leq c_2 2^{j+1}} \phi^p(\rho^*(u - a'))du = C \int_{c_1 2^j \leq \rho^*(u) \leq c_2 2^{j+1}} \phi^p(\rho^*(u))du,$$

or

$$C \int_{c_1 2^j \leq \rho^*(r\omega) \leq c_2 2^{j+1}} \phi^p(\rho^*(r\omega))r^{d-1} d\omega dr$$
$$= C \int_{c_1 2^j \leq r\rho^*(\omega) \leq c_2 2^{j+1}} \phi^p(r\rho^*(\omega))r^{d-1} d\omega dr,$$

where C is an independent constant.

Let $t = r\rho^*(\omega)$. We obtain

$$C \int_{c_1 2^j \le t \le c_2 2^{j+1}} \int_{S^{d-1}} \phi^p(t) t^{d-1} \frac{d\omega}{(\rho^*(\omega))^d} dt$$

$$= C \int_{c_1 2^j \le t \le c_2 2^{j+1}} \int_S \phi^p(t) t^{d-1} dt \frac{d\sigma_{\rho^*}(\eta)}{|\nabla \rho^*(\eta)|}, \tag{13}$$

where $d\sigma_{\rho^*}$ is the surface measure on

$$S = \{x : \rho^*(x) = 1\},$$

the ρ^*-unit sphere.

The expression (13) follows from the classical co-area formula. See, for example, [14], for a similar calculation.

Since $\rho^*(\eta) = \eta \cdot \nabla \rho^*(\eta)$, it follows that $|\nabla \rho^*(\eta)| \ge C > 0$, when η is on the ρ^*-unit sphere, for some different $C > 0$. It follows that (13) is bounded by

$$(13) \le C' 2^{j(d+1)} \cdot 2^{-j} \int_{c_1 2^j}^{c_2 2^{j+1}} \phi^p(t) t^{d-1} dt \le C'' c_j 2^{dj} \cdot 2^{-j \frac{d+1}{2} p},$$

where the first inequality above follows from the fact that all norms on finite dimensional normed spaces are comparable.

The conclusion follows after dividing both sides by 2^{dj} and taking p'th root. This recovers Theorem 1 from Theorem 3.

5 Sketch of Proof of a Slight Strengthening of Theorem 4

In this section we sketch the proof of the following result, which essentially follows from Bourgain's argument in [2].

Definition 2 We say that $E \subset \mathbb{R}^d$ has positive upper Lebesgue density if

$$\overline{dens_{\mathcal{L}}}(E) = \limsup_{R \to \infty} \frac{|E \cap B_R|}{|B_R|} > 0,$$

where B_R denote the ball of radius R.

Definition 3 Suppose that $E \subset \mathbb{R}^d$ with $\overline{dens_{\mathcal{L}}}(E) > 0$. We say that E' is a refinement of E of relative density r if $E' \subset E$ and

$$\overline{dens_{\mathcal{L}}}(E') \ge r \cdot \overline{dens_{\mathcal{L}}}(E).$$

Theorem 5 *Let $E \subset \mathbb{R}^d$ of positive upper Lebesgue density $\overline{dens_{\mathcal{L}}}(E)$. Then there exists a threshold $L_0 > 0$, a threshold $r \in (0, 1)$, and $x_0 \in E$, such that for any* *refinement E' of E of relative density r, and any $L > L_0$, there exists $x' \in E'$ such that $\rho^*(x_0 - x') = L$.*

Bourgain's proof of the case $d = 2$, $\rho^*(x) = |x|$, has two key ingredients. The first is the following maximal theorem which follows from Bourgain's proof of the circular maximal theorem.

Theorem 6 *Let σ be the arc-length measure on the circle S^1, and let σ_s be defined by the relation $\widehat{\sigma}_s(\xi) = \widehat{\sigma}(s\xi)$. Let P_t denote the Poisson semi-group kernel on $\mathbb{R}^2$, so $\widehat{P_t}(\xi) = e^{-t|\xi|}$. For $p > 2$, there exist constants $C(p) < \infty$ and $\alpha(p) > 0$, such that*

$$\left|\left|\sup_{s \geq t_0} |(f - (f * P_t) * \sigma_s)|\right|\right|_p \leq C(p)\left(\frac{t}{t_0}\right)^{\alpha(p)} ||f||_p.$$

The fact that the same conclusion is achieved if S^1 is replaced by any symmetric smooth convex curve with non-vanishing curvature follows from [18]. In higher dimensions, the following result follows from [6].

Theorem 7 *Let ρ be the Minkowski functional on the convex symmetric body K with a smooth boundary and everywhere non-vanishing curvature. Let ρ^* be as above. Let $S_\rho = \{x \in \mathbb{R}^d : \rho^*(x) = 1\}$, and let σ_ρ denotes the surface measure. Let $\sigma_{\rho,s}$ denote the dilated surface measure, as above. For $d > 2$, there exist $C < \infty$, such that*

$$\left|\left|\sup_{s \geq t_0} |(f - (f * P_t) * \sigma_{\rho,s})|\right|\right|_2 \leq C\left(\frac{t}{t_0}\right)^{\frac{d-2}{2}} ||f||_2.$$

The second ingredient in Bourgain's proof is the observation that if the conclusion of Theorem 4 is false in the case $d = 2$, $\rho^*(\xi) = |\xi|$, then there exists $U \subset [0, 1]^2$, $|U| > \epsilon$, and a sequence of positive numbers

$$s_1 > t_1 > s_2 > t_2 > \cdots > s_J > t_J,$$

J arbitrarily large, satisfying

$$s_{j+1} < \frac{1}{2} t_j,$$

and $x \in U \cap [s_j, 1-s_j]$, $s_j \in (0, 1/2)$, implies that $\sup_{s_j < t < s_{j+1}} |(1_{[0,1]^2} - f) * \sigma_t| = 1$. The same reduction can be performed in dimensions $d \geq 3$ with the Euclidean metric replaced by ρ^*. The contradiction is then obtained using Theorem 6. In dimensions three and higher, the contradiction is obtained in exactly the same way using Theorem 7.

Moreover, the same argument yields the slightly stronger conclusion, claimed in Theorem 5, that there exists a threshold $L_0 > 0$, a threshold $r \in (0, 1)$, and

$x_0 \in E$, such that for any refinement E' of E of relative density r, and any $L > L_0$, there exists $x' \in E'$ such that $\rho^*(x_0 - x') = L$. This is because assuming that this statement is false leads to exactly the same statement as above and the contradiction is obtained in the same way.

6 Proof of Theorem 2

Suppose that $\mathcal{E}(A)$ is an orthogonal collection in $L^2(K)$ in the sense that

$$\int_K e^{2\pi i x\cdot(a-a')}dx = 0 \text{ for all } a \neq a' \in A,$$

where K is as in Theorem 2.

The first listed author and Rudnev proved in [13] that if A is infinite, A is contained in a line. Moreover, they showed that if $d \neq 1 \mod 4$, then A must be finite.

The proof relies on two estimates that follow from the orthogonality assumption for $\mathcal{E}(A)$. The first says that if $\mathcal{E}(A)$ is an orthogonal set in $L^2(K)$, then for all $a, a' \in A$,

$$\rho^*(a-a') = \frac{k}{2} + \frac{d-1}{8} + O(|a-a'|^{-1}) \text{ for some integer } k \text{ as } |a-a'| \to \infty. \tag{14}$$

The second estimate says that if $a_0, a_1 \in A$, $a_0 \neq a_1$, then for all $a \in A$,

$$\rho^*(a_0 - a) - \rho^*(a_1 - a) = \frac{k}{2} + O(|a|^{-2}) \text{ for some integer } k \text{ as } |a| \to \infty. \tag{15}$$

The first estimate follows directly from Lemma 1. The second estimate uses more refined Bessel expansions (see Lemma 1.4 in [13]). Using these estimates, the authors established their claim.

Under the hypotheses of Theorem 2, the same arguments yield the following form of (14) and (15):

$$\rho^*(a-a') = \frac{k}{2} + \frac{d-1}{8} + O\left(\frac{\phi(|a-a'|)}{|a-a'|^{\frac{d+1}{2}}}\right)$$
$$\text{for some integer } k \text{ as } |a-a'| \to \infty, \tag{16}$$

where (5) holds, and (15) becomes

$$\rho^*(a_0 - a) - \rho^*(a_1 - a) = \frac{k}{2} + O\left(\frac{\phi(|a|)}{|a|^{\frac{d+1}{2}}} \cdot |a-a'|^{-1}\right)$$

$$\text{for some integer } k \text{ as } |a| \to \infty. \tag{17}$$

Observe that

$$O\left(\frac{\phi(|a-a'|)}{|a-a'|^{\frac{d+1}{2}}}\right) = o(1), \text{ as } |a-a'| \to \infty,$$

by assumption. Similarly,

$$O\left(\frac{\phi(|a|)}{|a|^{\frac{d+1}{2}}} \cdot |a-a'|^{-1}\right) = o(|a-a|^{-1}) \text{ as } |a-a'| \to \infty.$$

The geometric observation used to establish the conclusion of the main result in [13], namely Lemma 1.5, does not require anything stronger than (16) and (17) above. These are all that are needed to establish Lemma 1.5 and Proposition 1.6 in that paper. This results in the following version of the Erdős Integer Distance Principle.

Theorem 8 (Approximate Version of the Erdős Integer Distance Principle) *Let ρ^* be as above, corresponding to a symmetric convex body K with a smooth boundary and non-vanishing Gaussian curvature. Let $A \subset \mathbb{R}^d$, $d \geq 2$, and suppose that for $a, a_0, a_1 \in A$,*

$$\rho^*(a-a_0) - \rho^*(a-a_1) = o(|a|^{-1}), \tag{18}$$

and there exist constants c_1, c_2 such that for $a, a' \in A$,

$$\rho^*(a-a') = c_1 k + c_2 + o(1). \tag{19}$$

Then A is a subset of a line. If $d \neq 1 \mod 4$, then A is finite.

7 Remarks and Open Problems

The purpose of this section is to describe some open problems that may be accessible using the methods of this paper and related ideas.

(i) The authors of this paper previously proved [10] that if $g(x) = \chi_{B_d}(x)$, where B_d is the unit ball in dimensions two and higher, and $d \neq 1 \mod 4$, then there does not exist $S \subset \mathbb{R}^d \times \mathbb{R}^d$, such that $\{g(x-a)e^{2\pi i x \cdot b}\}_{(a,b)\in S}$ is an orthogonal basis for $L^2(\mathbb{R}^d)$. The case when $d = 1 \mod 4$ is work in progress. A reasonable question to ask is whether the same conclusion can be established under the assumption of approximate orthogonality.

(ii) In view of [12], it should be possible to show that the conclusion of Theorem 1 still holds if we assume that ∂K has one point, in the neighborhood of which it is smooth and has non-vanishing Gaussian curvature.

(iii) It is reasonable to conjecture that if K is a convex body in $\mathbb{R}^d$, $d \geq 2$, then there exists $S \subset \mathbb{R}^{2d}$ such that $\{\chi_K(x-a)e^{2\pi i x \cdot b}\}_{(a,b) \in S}$ is an orthogonal basis for $L^2(\mathbb{R}^d)$ if and only if K tiles $\mathbb{R}^d$ by translation, i.e iff K is a symmetric polytope of the type classified by Venkov [20] and McMullen [17]. It may be possible to approach this problem by combining the techniques of this paper and those of [16] who recently proved that if $K \subset \mathbb{R}^d$, $d \geq 2$, is convex, then $L^2(K)$ has an orthogonal basis of exponentials if and only if K tiles $\mathbb{R}^d$ by translation. The two dimensional case was previously established in [12].

(iv) The contrast between the proof of Theorem 1 and Theorem 2 raises the question of whether there is a strengthening of the approximate version of the Erdős Integer Distance Principle, proved in [13], where the point-wise (near) integer distance condition is replaced by an average version. For example, is it true that if A is an infinite separated subset of $\mathbb{R}^d$, $d \geq 2$, such that there exist constants c_1, c_2 and $p \in [1, \infty)$ so that

$$|a-a'| = c_1 k + c_2 + O(\mathcal{D}(|a-a'|)),$$

with

$$\limsup_{j \to \infty} \frac{1}{2^j} \int_{2^j}^{2^{j+1}} |\mathcal{D}(t)|^p dt = 0,$$

then A is a subset of a line?

If anything like this is true, it should lead to a considerable strengthening of Theorem 1. Heuristically, the idea (described in two dimensions for the sake of simplicity) is the following. Fix $a_0, a_1, a_0 \neq a_1$. Then 90% of the points in A are nearly $c_1 k + c_2$ distance away from a_0, and the same is true with respect to a_1. It follows that 81% of the points lie on finitely many hyperbolas with focal points at a_0 and a_1. Now choose a_2 that does not lie on a line that passes through a_0 and a_1 and play the same game with respect to a_1, a_2 and the points that are nearly shifted scaled integers with respect to a_2. The hyperbolas with focal points at a_0, a_1 intersect the hyperbolas with focal points at a_1, a_2 finitely many times (depending on the distance between those pairs of points) and shows that the resulting subset $A' \subset A$ (having at least 72.9% of the points) is a subset of a line. We now replay the same game, using the same points a_0, a_1, a_2 with respect to the complement of A'. This approach can probably be used to show that $A = A_{good} + A_{bad}$, where A_{good} is a subset of a line, and A_{bad} is arbitrarily sparse. This may lead to an alternate proof of Theorem 1. Whether one can conclude that A_{bad} is empty is an interesting question that should be investigated.

Acknowledgments The research of the first listed author was supported in part by the National Science Foundation under grant no. HDR TRIPODS—1934962 and by the NSF DMS—2154232

References

1. A. Beurling, *Local harmonic analysis with some applications to differential operators*, Some recent advances in the basic sciences, Academic press **1** (1966).
2. J. Bourgain, *Szem-eredi type theorem for sets of positive density in* $\mathbb{R}^k$, Israel J. Math. **54** (1986), no. 3, 307–316.
3. J. Bourgain, *Averages in the plane over convex curves and maximal operators*, J. Analyse Math. **47** (1986), 69–85.
4. H. Furstenberg, Y. Katznelson, and B. Weiss, *Ergodic theory and configurations in sets of positive density* Mathematics of Ramsey theory, 184–198, Algorithms Combin., 5, Springer, Berlin, (1990).
5. I. Gelfand and G. Shilov, *Generalized Functions*, Vol. 1, Academic Press, (1958).
6. A. Greenleaf, *Principal curvature and harmonic analysis*, Indiana Univ. Math. J. **30** (1981), no. 4, 519–537.
7. C. Herz, *Fourier transforms related to convex sets*, Ann. of Math. (2) **75** (1962) 81–92.
8. A. Iosevich and M. Kolountzakis, *A Weyl type formula for Fourier spectra and fraimes*, Proc. Amer. Math. Soc. **134** (2006), no. 11, 3267–3274.
9. A. Iosevich and A. Mayeli, *On complete and incomplete exponential systems*, to appear in the Proceeding of AMS.
10. A. Iosevich and A. Mayeli, *Gabor orthogonal bases on convexity*, Discrete Anal. (2018), Paper No. **19**, 11 pp.
11. A. Iosevich, N. H. Katz and S. Pedersen, *Fourier bases and a distance problem of Erdös. Math. Res. Lett.*, **6**, 251–255, (2001).
12. A. Iosevich, N. H. Katz and T. Tao, *Convex bodies with a point of curvature do not have Fourier bases*, Amer. J. Math., **123**, 115–120, (2001).
13. A. Iosevich and M. Rudnev, *A combinatorial approach to orthogonal exponentials*, Int. Math. Res. Not. (2003), no. 50, 2671–2685.
14. A. Iosevich and E. Sawyer, *Oscillatory integrals and maximal averages over homogeneous surfaces*, Duke Math. J. **82** (1996), no. 1, 103–141.
15. A. Iosevich and S. Pedersen, *How large are the spectral gaps?* Pacific J. Math. **192** (2000), no. 2, 307–314.
16. N. Lev, M. Matolcsi, *The Fuglede conjecture for convex domains is true in all dimensions*, Acta Mathematica, Volume 228 (2022), No. 2, Pages: 385 – 420.
17. P. McMullen, *Convex bodies which tile space by translation*, Mathematika, **27** (1):113–121, (1980).
18. G. Mockenhaupt, A. Seeger, and C. D. Sogge, *Local smoothing of Fourier integral operators and Carleson-Sjolin estimates*, J. Amer. Math. Soc. **6** (1993), no. 1, 65–130.
19. E. M. Stein, *Maximal functions. I. Spherical means*, Proc. Nat. Acad. Sci. U.S.A. **73** (1976), no. 7, 2174–2175.
20. B. A. Venkov, *On a class of Euclidean polyhedra*, Vestnik Leningrad. Univ. Ser. Mat. Fiz. Him., 9(2):11–31, (1954).

Application of Sparse Grids to Approximation and Optimization

Scott N. Kersey

1 Introduction

Sparse grids, introduced by Smolyak in 1963 [20], provide an efficient framework for addressing high-dimensional problems such as numerical integration, interpolation, and solving differential equations. Also known as discrete blending [11], Boolean interpolation [3, 4], hyperbolic crosses [5], and the combination technique [2, 9, 10, 17], sparse grids mitigate the curse of dimensionality by strategically selecting a sparse subset of grid points from a tensor product structure. This approach significantly reduces computational costs while maintaining high accuracy.

For example, a full grid on the hypercube $[0, 1]^d$ with n points per coordinate direction contains n^d grid points, whereas a sparse grid requires only $O(n \log(n)^{d-1})$ points. Despite the reduced number of grid points, cubic spline approximation on sparse grids achieves an accuracy of $O(h^4(\log h^{-1})^{d-1})$, where $h = \frac{1}{n}$, which is only slightly less than the $O(h^4)$ rate achieved by full grids [16].

Initially developed for numerical integration, sparse grids were later extended to solving partial differential equations by Zenger [22] and have since found applications in diverse fields such as numerical analysis, approximation theory, optimization, and machine learning. Significant advancements include the use of B-splines for interpolation, quasi-interpolation, and optimization [12, 13, 16], which form the foundation of this paper's contributions.

Specifically, we construct spline interpolants and quasi-interpolants on sparse grids using the combination technique. Gordon [7] first demonstrated this approach in the context of spline-blended surfaces, achieving approximation rates consistent

S. N. Kersey (✉)
Department of Mathematical Sciences, Georgia Southern University, Statesboro, GA, USA
e-mail: skersey@georgiasouthern.edu

D. Bilyk et al. (eds.), *Recent Advances in Approximation and Potential Theory*,
Applied and Numerical Harmonic Analysis,
https://doi.org/10.1007/978-3-031-95551-8_12

with the underlying spline network. Our work on spline interpolants follows the framework in [12], while the quasi-interpolants are based on univariate schemes described in [1, 8, 14, 15], with the specific quasi-interpolant detailed in [19].

Additionally, we explore the application of sparse grid methods based on quasi-interpolation to function optimization using B-splines. While previous studies have investigated hierarchical constructions [21] and gradient descent algorithms [6], our focus lies in leveraging sparse grid interpolation and quasi-interpolation to enhance the accuracy of function approximations in optimization tasks.

2 Sparse Grids

For this section, we restrict our grids to the hypercube $[0, 1]^d$, which is generalized to other 2-D rectangles for the computational portion of this paper. Let G_r denote the *regular grid* defined by the non-negative integer sequence $r := (r_1, \ldots, r_d)$ with $2^{r_i} + 1$ uniformly-spaced on [0, 1] in the i-th coordinate direction (including the end points). Our *sparse grid at level r in* $\mathbb{R}^d$ is defined as a union of regular grids:

$$G^{\alpha,\beta,d} := \bigcup_{\substack{|r|_1=\beta+d\alpha-2 \\ r_i \geq \alpha}} G_r.$$

For example, the following decomposition of the sparse grid $G^{2,4,2}$ into a union of regular grids

$$G^{2,4,2} = \bigcup_{\substack{|r|_1=6 \\ r_i \geq 2}} G_{(r_1,r_2)} = G_{(2,4)} \cup G_{(3,3)} \cup G_{(4,2)}$$

is illustrated in Fig. 1.

Note that the number of grid-points in $G^{2,4,2}$ is the sum of the grid points in each regular grid minus points counted more than once. This can be expressed by the following sum:

$$\begin{aligned} |G^{2,4,2}| &= (|G_{(2,4)}| + |G_{(3,3)}| + |G_{(4,2)}|) - (|G_{(2,3)}| + |G_{(3,2)}|) \\ &= (85 + 81 + 85) - (45 + 45) = 161. \end{aligned}$$

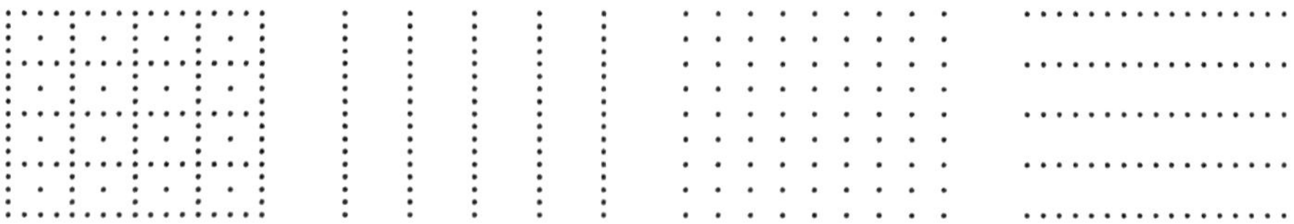

Fig. 1 Sparse Grid $G^{2,4,2}$ and regular subgrids $G_{(2,4)}$, $G_{(3,3)}$ and $G_{(4,2)}$

This reasoning of adding grid points and subtracting intersecting grid points leads to the following explicit expression for the number of grid points in a sparse grid in 2D

$$|G^{\alpha,\beta,2}| = \sum_{\substack{|r|_1=\beta+d\alpha-2 \\ r_i\geq\alpha}} |G_r| - \sum_{\substack{|r|_1=\beta+d\alpha-3 \\ r_i\geq\alpha}} |G_r|,$$

which generalizes in d-dimensions to

$$|G^{\alpha,\beta,d}| = \sum_{k=0}^{d-1}(-1)^k\binom{d-1}{k} \sum_{\substack{|r|_1=\beta+d\alpha-2-k \\ r_i\geq\alpha}} |G_r|$$

with

$$|G_r| = |G_{(r_1,\ldots,r_d)}| = \prod_{i=1}^{d}(2^{r_i}+1).$$

Table 1 contains the number of grid points needed for the sparse and full grids, as well as a percentage of the amount of data "compressed" in using the sparse grids instead of full grids. This percentage is calculated as follows:

$$\text{Compression} = 100 \times \frac{\text{Number of points in full grid} - \text{Number of points in sparse grid}}{\text{Number of points in full grid}}.$$

Table 1 Number of grid points in the sparse and full grids and the percent compression in using the sparse grid defined by lower level α and upper level β

Number of grid points				
α	β	Sparse	Full	Compression
2	2	25	25	0.0
2	3	65	81	19.8
2	4	161	289	44.3
2	5	385	1089	64.6
2	6	897	4225	78.8
2	7	2049	16641	87.7
2	8	4609	66049	93.0
2	9	10241	263169	96.1
2	10	22529	1050625	97.9
2	11	49153	4198401	98.8
2	12	106497	16785409	99.4

3 Splines on Sparse Grids

The B-Spline of order k (degree k-1) over the knots $t_0, \ldots, t_k$ is a divided difference of the truncated power function $(x)_+^{k-1}$:

$$B(x; t_0, \ldots, k) = (t_k - t_0)[t_0, \ldots, t_k](\cdot - x)_+^{k-1}$$

The Cardinal B-Spline basis function of order $k = 4$ (degree 3) with knots $\{-2, -1, 0, 1, 2\}$ is

$$B^k(x) := k[-2, -3, 0, 1, 2](\cdot - x)_+^{k-1}$$

and the Cardinal B-spline Series of order k with mesh spacing h is

$$\sum_{\alpha \in \mathbb{Z}} c_\alpha B^k(\frac{\cdot}{h} - \alpha)$$

It is well-known that the approximation order of cardinal order 4 spline interpolants to functions $f \in C^4(\mathbb{R})$ achieves order 4. Here, we solve

$$Sf := \sum_{\alpha \in \mathbb{Z}} c_\alpha \; B^k(\frac{\cdot}{h} - \alpha) \tag{1}$$

such that $Sf(\frac{\alpha}{h}) = f(\frac{\alpha}{h})$. We can also achieve approximation order $O(h^4)$ on $[0, 1]$ by the quasi-interpolant

$$Qf := \sum_{\alpha \in \mathbb{Z}} f\Big(\frac{\alpha}{h}\Big) \; \Psi^k\Big(\frac{\cdot}{h} - \alpha\Big) \tag{2}$$

with

$$\Psi^k(\cdot) := \frac{1}{6}\Big(- B^k(\cdot + 1) + 8\, B^k(\cdot) - B^k(\cdot - 1)\Big).$$

The advantage here is that no system of equation needs to be solved.

We now construct our spline approximations on sparse grids by the combination technique. To start, let $G_n = G_{n_1, \ldots, n_d}$ be regular grid with grid points

$$\frac{i}{n} = (\frac{i_1}{n_1}, \cdots, \frac{i_d}{n_d})$$

for $i_k \in \{0, \ldots, n_k\}$ and $h_k = \frac{1}{n_k}$. Then,

$$Q_n f(x_1, \ldots, x_d) = \sum_{i_1=0}^{n_1} \cdots \sum_{i_d=0}^{n_d} f(\frac{i_1}{n_1}, \ldots, \frac{i_d}{n_d}) \; \Psi^k(x_1) \cdots \Psi^k(x_d)$$

is a quasi-interpolant on G_n. Likewise, we solve the interpolation problem for the tensor product spline interpolant

$$S_n f(x_1, \dots, x_d) = \sum_{i_1=0}^{n_1} \cdots \sum_{i_d=0}^{n_d} c_{i_1,\dots,i_n} B^k(x_1) \cdots \Psi^k(x_d)$$

such that $S_n f(\frac{i_1}{n_1}, \dots, \frac{i_d}{n_d}) = f(\frac{i_1}{n_1}, \dots, \frac{i_d}{n_d})$. We can shorten our notation to

$$S_n f(x) = \sum_{x \in G_n} c_x B^k(\cdot - x) \quad \text{with} \quad S_n f\Big(\frac{i}{n}\Big) = f\Big(\frac{i}{n}\Big), \tag{3}$$

and

$$Q_n f(x) = \sum_{x \in G_n} f(x) \Psi^k(\cdot - x) \tag{4}$$

with

$$\Psi^k(x) = \prod_{k=1}^{d} \psi(2^{m_k} x_k).$$

Then using the combination technique we used to count grid points, we can formulate the quasi-interpolant on the sparse grid $G^{\alpha,\beta,d}$ as the linear combination of tensor product quasi-interpolants on regular grids as follows:

$$Q^{\alpha,\beta,d} f := \sum_{k=0}^{d-1} (-1)^k \binom{d-1}{k} \sum_{\substack{|r|_1 = \beta + d\alpha - 2 - k \\ r_i \geq \alpha}} Q_r f. \tag{5}$$

In the two-dimensional case, $d = 2$, the formula (5) reduces to

$$Q^{\alpha,\beta,2} f = \sum_{\substack{|r|_1 = \beta + d\alpha - 2 - k \\ r_i \geq \alpha}} Q_r f - \sum_{\substack{|r|_1 = \beta + d\alpha - 3 - k \\ r_i \geq \alpha}} Q_r f.$$

For the spline interpolant, we have

$$S^{\alpha,\beta,d} f := \sum_{k=0}^{d-1} (-1)^k \binom{d-1}{k} \sum_{\substack{|r|_1 = \beta + d\alpha - 2 - k \\ r_i \geq \alpha}} S_r f. \tag{6}$$

with $S_r f(G_r) = f(G_r)$.

4 Numerical Results

In this section, sparse grid approximations are compared to full grid approximations using spline interpolation and quasi-interpolation using Rosenbrock's "Banana" function [18]. Errors in the max norm (approximated by a uniform sampling on a 50×50 grid) are recorded as well as estimates for the rate of approximations. To estimate these rates, we assume the error at the i-th level is of the form $e_i = c h_i^\gamma$ for some constant c. With $h_{i-1} = 2h_i$, this leads to the estimate

$$\frac{e_{i-1}}{e_i} = \left(\frac{h_{i-1}^\gamma}{h_i^\gamma}\right) = 2^\gamma$$

from which we get the

$$\gamma = \ln\left(\frac{e_{i-1}}{e_i}\right) / \ln(2).$$

The spline interpolants and quasi-interpolants are used as surrogates for the original test functions, in optimizing for minima (Rosenbrock's function) and local maxima (Franke's). For the optimizer we use the interior point method in Matlab's built-in function fmincon(). Rates of convergence are estimated using the same formulation derived above for the rate of approximation.

Rosenbrock's function is well-known for testing optimization algorithms. It is a quartic polynomial, given explicitly by

$$f(x, y) = 100(y - x^2)^2 + (1 - x)^2, \tag{7}$$

and is very steep with a narrow curved valley. The function is plotted in Fig. 2, with a sparse grid super-imposed on the left image and contours on the right.

In Table 2, the approximation errors and estimated rates of approximation are given for full and sparse grids, using spline interpolation and quasi-interpolation. Our first observation is that the sparse and full errors are the same for each method. Due to the degree of the

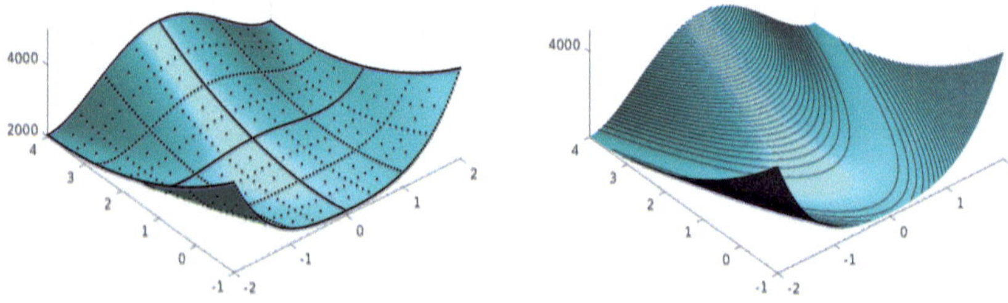

Fig. 2 Surface plot of the function in (7)

Table 2 Uniform Error in Sparse Grid and Full Grid Approximations to Rosenbrock's Banana Function on $[-2.5, 2.5]^2$ by spline interpolation and spline quasi-interpolation on grids defined by lower level α and upper level β, and on tensor product grids at grid level β with mesh spacing $h = 2^{-\beta}$

		Approximation error			
		Interpolation		Quasi-interpolation	
α	β	Sparse	Full	Sparse	Full
2	2	1.5e+02	1.5e+02	2.4e+02	2.4e+02
2	3	1.0e+01	1.0e+01	1.5e+01	1.5e+01
2	4	6.4e–01	6.4e–01	9.4e–01	9.4e–01
2	5	2.7e–02	2.7e–02	4.3e–02	4.3e–02
2	6	8.8e–04	8.8e–04	2.7e–03	2.7e–03
2	7	1.5e–05	1.5e–05	1.7e–04	1.7e–04
2	8	9.1e–07	9.1e–07	1.1e–05	1.1e–05
2	9	5.7e–08	5.7e–08	6.6e–07	6.6e–07
2	10	3.6e–09	3.6e–09	4.1e–08	4.1e–08
2	11	2.3e–10		2.6e–09	
2	12	3.0e–11		1.8e–10	

Rate of Approximation					
		Interpolation		Quasi-Interpolation	
α	β	Sparse	Full	Sparse	Full
2	2			0	0
2	3	3.9	3.9	4.0	4.0
2	4	4.0	4.0	4.0	4.0
2	5	4.6	4.6	4.4	4.4
2	6	5.0	5.0	4.0	4.0
2	7	5.9	5.9	4.0	4.0
2	8	4.0	4.0	4.0	4.0
2	9	4.0	4.0	4.0	4.0
2	10	4.0	4.0	4.0	4.0
2	11	4.0		4.0	
2	12	2.9		3.8	

polynomial being 2 in the y-direction, the full and sparse functions are actually the same. The estimates for the rate of approximation are consistently around 4.0, which is to be expected.

Rosenbrock's function is good for testing optimization algorithms due to it's very steep curved valley. The steepness can cause numerical issues, and the curved valley means that the optimization paths are curved, not straight. In Table 3, the error records the distance from the minimizer of the original function, which is (1,1), and the minimum of the surrogate functions at the given level of the sparse grid. The error converges to zero as the level of the grids increase (approximation improves), as we would expect. But, we also observe that because these errors are non-zero, the minimizer has shifted some by the approximations. This is not a surprise, given the steepness of the function.

The path taken by the interior point method in minimizing the surrogate function (at level $\beta = 12$) is plotted in Fig. 3. It is an interesting path, which starts moves from the corner to the boundary on the right, then moves towards and eventually follows along the valley to the minimum.

5 Conclusion

Sparse grids offer an efficient and robust framework for addressing high-dimensional problems, effectively mitigating the curse of dimensionality through the strategic selection of grid points. By leveraging the combination technique, sparse grids construct hierarchical operators that enable accurate and computationally efficient approximations, making them highly suitable for applications in numerical integration, interpolation, and optimization.

In this paper, we validated the theoretical foundations of sparse grids through numerical experiments on spline-based interpolation and quasi-interpolation. Our results demonstrated

Table 3 Optimization on Sparse Grid and Full Grid Approximations to Rosenbrock's Function on $[-2.5, 2.5]^2$ by spline interpolation and spline quasi-interpolation on grids defined by lower level α and upper level β, and on tensor product grids at grid level β with mesh spacing $h = 2^{-\beta}$. The initial point for the search is $(-2.5, 2.5)$, and the optimal point is $(1, 1)$

Optimization Error

		Interpolation		Quasi-Interpolation	
α	β	Sparse	Full	Sparse	Full
2	2	4.8e-01	4.8e-01	1.1e+00	1.1e+00
2	3	1.6e+00	1.6e+00	1.1e+00	1.1e+00
2	4	1.8e-01	1.8e-01	1.8e-01	1.8e-01
2	5	1.9e-02	1.9e-02	1.9e-02	1.9e-02
2	6	9.9e-03	9.9e-03	9.9e-03	9.9e-03
2	7	6.0e-04	6.0e-04	6.0e-04	6.0e-04
2	8	1.6e-04	1.6e-04	1.6e-04	1.6e-04
2	9	9.9e-06	1.0e-05	1.0e-05	1.0e-05
2	10	2.6e-06	2.5e-06	2.6e-06	2.5e-06
2	11	1.6e-07		2.6e-07	
2	12	3.9e-08		6.3e-08	

Optimization Rate

		Interpolation		Quasi-Interpolation	
α	β	Sparse	Full	Sparse	Full
2	2				
2	3	-1.7	-1.7	-0.0	-0.0
2	4	3.2	3.2	2.6	2.6
2	5	3.2	3.2	3.2	3.2
2	6	1.0	1.0	1.0	1.0
2	7	4.0	4.0	4.0	4.0
2	8	1.9	1.9	1.9	1.9
2	9	4.0	4.0	4.0	4.0
2	10	1.9	2.0	1.9	2.0
2	11	4.1		3.3	
2	12	2.0		2.0	

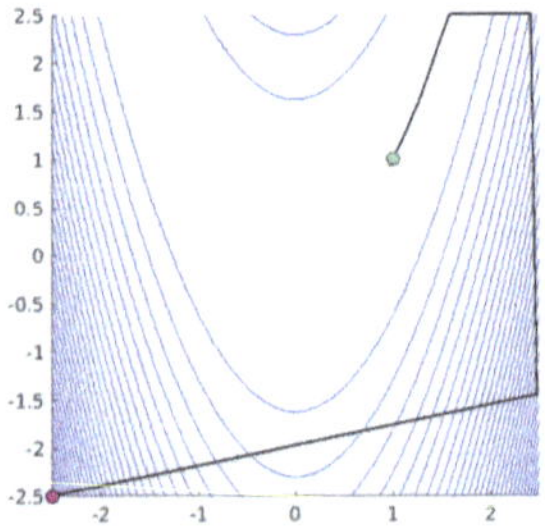

Fig. 3 Contour Plots and Optimization Path starting at the initial point $(-2.5, -2.5)$

that sparse grid operators can achieve approximation accuracy comparable to that of full grid operators while requiring significantly fewer grid points and storage. Specifically, the use of B-splines for interpolation and quasi-interpolation highlighted the advantages of sparse grids in achieving high accuracy with reduced computational costs.

Moreover, we extended the utility of sparse grid methods to function optimization tasks, showcasing their potential in high-dimensional settings. By constructing quasi-interpolants based on univariate schemes and leveraging the sparse grid framework, we provided efficient solutions for optimization problems without compromising approximation quality.

The findings presented here build upon prior advancements [12, 13, 16] and underscore the versatility of sparse grids in approximation and optimization tasks. Future work could explore the integration of sparse grids with machine learning techniques or investigate their applications to solving more complex differential equations. These extensions hold the promise of further broadening the impact of sparse grids across diverse scientific and engineering domains.

References

1. C. de Boor, *Splines as linear combinations of B-splines*, Approximation Theory II, G.G. Lorentz et al. (eds), Academic Press, New-York (1976), 1–47.
2. H.J. Bungartz and M. Griebel, *Sparse grids*, Acta Numerica **13** (2004), 147–269.
3. F. Delvos, *d-variate Boolean interpolation* J. Approx. Theory **34**(2) (1982), 99–114.
4. F. Delvos and H. Posdorf, *Generalized Biermann interpolation*, Resultate der Mathematik **5**(1-2) (1982), 6–18.
5. D. Dung, V. Temlyakov and T. Ullrich, *Hyperbolic Cross Approximation*, Advanced Courses in Mathematics – CRM Barcelona (2018), 1–218.
6. V. Ehirim, Spline interpolation and optimization on sparse grids, Master's Thesis, Georgia Southern University (2021), 1–53.
7. W. Gordon, *Distributive lattices and the approximation of multivariate functions*, in Approximations with Special Emphasis on Spline Functions, ed. I.J. Schoenberg, Academic Press (1969), 223–277.
8. W. Gao and Z. Wu, *Quasi-interpolation for linear functional data*, J. Comput. Appl. Math., **236**(13) (2012) 3256–3264.
9. J. Garcke, *Sparse grid tutorial*, https://www.researchgate.net/publication/228357801 (2006), 1–26.
10. M. Griebel, M. Schneider, and C. Zenger, *A combination technique for the solution of sparse grid problems*, in: R. Beauwens and P. De Groen, eds., Iterative Methods in Linear Algebra, North-Holland, Amsterdam, (1992), 263–281.
11. S. Kersey, *A dual basis for discrete polynomial blending*, Int. J. Numer. Meth. Appl. **15**(4) (2016), 267–303.
12. S. Kersey and C. Ehirim, Spline Approximation on Sparse Grids, *Mathematical Analysis and Computation*, Springer (2023), 1–13.
13. S. Kersey, B-Spline Quasi-Interpolation Applied to Sparse Grids, In progress (2024).
14. T. Lyche and L. Schumaker, *Local spline approximation*, J. Approx. Theory **15** (1975), 294–325.
15. B. Lee, T. Lyche and K. Morken, *Some examples of quasi-interpolants constructed from local spline projectors*, Mathematics Methods for Curves and Surfaces, Oslo, (2000), 243–252.
16. B. Jeong, S. Kersey and J. Yoon (2021), *Approximation of Multivariate Functions on Sparse Grids by Kernel-Based Quasi-Interpolation*, SIAM J. on Scientific Computing (SISC), 42(3), 953–979.
17. C. Pflaum and A. Zhou, *Error analysis of the combination technique*, Numer. Math. **84**(2) (1999), 327–350.
18. H. Rosenbrock, An automatic method for finding the greatest of least value of a function, The Computer Journal **3**(3) (1960), 175–184.
19. P. Sablonniére, *Univariate spline quasi-interpolation and applications to numerical analysis*, arXiv:math/0504022v1 (2005).
20. S.A. Smoljak, *Quadrature and interpolation formulas for tensor products of certain classes of functions*, Soviet Math. Dokl., **4** (1963), 240–243.
21. J. Valentin, *B-Splines for Sparse Grids: Algorithms and Applications to Higher-Dimensional Optimization*, Dissertation, Institute f ur Parallele und Verteilte Systeme der Universita at Stuttgart (2019), 1–258.
22. C. Zenger, *Sparse grids*, in Parallel Algorithms for Partial Differential Equations (Kiel, 1990), Notes on Numerical Fluid Mechanics, vol. 31 (1991), 241–251.

Shape Preserving Approximation of Periodic Functions–A Survey

Dany Leviatan and Igor O. Shevchuk

1 Introduction

1.1 In the Beginning

For many years the cover of "Constructive Approximation", edited by Edward Saff, contained a picture of the celebrated Chebyshev polynomials $T_n(x) := \cos(n \arccos x)$, having the least uniform norm on $[-1, 1]$ among all algebraic polynomials of degree n with the same leading coefficient (Fig. 1).

Fewer people know that Chebyshev [20] also had monotone polynomials having the least uniform norm on $[-1, 1]$ among all algebraic monotone polynomials of degree $n > 1$, having the same monotonicity, with the same leading coefficient. For example, for $n = 5$ they have the form (Fig. 2).

Chebyshev [20] also proved that *if*

$$P_n(x) = x^n + a_1 x^{n-1} + \cdots + a_{n-1} x + a_n$$

D. Leviatan (✉)
Raymond and Beverly Sackler School of Mathematical Sciences, Tel Aviv University, Tel Aviv, Israel
e-mail: leviatan@tauex.tau.ac.il

I. O. Shevchuk
Faculty of Mechanics and Mathematics, Taras Shevchenko National University of Kyiv, Kyiv, Ukraine
e-mail: shevchukio@knu.ua

D. Bilyk et al. (eds.), *Recent Advances in Approximation and Potential Theory*, Applied and Numerical Harmonic Analysis,
https://doi.org/10.1007/978-3-031-95551-8_13

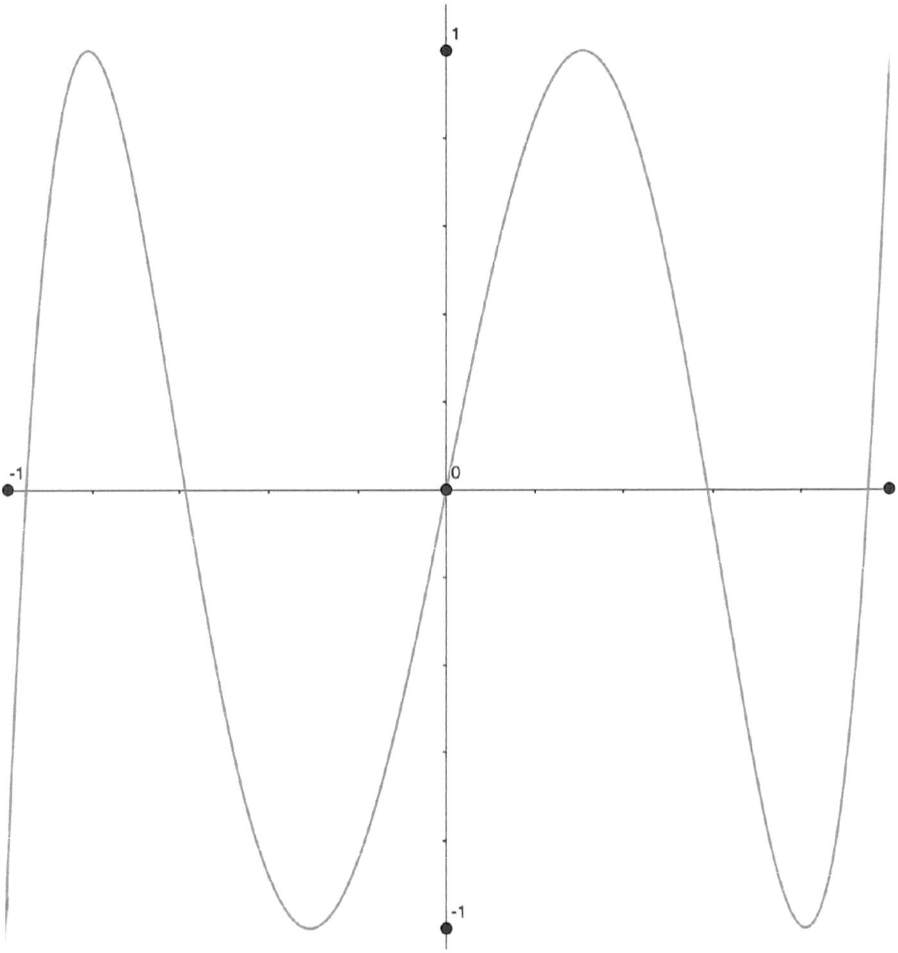

Fig. 1 The Chebyshev polynomial $T_n(x) = \cos(n \arccos x)$, $n = 5$, having the least uniform norm on $[-1, 1]$ among all algebraic polynomials $2^{n-1}x^n + a_1 x^{n-1} + \cdots + a_{n-1}x + a_n$

is monotone on $[-1, 1]$, *then, for even* n,

$$\|P_n\|_{C[-1,1]} \ge 2\left(\frac{(\frac{n}{2})!}{(n-1)!!}\right)^2,$$

and if $n \ge 3$ *is odd, then*

$$\|P_n\|_{C[-1,1]} \ge \left(\frac{\left(\frac{n-1}{2}\right)!}{(n-2)!!}\right)^2 \frac{n+1}{n \pm 1},$$

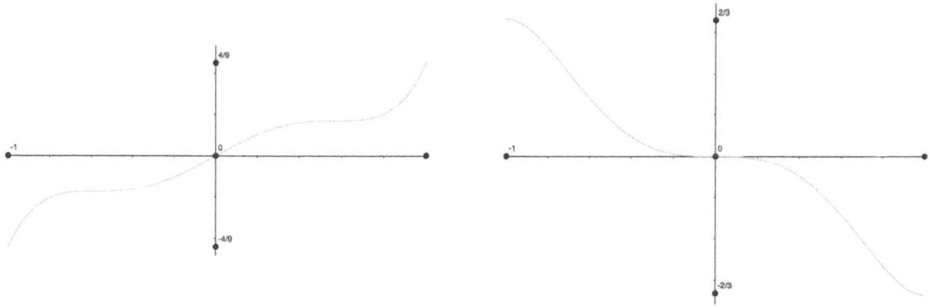

Fig. 2 The Chebyshev monotone on $[-1, 1]$ polynomials, having the least uniform norm on $[-1, 1]$ among all algebraic polynomials $x^n + a_1x^{n-1} + \cdots + a_{n-1}x + a_n$, monotone on $[-1, 1]$. Left $P_5^{\uparrow}(x) = x^5 - \frac{10}{9}x^3 + \frac{5}{9}x$, right $P_5^{\downarrow}(x) = x^5 - \frac{5}{3}x^3$

where $\pm$ *means* $+$ *for increasing polynomials and* $-$ *for decreasing ones. Moreover, the estimates are exact.*

Bernstein [2] generalized the problem of Chebyshev (that is, find a polynomial that deviates least from zero when the leading coefficient is given) in the case of multiply monotone polynomials. (A function $f \in C^q[-1, 1]$ is called *multiply monotone* if $f^{(r)}(x) \geq 0$, $x \in [-1, 1]$, for all $1 \leq r \leq q$.)

Moving to higher order of shape, namely, convex, 3-monotone, etc., we define

Definition 1 A function f is called q-monotone, $q \in \mathbb{N}$, on $[a, b]$, if for any collection $\{x_j\}_{j=0}^q$ of distinct points $x_j \in [a, b]$ the divided difference

$$[x_0, x_1, \ldots, x_q; f] \geq 0.$$

I.e., a 1-monotone function is a nondecreasing function, and a 2-monotone function is a convex one.

Clearly, if $f \in C[a, b] \cap C^q(a, b)$, then f is q-monotone on $[a, b]$, if and only if $f^{(q)}(x) \geq 0$, $x \in (a, b)$.

The next step in Shape Preserving Approximation (SPA) is due to Lorentz [15] who observed that the Bernstein polynomials preserve q-monotonicity, *i.e., if a function* f *is* q*-monotone on* $[0, 1]$, *then its Bernstein polynomials*

$$B_n(f, x) = \sum_{j=0}^{n} \binom{n}{j} f\left(\frac{i}{n}\right) x^j(1-x)^{n-j}$$

are q*-monotone on* $[0, 1]$ *as well.*

Since the Bernstein polynomials associated with $f \in C[a, b]$, uniformly approximate f on $[a, b]$, an immediate consequence is a "q-monotone Weierstrass type theorem", namely, *the* q*-monotone polynomials are dense in the continuous* q*-monotone functions on* $[a, b]$, *in the uniform norm.*

1.2 SPA by Algebraic Polynomials

It is well known that *for each function* $f \in C[-1, 1]$, *there is a sequence* $\{P_n\}_{n=1}^{\infty}$ *of algebraic polynomials of degree* $< n$, *such that*

$$|f(x) - P_n(x)| \le c\omega\left(f, \frac{\sqrt{1-x^2}}{n} + \frac{1}{n^2}; [-1, 1]\right), \quad x \in [-1, 1], \tag{1}$$

where c is an absolute constant, and $\omega(f, \cdot; [-1, 1])$ *is the modulus of continuity of* f *on* $[-1, 1]$.

In 1968, Lorentz and Zeller [16] proved the following significant result in SPA of functions $f \in C[a, b]$ by algebraic polynomials. They proved that *for each increasing function* $f \in C[-1, 1]$, *there is a sequence* $\{P_n\}_{n=1}^{\infty}$ *of increasing algebraic polynomials of degree* $< n$, *such that (1) holds.*

DeVore [3] obtained Jackson-type estimates in SPA of r times continuously differentiable monotone functions.

Newman [17] obtained the first estimate in comonotone SPA, by algebraic polynomials, of piecewise-monotone functions $f \in C[a, b]$.

Leviatan [10] obtained Jackson type estimates for convex approximation. See also Beatson [1] for q-monotone approximation.

Dozens of authors have investigated SPA by algebraic polynomials and have obtained various pointwise and uniform estimates, analogous to known ones in unconstrained approximation. The interested reader should consult, e.g., the surveys [8] and [9].

1.3 SPA by Trigonometric Polynomials, Preliminaries

The theory of SPA of periodic functions by trigonometric polynomials had to wait for the Newman [17] comonotone result, as the only monotone periodic functions are constants.

One may consider the first result in comonotone approximation of 2π-periodic functions by trigonometric polynomials as the one obtained by Lorentz and Zeller [16, Theorem 1].

Theorem 1 ([16]) *Let* $f \in C(\mathbb{R})$ *be an even* 2π*-periodic function, decreasing in* $[0, \pi]$. *Then for each* $n \in \mathbb{N}$ *there is an even trigonometric polynomial* T_n *of degree* $< n$, *decreasing in* $[0, \pi]$, *such that*

$$\|f - T_n\|_{C(\mathbb{R})} \le c\omega\left(f, \frac{1}{n}\right), \tag{2}$$

where c is an absolute constant and $\omega(f, \cdot)$ *is the modulus of continuity of* f *on* $\mathbb{R}$.

As mentioned above, Lorentz and Zeller were interested in monotone approximation of an increasing $f \in C[-1, 1]$, by increasing algebraic polynomials. They substituted $x = \cos t$ and obtained an even 2π-periodic continuous function $f(\cos t)$, $t \in \mathbb{R}$, decreasing on $[0, \pi]$, which they called bell shaped. They approximated it by bell shaped trigonometric polynomials, and by the same substitution obtained the monotone algebraic polynomials. We observe that such a bell-shaped function has 2 changes of monotonicity in each period.

Furthermore, for a differentiable function, we have

Theorem 2 ([12]) *Let $r \in \mathbb{N}$ and let $f \in C^r$ have $2s$ changes of monotonicity in each period. Then for each $n \in \mathbb{N}$ there is a trigonometric polynomial T_n of degree $< n$, such that*

$$f'(x)T_n'(x) \geq 0, \quad x \in \mathbb{R}, \tag{3}$$

and

$$\|f - T_n\|_{C(\mathbb{R})} \leq \frac{c(r, s)}{n^r}\omega\left(f^{(r)}, \frac{1}{n}\right), \tag{4}$$

where $c = c(r, s)$ is a constant, depending only on r and s.

Remark 1 Theorem 2 is valid also for $r = 0$ (Pleshakov [18]), when the inequality in (3) is assumed only a.e..

If we denote by W^r, $r \in \mathbb{N}$, the class of 2π-periodic functions f, having an absolutely continuous $(r-1)st$ derivative in $\mathbb{R}$, and such that $\|f^{(r)}\|_{L_\infty(\mathbb{R})} \leq 1$, then an immediate consequence is

Corollary 1 *Let $r \in \mathbb{N}$ and let $f \in W^r$ have $2s$ changes of monotonicity in each period. Then for each $n \in \mathbb{N}$ there is a trigonometric polynomial T_n of degree $< n$, such that (3) holds (a.e. if $r = 1$), and*

$$\|f - T_n\|_{C(\mathbb{R})} \leq \frac{c(r, s)}{n^r}.$$

We survey here the validity of Jackson-type estimates in Shape Preserving Approximation (SPA) of 2π-periodic functions by trigonometric polynomials. We concentrate on comonotone (Sect. 2) and coconvex (Sect. 3) approximation (see Definition 2 below), since for the so called co-q-monotone, $q > 2$, approximation there are no Jackson type estimates (see details in Sect. 4). Section 5 is devoted to comparing the degrees of unconstrained approximation by trigonometric polynomials and of SPA.

Let $C = C^0$ be the space of 2π-periodic continuous functions f on $\mathbb{R}$, endowed with the uniform norm

$$\|f\| := \|f\|_{C(\mathbb{R})} = \max_{x \in \mathbb{R}} |f(x)|.$$

For $r \in \mathbb{N}$, denote by C^r the subspace of r times continuously differentiable functions $f \in C$.

For $f \in C$ and $k \in \mathbb{N}$, we denote by

$$\omega_k(f,t) = \sup_{x\in\mathbb{R}, 0<h\le t} \left| \sum_{i=0}^{k} (-1)^i \binom{k}{i} f\big(x + (k/2 - i)h\big) \right|,$$

the kth modulus of smoothness of f.

Finally, denote by

$$E_n(f) := \inf \|f - T_n\|,$$

where the infimum is taken over all trigonometric polynomials T_n of degree $< n$, the degree of approximation by these polynomials.

In this paper we discuss the approximation of $f \in C^r$, $r \in \mathbb{N}_0$, that changes its monotonicity or its convexity finitely many times in its period $(-\pi, \pi]$, by trigonometric polynomials that follow the monotonicity or convexity of f, respectively (i.e., comonotone, respectively, coconvex approximation). Note that a 2π-periodic function may not be monotone or convex unless it is a constant. A nonconstant trigonometric polynomial may only have a finite even number of changes of monotonicity or convexity in a period. Hence, we are limited to discussion of comonotone and coconvex approximation of f that has $2s$, $s \ge 1$, changes of monotonicity in each period. We also look at functions that change their q-monotonicity, $q > 2$, finitely many, that is, $2s$ times in each period.

Let $Y_s := \{y_i\}_{i\in\mathbb{Z}}$, $s \in \mathbb{N}$, $y_i \in \mathbb{R}$, be such that $y_i < y_{i+1}$ and $y_{i+2s} = y_i + 2\pi$, $i \in \mathbb{Z}$. Denote by $\Delta^{(q)}(Y_s)$, $q \in \mathbb{N}$, the collection of functions $f \in C$, such that $(-1)^{i-1} f$ is q-monotone on the intervals $[y_{i-1}, y_i]$.

Note that if we denote

$$\Pi(x) := \prod_{i=1}^{2s} \sin \frac{x - y_i}{2}, \tag{5}$$

then for $f \in C^q$, $q \in \mathbb{N}$, $f \in \Delta^{(q)}(Y_s)$, if and only if

$$f^{(q)}(x)\Pi(x) \ge 0, \quad x \in \mathbb{R}.$$

Definition 2 For $f \in \Delta^{(q)}(Y_s)$, let

$$E_n^{(q)}(f, Y_s) := \inf_{T_n \in \Delta^{(q)}(Y_s)} \|f - T_n\|,$$

denote the error of best approximation of f by co-q-monotone trigonometric polynomials $T_n \in \Delta^{(q)}(Y_s)$ of degree $< n$. In particular, if $q = 1$, then we write comonotone instead of co-1-monotone, and if $q = 2$, then we write coconvex instead of co-2-monotone.

For $f \in C^r \cap \Delta^{(q)}(Y_s)$, we are interested in the validity of SPA Jackson-type estimates,

$$E_n^{(q)}(f, Y_s) \le \frac{c(k, r, s, q)}{n^r} \omega_k \left(f^{(r)}, \frac{1}{n}\right), \quad n \ge 1. \tag{6}$$

If SPA Jackson-type estimates are invalid, we will discuss the validity of SPA weak Jackson-type estimates,

$$E_n^{(q)}(f, Y_s) \le \frac{c(r, k, s, q)}{n^r} \omega_k \left(f^{(r)}, \frac{1}{n}\right), \quad n \ge N(r, k, q, Y_s), \tag{7}$$

namely, in the latter case, the inequality in (7) is valid only for N which depends not only on the (r, k, s, q), rather, it depends also on the disposition of the points Y_s.

Thus, we will present all quadruplets (k, r, s, q), for which SPA Jackson-type estimates hold with constant $c = c(k, r, s, q)$, depending on these parameters only. For all other quadruplets we have counterexamples. Moreover, we will present all quadruplets, for which SPA Jackson-type estimate is invalid, but SPA weak Jackson-type estimates are valid, as well as all quadruplets, for which even the latter is invalid.

Note that since f is 2π-periodic, Whitney's inequality implies that for any $k \ge 1$,

$$\|f - f(0)\| \le c(k)\omega_k(f, 2\pi),$$

that yields, for any N and all $1 \le n \le N$, and $f \in \Delta^{(q)}(Y_s)$,

$$E_n^{(q)}(f, Y_s) \le E_1^{(q)}(f, Y_s) \le \|f - f(0)\| \le c(k, N)\omega_k(f, 1/n). \tag{8}$$

Hence, in view of the properties of the moduli of smoothness, if we are able to establish the inequality in (6) for $n \ge N(r, k, s, q)$, then it is valid for $n \ge 1$, and whenever (7) is valid, we may obtain (6) provided we are willing to allow the constant $c = c(r, k, q, Y_s)$, that is, to depend also on Y_s.

2 Comonotone Approximation of Periodic Functions

In this section we deal with functions $f \in C^r \cap \Delta^{(1)}(Y_s)$, $r \in \mathbb{N}_0$, $s \in \mathbb{N}$.

As already mentioned, (6) with $q = 1$, holds for the pairs (r, k), $k = 1$, $r \in \mathbb{N}_0$ (see Theorems 1, 2, and Remark 1).

We write $(r, k) \in A_s$, if either $k = 1$, or $k = 2$ and $r \ge 2s - 2$, or $k = 3$ and $r \ge 2s - 1$, or $k \in \mathbb{N}$ and $r \ge 2s$. Then, we have extended the above, proving

Theorem 3 ([12] and [14]) *For each $s \in \mathbb{N}$ and $(r, k) \in A_s$ there is a constant $c(k, r, s)$, such that if a function $f \in C^r \cap \Delta^{(1)}(Y_s)$, then*

$$E_n^{(1)}(f, Y_s) \leq \frac{c(k, r, s)}{n^r} \omega_k \left(f^{(r)}, \frac{1}{n} \right), \quad n \geq 1. \tag{9}$$

However, there is no other pair (r, k) for which (9) is valid [14], as we explain below.

Perhaps, the most interesting case is,

Theorem 4 *Let $0 \leq r < 2s - 2$. There is a constant $c(r) > 0$, such that for each $n \geq 1$, there exist a collection Y_s and a function $f \in C^r \cap \Delta^{(1)}(Y_s)$, such that*

$$E_n^{(1)}(f, Y_s) > n \frac{c(r)}{n^r} \omega_2 \left(f^{(r)}, \frac{1}{n} \right).$$

Corollary 2 *Let $0 \leq r < 2s - 2$ and $k \geq 2$. For every $A > 0$ and $N > 0$, there exist $n > N$, Y_s and an $f \in C^r \cap \Delta^{(1)}(Y_s)$, such that*

$$E_n^{(1)}(f, Y_s) > \frac{A}{n^r} \omega_k \left(f^{(r)}, \frac{1}{n} \right). \tag{10}$$

But, we also have

Theorem 5 *Let $r = 2s - 2$. There is a constant $c(r) > 0$, such that for each $n \geq 1$, there exist a collection Y_s and a function $f \in C^r \cap \Delta^{(1)}(Y_s)$, such that*

$$E_n^{(1)}(f, Y_s) > \sqrt{n} \frac{c(r)}{n^r} \omega_3 \left(f^{(r)}, \frac{1}{n} \right),$$

and

Theorem 6 *Let $r = 2s - 1$. There is a constant $c(r) > 0$, such that for each $n \geq 1$, there exist a collection Y_s and a function $f \in C^r \cap \Delta^{(1)}(Y_s)$, such that*

$$E_n^{(1)}(f, Y_s) > \sqrt[3]{n} \frac{c(r)}{n^r} \omega_4 \left(f^{(r)}, \frac{1}{n} \right).$$

Hence, we conclude that

Corollary 3 *Let $r = 2s - 2$ and $k \geq 3$ or $r = 2s - 1$ and $k \geq 4$. For every $A > 0$ and $N > 0$, there exist $n > N$, Y_s and an $f \in C^r \cap \Delta^{(1)}(Y_s)$, such that (10) holds for the appropriate pairs.*

If (9) is invalid and $r \geq 2$, then all is not lost, as it still follows by Dzyubenko [5], that the estimate (7) with $q = 1$, holds. In addition, for $n > N(Y_s)$ we may guarantee that the distance between any two consecutive y_i's is bigger than $\frac{5\pi}{n}$. Therefore, if we take $n > N(Y_s)$, then [14] yields that (7) with $q = 1$ is valid for

$r = 1$ and $k = 2, 3$, as well. Finally, the validity of (7) with $q = 1$, when $r = 0$ and $k = 2$, is proved in [6].

However, if $r = 0$ and $k \geq 3$, and if $r = 1$ and $k \geq 4$, then even (7) with $q = 1$, is invalid, since (see [7]), for each Y_s there is a function $f \in C^r \cap \Delta^{(1)}(Y_s)$, $r = 0, 1$, such that for these pairs (r, k),

$$\limsup_{n\to\infty} \frac{n^r}{\omega_k(f^{(r)}, 1/n)} E_n^{(1)}(f, Y_s) = \infty.$$

The case $r = 0$ and $k \geq 3$ originally appeared in [4].

We can now summarize the results. In the tables below we denote by "+" the pairs (r, k) for which (9) is valid. (The column $(r, 0)$ corresponds to Corollary 1.) We denote by $\oplus$ the pairs (r, k) for which (9) is invalid, but (7) with $q = 1$, is valid, and finally, – denotes that even (7) with $q = 1$, is invalid.

r	⋮	⋮	⋮	⋮	⋮	⋮	⋰
$2s+1$	+	+	+	+	+	+	⋯
$2s$	+	+	+	+	+	+	⋯
$2s-1$	+	+	+	+	$\oplus$	$\oplus$	⋯
$2s-2$	+	+	+	$\oplus$	$\oplus$	$\oplus$	⋯
$2s-3$	+	+	$\oplus$	$\oplus$	$\oplus$	$\oplus$	⋯
⋮	⋮	⋮	⋮	⋮	⋮	⋮	⋮
2	+	+	$\oplus$	$\oplus$	$\oplus$	$\oplus$	⋯
1	+	+	$\oplus$	$\oplus$	−	−	⋯
0		+	$\oplus$	−	−	−	⋯
	0	1	2	3	4	5	k

$s > 2$

r	⋮	⋮	⋮	⋮	⋮	⋮	⋰
5	+	+	+	+	+	+	⋯
4	+	+	+	+	+	+	⋯
3	+	+	+	+	$\oplus$	$\oplus$	⋯
2	+	+	+	$\oplus$	$\oplus$	$\oplus$	⋯
1	+	+	$\oplus$	$\oplus$	−	−	⋯
0		+	$\oplus$	−	−	−	⋯
	0	1	2	3	4	5	k

$s = 2$

r	⋮	⋮	⋮	⋮	⋮	⋮	⋰
3	+	+	+	+	+	+	⋯
2	+	+	+	+	+	+	⋯
1	+	+	+	+	−	−	⋯
0		+	+	−	−	−	⋯
	0	1	2	3	4	5	k

$s = 1$

We wish to emphasize that these tables differ substantially from the corresponding tables for comonotone approximation, by algebraic polynomials, of continuous functions on a finite interval (see, [8, Tables 16 and 17]). We note that when one discusses functions on a finite interval, they may have any (odd or even) number of extremum points, and we should limit ourselves to comparing only to the case of an even number of such points. Thus, adapting to our notation, we have $2s$ extremum points, $s \ge 1$.

$$
\begin{array}{c|cccccccc}
r & \vdots & \vdots & \vdots & \vdots & \vdots & \vdots & \iddots \\
2s+2 & + & + & + & + & + & + & \cdots \\
2s+1 & + & + & + & + & + & + & \cdots \\
2s & + & + & + & \oplus & \oplus & \oplus & \cdots \\
2s-1 & + & + & \oplus & \oplus & \oplus & \oplus & \cdots \\
\vdots & \vdots & \vdots & \vdots & \vdots & \vdots & \vdots & \cdots \\
2 & + & + & \oplus & \oplus & \oplus & \oplus & \cdots \\
1 & + & + & \oplus & \oplus & - & - & \cdots \\
0 & & + & \oplus & - & - & - & \cdots \\
 & 0 & 1 & 2 & 3 & 4 & \cdots\cdots & k
\end{array}
\qquad
\begin{array}{c|cccccccc}
r & \vdots & \vdots & \vdots & \vdots & \vdots & \vdots & \iddots \\
4 & + & + & + & + & + & + & \cdots \\
3 & + & + & + & + & + & + & \cdots \\
2 & + & + & + & \oplus & \oplus & \oplus & \cdots \\
1 & + & + & \oplus & \oplus & - & - & \cdots \\
0 & & + & \oplus & - & - & - & \cdots \\
 & 0 & 1 & 2 & 3 & 4 & 5 & k
\end{array}
$$

$s > 1$ $\qquad\qquad$ $s = 1$

3 Coconvex Approximation of Periodic Functions

In this section we deal with functions $f \in C^r \cap \Delta^{(2)}(Y_s)$, $r \in \mathbb{N}_0$, $s \in \mathbb{N}$.

The first results in coconvex approximation of 2π-periodic functions by trigonometric polynomials are due to Popov [19] who proved that (7) with $q = 2$, is valid for $(r, k) = (0, 2)$, which readily implies that it is also valid for the pairs $(1, 1)$ and $(0, 1)$. Later, Zalizko [21] proved the validity of (7) with $q = 2$, for $(r, k) = (0, 3)$, which, in turn, readily implies its validity for the pairs $(0, 2)$, $(1, 2)$ and $(2, 1)$, as well as all those claimed by Popov.

Recently [13], we have shown that if $f \in C^2 \cap \Delta^{(2)}(Y_s)$, then (7) with $q = 2$ is valid for $(r, k) = (2, 3)$, so, in turn, also for the pair $(2, 2)$ that was missing above.

Finally, we have proved

Theorem 7 ([13]) *For each $k \in \mathbb{N}$ and $s \in \mathbb{N}$, there is a constant $c(k, s)$, such that if a function $f \in C^3 \cap \Delta^{(2)}(Y_s)$, then*

$$E_n^{(2)}(f, Y_s) \le \frac{c(k, s)}{n^3}\omega_k\left(f^{(3)}, \frac{1}{n}\right), \quad n \ge N(k, Y_s).$$

And as a consequence we get a general result.

Corollary 4 *For each $r \ge 3$, $k \in \mathbb{N}$ and $s \in \mathbb{N}$, there is a constant $c(r, k, s)$, such that if a function $f \in C^r \cap \Delta^{(2)}(Y_s)$, then*

$$E_n^{(2)}(f, Y_s) \le \frac{c(k, r, s)}{n^r}\omega_k\left(f^{(r)}, \frac{1}{n}\right), \quad n \ge N(r, k, Y_s). \tag{11}$$

Recall that W^r, $r \in \mathbb{N}$, is the class of 2π-periodic functions $f \in AC^{(r-1)}(\mathbb{R})$ such that $\|f^{(r)}\|_{L_\infty(\mathbb{R})} \le 1$. Analogous to Corollary 1, an immediate consequence of the above is

Corollary 5 *Let $r \in \mathbb{N}$ and let $f \in W^r \cap \Delta^{(2)}(Y_s)$. Then*

$$E_n^{(2)}(f, Y_s) \le \frac{c(r, s)}{n^r}, \quad n \ge N(r, Y_s).$$

A natural question, of course, is what about (6) with $q = 2$.

Alas, (6) with $q = 2$ cannot be had. A recent paper [7, Theorem 1.1] yields that for each pair $r \in \mathbb{N}$ and $s \in \mathbb{N}$, there is a constant $c(r, s) > 0$, such that for any $n \in \mathbb{N}$, there exist a collection Y_s and a function $f \in W^r \cap \Delta^{(2)}(Y_s)$, for which

$$E_n^{(2)}(f, Y_s) > \frac{c(r, s)}{n^{r-1}}. \tag{12}$$

This, in turn, implies that for each triple of integers $r \in \mathbb{N}_0$, $k \in \mathbb{N}$ and $s \in \mathbb{N}$, there is a constant $c(k, r, s) > 0$, such that for any $n \in \mathbb{N}$, there exist a collection Y_s and a function $f \in C^r \cap \Delta^{(2)}(Y_s)$, for which

$$E_n^{(2)}(f, Y_s) > \frac{c(k, r, s)}{n^{r-1}} \omega_k\left(f^{(r)}, \frac{1}{n}\right).$$

Moreover, if $k \ge 4$ and $0 \le r \le 2$, then (11) is invalid even if we allow N to depend on f. Indeed, [7, Theorems 1.2 and 1.3] yields that for each Y_s and $0 \le r \le 2$, there is a function $f \in C^r \cap \Delta^{(2)}(Y_s)$, such that

$$\limsup_{n\to\infty} \frac{n^r E_n^{(2)}(f, Y_s)}{\omega_4(f^{(r)}, 1/n)} = +\infty.$$

In fact, for $r = 1$ this holds also with ω_3.

We summarize the above in the following table, where "$\oplus$" denotes pairs (r, k) for which only (11) is valid, so that, in particular, (6) with $q = 2$, is invalid, and "$-$" denotes pairs (r, k) for which (11) is not valid even if we allow $N = N(f)$.

r	$\vdots$	$\vdots$	$\vdots$	$\vdots$	$\vdots$	$\vdots$	$\iddots$
4	$\oplus$	$\oplus$	$\oplus$	$\oplus$	$\oplus$	$\oplus$	$\cdots$
3	$\oplus$	$\oplus$	$\oplus$	$\oplus$	$\oplus$	$\oplus$	$\cdots$
2	$\oplus$	$\oplus$	$\oplus$	$\oplus$	$-$	$-$	$\cdots$
1	$\oplus$	$\oplus$	$\oplus$	$-$	$-$	$-$	$\cdots$
0		$\oplus$	$\oplus$	$\oplus$	$-$	$-$	$\cdots$
	0	1	2	3	4	5	k

It is interesting to point out that the corresponding table for coconvex approximation, by algebraic polynomials, of continuous functions on a finite interval, [8, p. 39, Table 23] is exactly the same. Again we note that algebraic polynomials may have any (odd or even) finite number of inflection points in a finite interval, and we should limit ourselves to comparing only with the table for an even number of inflection points. Surprisingly, if the function has a single inflection point, then the truth table for the algebraic case is completely different (see [8, p. 38, Table 22]).

4 No Jackson Estimates for $q \geq 3$

Our main result is

Theorem 8 ([11]) *For each $q \geq 3$, $r \in \mathbb{N}$, $s \in \mathbb{N}$ and any Y_s, there exists a function $f \in \Delta^{(q)}(Y_s) \cap W^r$, such that*

$$\limsup_{n\to\infty} n^r E_n^{(q)}(f, Y_s) = \infty.$$

Corollary 6 *For each $q \geq 3$, $k \in \mathbb{N}$, $r \in \mathbb{N}_0$, $s \in \mathbb{N}$ and any Y_s, there exists a function $f \in \Delta^{(q)}(Y_s) \cap C^r$, such that*

$$\limsup_{n\to\infty} \frac{n^r E_n^{(q)}(f, Y_s)}{\omega_k(f^{(r)}, 1/n)} = \infty.$$

The following are less general but more precise statements (see [11]).

Theorem 9 *For each $q \geq 3$, $r \leq q-2$, $s \in \mathbb{N}$ and any Y_s, there exists a function $f \in \Delta^{(q)}(Y_s) \cap W^r$, such that*

$$E_n^{(q)}(f, Y_s) \geq C(q, Y_s), \quad n \in \mathbb{N},$$

where $C(q, Y_s) > 0$ depends only on q and Y_s.

We also have,

Theorem 10 *For each $q \geq 3$, $s \in \mathbb{N}$ and any Y_s, there exists a function $f \in \Delta^{(q)}(Y_s) \cap W^{q-1}$, such that*

$$nE_n^{(q)}(f, Y_s) \geq C(q, Y_s), \quad n \in \mathbb{N},$$

where $C(q, Y_s) > 0$ depends only on q and Y_s.

Our final result is

Theorem 11 *Let $q \geq 3$, $r \geq q$ and $s \in \mathbb{N}$. For each sequence $\{\varepsilon_n\}_{n=1}^{\infty}$ of positive numbers, tending to infinity, there is a function $f \in \Delta^{(q)}(Y_s) \cap W^r$, such that*

$$\limsup_{n\to\infty} \varepsilon_n n^{r-q+2} E_n^{(q)}(f, Y_s) = \infty.$$

5 Comparing to Unconstrained Approximation

While it is clear that one always has $E_n(f) \leq E_n^{(q)}(f, Y_s)$ for any $f \in \Delta^{(q)}(Y_s)$, there are cases where the two degrees are of the same order.

In particular, we have the following two results.

Theorem 12 *Let $f \in \Delta^{(1)}(Y_s)$. Given $\alpha > 0$. If*

$$n^{\alpha} E_n(f) \leq 1, \quad n \geq 1, \tag{13}$$

then for $\alpha \notin \mathbb{N}$, or $\alpha \in \mathbb{N}$ and $\alpha \geq 2s - 1$, we have

$$n^{\alpha} E_n^{(1)}(f, Y_s) \leq c(s, \alpha), \quad n \geq 1,$$

and if $\alpha \in \mathbb{N}$ and $\alpha < 2s - 1$, then

$$n^{\alpha} E_n^{(1)}(f, Y_s) \leq c(s, \alpha), \quad n \geq N(Y_s). \tag{14}$$

Theorem 13 *Let $f \in \Delta^{(2)}(Y_s)$. If (13) holds for some $\alpha > 0$, then*

$$n^{\alpha} E_n^{(2)}(f, Y_s) \leq c(s, \alpha), \quad n \geq N(Y_s),$$

and the inequality is, in general, invalid with N independent of Y_s.

Indeed, the above theorems follow by the facts that if $r < \alpha < r + 1$, $r \in \mathbb{N}_0$, then

$$E_n(f) \leq n^{-\alpha}, \quad n \geq 1, \quad \Rightarrow \quad f \in C^r \quad \text{and} \quad \omega(f^{(r)}, t) \leq c(\alpha) t^{\alpha - r},$$

and if $\alpha = r \in \mathbb{N}$, then

$$E_n(f) \leq n^{-r}, \quad n \geq 1, \quad \Rightarrow \quad f \in C^{r-1} \quad \text{and} \quad \omega_2(f^{(r-1)}, t) \leq c(r) t.$$

For Theorem 12 we apply these facts to Theorem 3 and the paragraph under Corollary 3, and for Theorem 13 we apply them to (11) and the earlier known estimates above Theorem 7. The last assertion of Theorem 13 follows by (12). Also, (14) is sharp in the same sense.

References

1. R. K. Beatson, The degree of monotone approximation, *Pacific J. Math.* 74(1978), 5–14.
2. S. N. Bernstein, Sur les polynômes multiplement monotones, *Soobshch. Kharkov Soc. Math.*, (4) 1 (1927).
3. R. A. DeVore, Monotone approximation by polynomials, *SIAM J. Math. Anal.* 8(1977), 906–921.
4. G. Dzyubenko, Counterexample in comonotone approximation of periodic functions, *Trans. Inst. Math. NAS Ukraine* 5 (2008), 113–123 (in Ukrainian).
5. G. Dzyubenko, Comonotone approximation of twice differentiable periodic functions, *Ukrainian Mat. J.*, 61 (2009), 519–540.
6. G. Dzyubenko and M. G. Pleshakov, Comonotone approximation of periodic functions, *Math. Notes*, 83 (2008), 180–189.
7. G. Dzyubenko, V. Voloshyna and L. Yushchenko, Negative results in coconvex approximation of periodic functions, *J. Approx. Theory*, 267:105582 (2021), 1–11.
8. K. A. Kopotun, D. Leviatan, A. Prymak and I. A. Shevchuk, Uniform and pointwise shape preserving approximation (SPA) by algebraic polynomials, *Surveys in Approx. Theory* 6 (2011), 24–74. (see https://www.surveys-inapproximationtheory.com/papers/16/)
9. K. A. Kopotun, D. Leviatan and I. A. Shevchuk, Uniform and pointwise shape preserving approximation (SPA) by algebraic polynomials: an update. *The SMAI journal of computational mathematics*, **5** (2019) , 99–108.
10. D. Leviatan, Pointwise estimates for convex polynomial approximation, *Proc. Amer. Math. Soc.*, 98 (1986), 471–474.
11. D. Leviatan, O. V. Motorna and I. A. Shevchuk, No Jackson-type estimates for piecewise q-monotone, $q \geq 3$, trigonometric approximation, *Ukrainian Math. J.* 74 (2022), 662–675.
12. D. Leviatan, M. V. Shchehlov and I. O. Shevchuk, Comonotone approximation of periodic functions *J. Approx. Theory* 299:106015 (2024), 1–16.
13. D. Leviatan and I. A. Shevchuk, Coconvex approximation of periodic functions, *Constr. Approx.* 57 (2023), 695–726.
14. D. Leviatan and I. O. Shevchuk, Comonotone approximation of periodic functions: Conclusion, *Constr. Approx.* 62 (2025).
15. G. G. Lorentz, *Bernstein polynomials*. Univ. of Toronto Press. Toronto, 1953, 130pg.
16. G. G. Lorentz and K. L. Zeller, Degree of approximation by monotone polynomials I, *J. Approx. Theorey* 1 (1968), 501–504.
17. D. J. Newman, Efficient co-monotone approximation,*J. Approx. Theory*, 25 (1979), 189–192.
18. M. G. Pleshakov, Comonotone Jackson's inequality, *J. Approximation Theory*, 99 (1999), 409–421.
19. P. A. Popov, An analog of the Jackson inequality for coconvex approximation of periodic functions, *Ukrainian Math. J.*, 53 (2001), 1093–1105.
20. P. L. Tchebyshev, Sur les functions qui diffèrent le moins possible de zéro, *J. de math. pures et appl.*, II série, XIX (1874), 319-346.
21. V. D. Zalizko, Coconvex approximation of periodic functions, *Ukrainian Math. J.* 59 (2007) 28–44.

Spaces of Trigonometric Series and the Wiener Algebra

Elijah Liflyand

Dedicated to E. Saff on the occasion of his 80th birthday.

1 Introduction

The classical problem whether a trigonometric series is the Fourier series of an integrable function has more than a century long and sometimes intriguing history. One cannot say that this story is over, especially in the multivariate case but it has recently become more or less completed and transparent. To a large extent, it had happened due to a parallel study of Fourier transforms. Here we outline this material in a concentrated form.

The following observation is worth mentioning. Many professionals never coming across this problem directly do not distinguish between two types of series, mainly having Fourier series in mind when saying or writing trigonometric series. We emphasize that by trigonometric series we mean an infinite linear combination of harmonics with that or another sequence of coefficients only. On the other hand, saying Fourier series, we firmly mean a very similar infinite sum but with coefficients being the Fourier coefficients of an integrable function. We add that this is not the question of convergence: on the one hand, famous Kolmogorov's example [20] shows that the Fourier series of an integrable function may be divergent even at *each point*; on the other hand, a trigonometric series may be convergent almost everywhere or even everywhere but not be a Fourier series.

The structure of the paper is as follows. In the following section, the period from the first result in 1913 till the results in mid 1950s is surveyed. Then, in Sect. 3, further results are described. Section 4 presents the Fourier transform approach due to the author. Numerous results on the integrability of the Fourier transform of a function of bounded variation are formulated, from which most of the known

---

E. Liflyand (✉)
Department of Mathematics, Bar-Ilan University, Ramat-Gan, Israel
e-mail: liflyand@math.biu.ac.il

D. Bilyk et al. (eds.), *Recent Advances in Approximation and Potential Theory*,
Applied and Numerical Harmonic Analysis,
https://doi.org/10.1007/978-3-031-95551-8_14

conditions for a trigonometric series to be a Fourier series are derived. In Sect. 5, recent results by Trigub and the author are given. These results are necessary and sufficient and, in a sense, lead to, if not all the results but to the ways of finding them without any preliminary restrictions on the sequence of the coefficients of a trigonometric series. In the last section, multidimensional extensions of these recent results are presented. Though the list of references is long enough containing 45 items, it by no means exhausts the stock of the relevant sources; this was an initial idea to present only substantial results and references.

Before proceeding to the presentation in essence, we mention that notation " $\lesssim$" will be used as abbreviation for " $\le C$ ", with C being an absolute positive constant.

2 Trigonometric Series Versus Fourier Series: First Steps

The problem whether a trigonometric series is the Fourier series of an integrable function in symbols reads as follows. Given a trigonometric series

$$\frac{a_0}{2}+\sum_{k=1}^{\infty}(a_k\cos kx+b_k\sin kx),$$

under what assumptions on its coefficients this series is the Fourier series of an integrable function. Sometimes, it is briefly said "a Fourier series" in place of "the Fourier series of an integrable function". Frequently, the cosine series

$$\frac{a_0}{2}+\sum_{k=1}^{\infty}a_k\cos kx \tag{1}$$

and the sine series

$$\sum_{k=1}^{\infty}b_k\sin kx \tag{2}$$

are investigated separately, since there is a difference in their behavior. Usually, solutions for the latter requires additional assumptions. In a different way, we shall say in the cases where the series is a Fourier series that the sequence of coefficients belongs to $\widehat{L^1}$.

The corresponding trigonometric series can be rewritten in the complex form

$$\sum_{k=-\infty}^{\infty}c_ke^{ikt}, \tag{3}$$

where in the case of (3) being the Fourier series of $f \in L^1(\mathbb{T})$, $\mathbb{T} = [-\pi, \pi)$, the numbers c_k are the Fourier coefficients of f:

$$c_k = \frac{1}{2\pi} \int_{-\pi}^{\pi} f(t) e^{-ikt} \, dt.$$

There is a common opinion that this study began in 1913, with the work of W.H. Young [45]. One can hardly doubt in this, just because earlier works have never been found nor mentioned. What we wish to do shortly after is to present more or less original statements of this paper. In fact, the paper is rather long, contains many results and mainly devoted to the study of bounded functions. The latter issue did not find many followers. We will formulate two assertions in this work, with a slight alteration of notation and use of today terminology.

Theorem 1 *If $a_0 \ge a_1 \ge \ldots$ form a monotone decreasing sequence with zero as limit, and their differences also form a monotone decreasing sequence, that is,*

$$a_0 - a_1 \ge a_1 - a_2 \ge \ldots,$$

then the trigonometric series (1) is the Fourier series of a positive summable function.

Theorem 2 *If $b_1 \ge b_2 \ge \ldots$ form a monotone decreasing sequence of constants with zero as limit and $\sum\limits_{k=1}^{\infty} \frac{b_k}{k}$ converges, then (2) is the Fourier series of a summable function which, for positive values of x, is bounded below, and, for negative values of x, is bounded above.*

Theorem 1 is considered as the first step in the theory. Of course, later authors equivalently formulated its main assumption as that the sequence $\{a_k\}$ is *convex*. Young himself gives a footnote that the assumption of monotonicity of $\{a_k\}$ is somewhat redundant. On the other hand, he mentions an important property of positivity of the sum of the series.

In a while, Kolmogorov [19] showed that the same is true for the so-called quasi-convex sequences $\{a_k\}$, namely, those satisfying

$$\sum_{k=0}^{\infty} (k+1) |\Delta^2 a_k| < \infty.$$

In 1934, Pfleger [32] proved that similar to the fact that every real sequence of bounded variation is a difference of two monotone sequences, every real quasi-convex sequence is a difference of two convex sequences.

Concerning Theorem 2, it is of interest for us that Young also dealt with sine series and understood that mere monotonicity of the coefficients is not enough for a sine series to be a Fourier series. It is not thus difficult to construct a sine series that is not a Fourier series. It is clear that in view of the Riemann-Lebesgue lemma, $\widehat{L^1}$ itself is a subspace of c_0, the space of *null sequences* (vanishing at infinity).

A much more challenging problem was to construct a cosine series with monotone coefficients that is not a Fourier series. This was done by Sidon [37] (see also [3, Volume I]). Taking into account that every sequence of bounded variation can be represented as a difference of two monotone sequences, this means that also the space of sequences of bounded variation is not a subspace of $\widehat{L^1}$. From that time, activity in the topic became more or less typical in the sense that people started from the null sequence of coefficients with *bounded variation*, written

$$bv = \left\{ d = \{d_k\} : \|d\|_{bv} = \sum_{k=0}^{\infty} |\Delta d_k| < \infty \right\},$$

where $\Delta d_k = d_k - d_{k+1}$, and searched for additional conditions providing the property of the trigonometric series to be a Fourier series. It is not our task to mention all the contributions, we are going to give or mention only the most notable among them. Before proceeding to concrete results, we observe one important and very attractive property of trigonometric series with coefficients of bounded variation, or just monotone. Such a cosine series converges for every $x \neq 0 (\mathrm{mod}\, 2\pi)$, while the sine series converges everywhere. This is not difficult to see by applying one very important and popular expedient in analysis, the so-called Abel transformation:

$$\sum_{k=m}^{n} a_k b_k = \sum_{k=m}^{n-1} A_m^k (b_k - b_{k+1}) + A_m^n b_n, \tag{4}$$

where $A_m^k = a_m + \ldots + a_k$. Indeed, applying (4) to (1) and (2), respectively, we have

$$\frac{a_0}{2} + \sum_{k=1}^{\infty} a_k \cos kx = \sum_{k=0}^{\infty} D_k^c(x) \Delta a_k \tag{5}$$

and

$$\sum_{k=1}^{\infty} b_k \sin kx = \sum_{k=1}^{\infty} D_k^s(x) \Delta b_k, \tag{6}$$

where D_k^c and D_k^s are the Dirichlet kernels, the sums of cosines and sines, respectively. By the boundedness of the kernel $D_k^c(x)$ for every $x \neq 0 (\mathrm{mod}\, 2\pi)$, and of the kernel $D_k^s(x)$ for all x, and by the fact that both sequences of coefficients are bv sequences, we conclude that both series converge absolutely, which yields the desired properties.

3 Trigonometric Series Versus Fourier Series: Continued

Despite of Sidon's discovery, trigonometric series with monotone coefficients continued to attract the interest of researchers. They possess many nice properties, and every classical source contains a chapter or section on this topic; see, e.g., [46, Chapter V]. Important additional details can be found in [31].

To the best of our knowledge, a convenient description of $\widehat{L^1}$ in terms of a given sequence alone did not exist till recently. Actually, there were some characterizations, e.g., [10, 33–36], but they were too complicated to be applied to concrete problems and they involved properties of functions to be restored from the given expansion. Hence, certain subspaces of $\widehat{L^1}$ have been studied so that they are both as wide as possible and described in terms convenient for applications. Because of convergence, a part of the problem concerning trigonometric series with coefficients of bounded variation is widely known, with slight abuse of terminology, as integrability of trigonometric series.

The first period of investigation ended, in a sense, in 1956 when Boas generalized all previous results in [7]. We do not reproduce his result but explain when formulating Telyakovskii's ones, which became the strongest known for a long time. In the following years, extensions of Boas' results were studied.

Let us give a corresponding list of sequence spaces all of which are subspaces of $\widehat{L^1}$ and ensure the integrability of the corresponding trigonometric series. These results were obtained in the second half of the twentieth century. This list does not pretend to be comprehensive. Though most of the strongest known conditions are in this list, the selection is partly a matter of taste. Here, by $\{d_k\}$ we mean either $\{a_k\}$ or $\{b_k\}$.

(1) The so-called Boas-Telyakovskii condition (see, e.g., [40] or [41]):

$$s_d = \sum_{n=2}^{\infty} \left| \sum_{k=1}^{[\frac{n}{2}]} \frac{\Delta d_{n-k} - \Delta d_{n+k}}{k} \right| < \infty, \tag{7}$$

where $[\alpha]$ means the integer part of the number α. This is a generalization of Boas' result in the way that in [7] the sign of absolute value in the representation for s_d was inside the second sum.

(2) Fomin's condition [11]:

$$\|d\|_{o_p} = \sum_{n=0}^{\infty} 2^{\frac{n}{p'}} \left\{ \sum_{k=2^n}^{2^{n+1}-1} |\Delta d_k|^p \right\}^{\frac{1}{p}} < \infty, \qquad 1 < p < \infty, \quad \frac{1}{p} + \frac{1}{p'} = 1. \tag{8}$$

(3) The Sidon-Telyakovskii condition [42]:
There exists a sequence $\{A_k\}$ such that

$$A_k \downarrow 0 \ (k \to \infty), \quad \sum_{k=0}^{\infty} A_k < \infty \qquad \text{and} \qquad |\Delta d_k| < A_k. \tag{9}$$

Here the space in **(3)** is a subspace of any sequence space in **(2)**. In turn, all of them are subspaces of the space in **(1)**. The most natural and effective choice of $\{A_n\}$ is

$$A_n = \max_{n \ge k} |\Delta d_k|.$$

(4) The Buntinas–Tanovic-Miller condition (see, e.g., [9]):

Let $\{k_n\}$ be an increasing sequence and $\{m_n\}$, $1 \le m_n \le k_{n+1}$, a non-decreasing sequence. Then $d \in hv^p$ if

$$\sum_{n=0}^{\infty} m_n^{\frac{1}{p'}} \left\{ \sum_{k=k_n}^{k_{n+1}-1} |\Delta d_k|^p \right\}^{\frac{1}{p}} + \sum_{n=0}^{\infty} \ln\left(\frac{k_{n+1}}{m_n}\right) \sum_{k=k_n}^{k_{n+1}-1} |\Delta d_k| < \infty. \tag{10}$$

If $k_n = 2^n$ and $m_n = 2^{n+1}$, we get o_p. If $m_n = 1$, then (10) reduces to

$$\sum_{n=1}^{\infty} |\Delta d_n| \ln n < \infty.$$

Buntinas and Tanovic-Miller also introduced a scale of HV^p spaces each of which is a linearization of hv^p.

(5) More recent is an amalgam type space (see [2, 9, 24]), in which the condition for a sequence d is

$$\sum_{n=0}^{\infty} \left\{ \sum_{m=1}^{\infty} \left[\sum_{k=m2^n}^{(m+1)2^n-1} |\Delta d_k| \right]^2 \right\}^{\frac{1}{2}} < \infty, \tag{11}$$

was involved in these problems; for further analysis see [12], while for functions, analogous spaces will be discussed in the next section.

The traditional way to prove such conditions (or to put these sequence spaces into play) is the same application of Abel's transformation as above. This is not a surprise if we observe that all the conditions are given in terms of the differences of the coefficients (again, if one understands $\{d_k\}$ as one of the sequences of coefficients, either $\{a_k\}$ or $\{b_k\}$. More precisely, the absolute values of the corresponding series are integrated over $(0, \pi)$ or $(\frac{\pi}{N}, \pi)$, with appropriate bounds in terms of the coefficients. Such estimates are traditionally called *Sidon inequalities*. This name goes back to one of such inequalities obtained by Sidon [38]:

$$\frac{1}{N+1} \left\| \sum_{k=0}^{N} c_k D_k \right\|_{L^1(\mathbb{T})} \le \max_{0 \le k \le N} |c_k|,$$

where c_k are arbitrary numbers. For generalizations and applications, see, e.g., a survey by Fridli [12].

Such was a state of art in the problem in question at the beginning of the 1990s.

4 A Fourier Transform Approach

In this section, we return to the topic of trigonometric series not because of different results (though some of them, very recent, will be given) but more because of different methods. This became possible after the introduction of new notions and terminology. To start with, recall the so-called Boas-Telyakovskii condition (7). It can be replaced by the assumption that the sequence of differences of the coefficients belongs to the discrete Hardy space h_o. Without going into details of the discrete Hardy space h and its subspaces h_e and h_o for even and odd sequences respectively, as well as for the non-periodic prototypes H^1, H_e^1 and H_o^1 (their definitions and properties appropriate for the problems in question are collected in [26] and [27]), we mention that it turns out to be not only intrinsic language for important results in the topic but a way to indicate what kind of tools should be involved in this study.

Instead of the traditional way to prove such results by means of Sidon type inequalities mentioned above, a new approach to these problems was introduced in [21]. The idea was to switch the study from the periodic object trigonometric series to their non-periodic analog the Fourier transform. As in the case of trigonometric series, an initial result in this topic was also concerned with convexity. It was due to Trigub (see, e.g., [44, 6.4.7]) and gave a general formula, sometimes of asymptotic nature, for the Fourier transform of a convex function.

In general, let us consider a locally absolutely continuous function f on $(0, \infty)$ such that

$$\lim_{x\to\infty} f(x) = 0$$

and $f' \in X$, where X is a subspace of the space of integrable functions. For example, the derivative of a function of bounded variation, locally absolutely continuous on $(0, \infty)$ and vanishing at infinity is Lebesgue integrable; we can denote

$$\|f\|_{BV} = \int_0^\infty |f'(x)|\,dx < \infty.$$

Various spaces X are considered in detail in [26] (see also [16] and [27]); roughly speaking, most of them are non-periodic analogs of the sequence spaces considered earlier, in the sense that the sequences are replaced by functions and the sums by integrals. One of these spaces is the real Hardy space $H^1(\mathbb{R})$. We describe how it comes into play in a natural way. For the given class of functions, we integrate by parts in the formula for the Fourier transform

$$\widehat{f}(x) = \int_{\mathbb{R}} f(t)e^{-ixt}\,dt, \tag{12}$$

understood in the improper sense. What we arrive at instead is

$$\frac{\widehat{f'}(x)}{ix}. \tag{13}$$

Integrating $|\widehat{f}|$ over $\mathbb{R}$ is the same as integrating over $\mathbb{R}$ the absolute value of (13). This leads to a single necessary and sufficient condition for the integrability of the Fourier transform of functions in the considered class. Following [17], we introduce the space

$$Q = \{g \in L^1(\mathbb{R}) : \int_{\mathbb{R}} \frac{|\widehat{g}(x)|}{|x|}\,dx < \infty\},$$

a proper subspace of $L^1(\mathbb{R})$, or, more precisely, of $L^1_0(\mathbb{R})$—the space of $L^1(\mathbb{R})$ functions with mean zero; we note that frequently subspaces of this space are used when only even or odd functions are considered: Q_e and Q_o, respectively. The mentioned criterion is given by

Theorem 3 *Let f be absolutely continuous on $\mathbb{R}$ and vanishing at infinity. Then $\widehat{f} \in L^1(\mathbb{R})$ if and only if $f' \in Q$.*

In fact, the result itself is just integration by parts, while numerous delicate particular conditions, sometimes sophisticated, can be found in [26]. What makes it essential is that Q is an interesting space, wider than all the other spaces of integrability of the Fourier transform of a function of bounded variation. As mentioned, for f with bounded variation, its derivative f' exists almost everywhere and is Lebesgue integrable on $\mathbb{R}$.

The Hilbert transform of a function $g \in L^1(\mathbb{R})$ is

$$\mathcal{H}g(x) = \frac{1}{\pi}\int_{\mathbb{R}} \frac{g(t)}{x-t}\,dt, \tag{14}$$

where the integral is understood in the improper (Principal Value) sense, as

$$\mathcal{H}g(x) = \lim_{\delta\to 0+} \mathcal{H}_\delta g(x),$$

with

$$\mathcal{H}_\delta g(x) = \frac{1}{\pi}\int_{|t-x|>\delta} \frac{g(t)}{x-t}\,dt$$

being the *truncated Hilbert transform*.

If g is integrable, its Hilbert transform exists almost everywhere but is not necessarily integrable. Moreover, it can be even not locally integrable. When the Hilbert transform is integrable, we say that g is in *the (real) Hardy space* $H^1 := H^1(\mathbb{R})$. Applying the well-known Hardy inequality (see, e.g., [14, (7.24)]; in [26] the author suggests to call it the Fourier-Hardy inequality)

$$\int_{\mathbb{R}} \frac{|\widehat{g}(x)|}{|x|}\,dx \lesssim \|g\|_{H^1(\mathbb{R})} \tag{15}$$

to (13), we derive the following strong condition for the integrability of the Fourier transform.

Theorem 4 *If a function f is locally absolutely continuous on $\mathbb{R}\setminus\{0\}$,*

$$\lim_{|t|\to\infty} f(t) = 0,$$

and $f' \in H^1(\mathbb{R})$, then the Fourier transform $\widehat{f}$ is Lebesgue integrable on $\mathbb{R}$.

More sophisticated results, including asymptotic ones, are collected, as mentioned, in [26]; see also references therein. It is of considerable interest that in Theorem 4 there is no need to assume that f is of bounded variation. Indeed, what we need the boundedness of variation for is the differentiability of the derivative. But here we get this from the assumption $f' \in H^1(\mathbb{R})$.

Let us come nearer to the Fourier transform approach. In order to present an example of a theorem on the behavior of the Fourier transform, we consider $\widehat{f_c}$ and $\widehat{f_s}$ on the half-axis in place of $\widehat{f}$ on $\mathbb{R}$, where

$$\widehat{f_c}(x) = \int_0^\infty f(t)\cos xt\,dt \tag{16}$$

is the cosine Fourier transform and

$$\widehat{f_s}(x) = \int_0^\infty f(t)\sin xt\,dt \tag{17}$$

is the sine Fourier transform. The right-hand sides of (16) and (17), with f in question, are natural analogs of trigonometric series (1) and (2), with bv coefficients, respectively.

Theorem 5 *Let f be as above. Then for any $y > 0$ we have*

$$\widehat{f_c}(y) = \theta\gamma_1(y),$$

and

$$\widehat{f_s}(y) = \frac{1}{y} f\left(\frac{\pi}{2y}\right) + \theta\gamma_2(y),$$

where $|\theta| \le C$ *and*

$$\int_0^\infty |\gamma_j(y)|\,dy \le \|f'\|_X, \qquad j = 1, 2.$$

With this or similar theorems in hand (again, they can be found first of all in [26]; see also [13] and [21]), we are now able to show a way how to strengthen the known integrability results for trigonometric series. Given cosine and sine series (1) and (2), with the null sequence of coefficients $\{a_k\}$ and $\{b_k\}$ such that their difference sequences $\{\Delta a_k\}$ and $\{\Delta b_k\}$, where

$$\Delta a_k = a_k - a_{k+1} \qquad \text{and} \qquad \Delta b_k = b_k - b_{k+1},$$

are in a sequence space χ (acting in the same sense for sequences as X above for functions). A "bridge" for passing from Fourier integrals to trigonometric series (and vice versa) is now in order. In the first one by Belinsky, in [4], the function f was assumed to be of bounded variation (and of compact support which is an unnecessary restriction). Further results are due to Trigub, see, e.g., [44, Theorem 4.1.2]. For the function f of bounded total variation and vanishing at infinity, it reads as follows:

$$\sup_{|x|\le\pi} \left| \int_{\mathbb{R}} f(t)e^{ixt}dt - \sum_{k\in\mathbb{Z}} f(k)e^{ixk} \right| \lesssim \|f\|_{BV}. \tag{18}$$

With this in hand, we set for $x \in [k, k+1]$,

$$a(x) = a_k + (k - x)\Delta a_k, \qquad a_0 = 0, \tag{19}$$

$$b(x) = b_k + (k - x)\Delta b_k. \tag{20}$$

So, we construct a corresponding function by means of linear interpolation of the sequence of coefficients.

Starting with assumptions on the coefficients as belonging to χ and then transferring its properties to a or b as belonging to X, we arrive at

Theorem 6 *For each* y*, where* $0 < y \le \pi$*, we have*

$$\sum_{k=1}^\infty a_k \cos ky = \theta\gamma_1(y)$$

and

$$\sum_{k=1}^\infty b_k \sin ky = \frac{1}{y} b\Big(\frac{\pi}{2y}\Big) + \theta\gamma_2(y),$$

where $|\theta| \le C$ *and*

$$\int_0^{\pi} |\gamma_j(y)|\, dy \le \|\{\Delta a_k\}\|_X$$

for the first relation ($j = 1$) *and similarly for* b ($j = 2$).

Making this theoretical scheme more realistic, in place of abstract X we discuss particular but more practical spaces than Q (or Q_e and Q_o) which are involved in the study of the integrability of the Fourier transform of a function of bounded variation.

One of them is the scale of subspaces of Q (more precisely, of Q_e and Q_o) proved to be convenient in applications.

Definition 1 For $1 < q < \infty$, the space $O_q := O_q(\mathbb{R}_+)$ is the space of functions g with finite norm

$$\|g\|_{O_q} = \int_0^{\infty} \left(\frac{1}{x}\int_x^{2x} |g(t)|^q dt\right)^{\frac{1}{q}} dx. \tag{21}$$

All these spaces and their sequence analogs first appeared in the paper by D. Borwein [8], but became - for sequences - widely known after the above mentioned paper by G. A. Fomin [11].

Further, for $q = \infty$, we define the corresponding space as follows.

Definition 2 The space O_∞ is the space of functions g with finite norm

$$\|g\|_{O_\infty} = \int_0^{\infty} \operatorname*{ess\,sup}_{x \le t \le 2x} |g(t)|\, dx. \tag{22}$$

The role of the integrable monotone majorant for problems of almost everywhere convergence of singular integrals is known from the work of D.K. Faddeev (see, e.g., [1, Ch. IV, §4]; also [39, Ch. I]); for spectral synthesis problems it was used by Beurling [6], for more details see [5].

In parallel to O_q, we can introduce the spaces E_q as

$$E_q := E_q(\mathbb{R}_+) = \left\{g : g \in O_q \quad \text{and} \quad \int_0^{\infty} \frac{1}{x}\left|\int_0^x g(t)\, dt\right| dx < \infty\right\};$$

cf. [26, Ch. 2, Section 2.4] or [27, Ch. 6, §6.4]. Naturally, these spaces are endowed with the norm

$$\|g\|_{E_q} = \|g\|_{O_q} + \int_0^{\infty} \frac{1}{x}\left|\int_0^x g(t)\, dt\right| dx.$$

If we, as usual, consider the affiliation of f' to one of these spaces, the last integral turns into

$$\int_0^\infty \frac{|f(t)|}{t}\,dt. \tag{23}$$

A more general case is that where f' belongs to the real Hardy space $H^1(\mathbb{R})$ (or its even or odd subspaces H_e^1 and H_o^1 naturally considered on $\mathbb{R}_+$). Theorems 4 and 5 above are examples of such results.

There are certain interrelations between these spaces. For example,

$$O_\infty \hookrightarrow O_{p_1} \hookrightarrow O_{p_2} \hookrightarrow H_o^1 \hookrightarrow Q_o \hookrightarrow L^1 \ (p_1 > p_2 > 1), \tag{24}$$

which can be found in [21] or [16, Ch. 3]. Similar relations are true in the case of Q_e and its subspaces.

One more chain of subspaces of Q, also wider than the above O_q but incomparable with H^1, proved to be of importance in the considered problems (see [24] or [16]). We start with the definition of these classes.

Definition 3 A locally integrable function g defined on $\mathbb{R}_+$ belongs to $A_{1,p}$ if

$$\|g\|_{A_{1,p}} = \sum_{m=-\infty}^{\infty} \left\{ \sum_{j=1}^{\infty} \left[\int_{j2^m}^{(j+1)2^m} |g(t)|\,dt \right]^p \right\}^{\frac{1}{p}} dx < \infty. \tag{25}$$

Such a space is *amalgam* in nature, since each of the m-th summands is the norm in the Wiener amalgam space $W(L^1, \ell^p)$ for functions

$$G_m(t) = \begin{cases} 2^m g(2^m t), & if \quad t \ge 1, \\ 0, & \text{otherwise}, \end{cases}$$

where ℓ^p, $1 \le p < \infty$, is a space of sequences $\{d_j\}$ endowed with the norm

$$\|\{d_j\}\|_{\ell^p} = \left(\sum_{j=-\infty}^{\infty} |d_j|^p \right)^{\frac{1}{p}}. \tag{26}$$

The norm of a function $h : \mathbb{R} \to \mathbb{C}$ in the amalgam space $W(L^1, \ell^p)$ is taken as

$$\left\| \left\{ \int_j^{j+1} |h(t)|\,dt \right\} \right\|_{\ell^p}.$$

We only mention that the greater is p the wider is $A_{1,p}$. Therefore, Theorem 5 is the most general for $X = A_{1,2}$ rather than with $X = A_{1,p}$, $1 < p < 2$.

Recently [25] (see also [27]), a counterpart of Theorem 5 has been obtained, in the sense that instead of H_o^1 as X the space H_e^1 came into play. This resulted in an asymptotic formula for $\widehat{f_c}$ rather than for $\widehat{f_s}$. This result was completely new, that is, never obtained before directly for sequence analogs. The corresponding theorem for sequences is a counterpart of Theorem 6 in the case where in the latter h_o is taken as χ. The discrete Hardy space h can be defined in the same way as H^1 above. For the sequence $d = \{d_k\} \in \ell^1$, the discrete Hilbert transform is defined for $m \in \mathbb{Z}$ as (see, e.g., [18, (13.127)])

$$\hbar d(m) = \sum_{\substack{k=-\infty \\ k \neq m}}^{\infty} \frac{d_k}{m-k}. \tag{27}$$

Correspondingly, h is the space of summable sequences with summable discrete Hilbert transform. The cases and notation for even and odd sequences are obvious. So, the mentioned new result, with the space h_e taken instead of h_o reads as follows (see also [26, Chapter 8, Theorem 8.8]).

Theorem 7 *If the coefficients $\{a_k\}$ in (1) and $\{b_k\}$ in (2) tend to* 0 *as $k \to \infty$, and the sequences $\{\Delta a_k\}$ and $\{\Delta b_k\}$ are in h_e, then (2) represents an integrable function on $[0, \pi]$, and, for $N \le \frac{\pi}{2x} < N+1$,*

$$\sum_{k=1}^{\infty} a_k \cos kx = \frac{A}{x} a\left(\frac{\pi}{2x}\right) + \frac{\pi}{2x} \sum_{k=1}^{N-1} \Delta a_k \ln \frac{e\pi}{2x(k+1)(1+\frac{1}{k})^k} + G(x),$$

where

$$A = \frac{2}{\pi}\left(\int_0^{\frac{\pi}{2}} \frac{1-\cos t}{t}\, dt - \int_{\frac{\pi}{2}}^{\infty} \frac{\cos t}{t}\, dt\right)$$

and

$$\int_0^{\pi} |G(x)|\, dx \lesssim \|\{\Delta a_k\}\|_{h_e}.$$

In fact, all the above results for trigonometric series, except that by Buntinas and Tanovic-Miller (merely not studied in this respect yet), are easily derived from the relevant results for the Fourier transforms.

5 Wiener Algebras Come to Play in Full Extent

The above approach does not cover all the options because of the reduction to functions and sequences of bounded variation or those related to them. One may imagine the following hypothetic situation: the Fourier coefficients of the integrable function in the above mentioned Kolmogorov's counterexample are given as the sequence of coefficients of a trigonometric series. How can one even try to figure out whether this series is a Fourier series or not? We know from [20] that neither bounded variation nor convergence can be expected.

Recently, mentioned restrictions has been removed (or, more precisely, the range of the Fourier transform approach has been extended) by finding in [29] an optimal necessary and sufficient condition for a trigonometric series to be a Fourier series. In fact, the same can be found in an earlier work [15]; the reason that this and similar results in [15] have gone unnoticed by the researchers of trigonometric series for a long time is probably due to the absence of applications in that paper. The authors of [29] succeeded to make out this fact at the very last moment before publication. Luckily, this did not destroy [29] since there were many other results of this kind there formulated and proven in a more general way as well as many applications were given completely absent in [15]. In the comprehensive case, there is less reason to distinguish between the cosine and sine series, and the general complex form (3) is more appropriate. Similarly to above, the answer is given in terms of a single, quite simple function ℓ_c, piecewise linear, continuous and such that $\ell_c(k) = c_k$, $k \in \mathbb{Z}$. We follow Lemma D in [15]. Let

$$\Delta(x) = \begin{cases} 1 - |x|, & |x| \le 1, \\ 0, & |x| > 1. \end{cases}$$

Lemma 1 *Let $\{c_k\}$ be a sequence of complex numbers and let*

$$\ell_c(x) = \sum_{k=-\infty}^{\infty} c_k \Delta(x-k), \qquad x \in \mathbb{R}. \tag{28}$$

Then $\ell_c(k) = c_k$, $k \in \mathbb{Z}$, and ℓ_c is linear on each interval $[k, k+1]$.

In other words, this lemma says that the function whose graph consists of the line segments joining (k, c_k) to $(k+1, c_{k+1})$ for $k = 0, \pm 1, \pm 2, \dots$ is given by the series (28).

The answer in [29] is given in terms of the Wiener algebra

$$W_0 = W_0(\mathbb{R}) := \left\{ f(x) : \, f(x) = \int_{\mathbb{R}} e^{-ixy} g(y)\, dy, \; g \in L^1(\mathbb{R}) \right\}.$$

Criterion *For series (3) to be a Fourier series, it is necessary and sufficient that* $\ell_c \in W_0(\mathbb{R})$.

It is hardly believable that a necessary and sufficient condition closer to the initial form of the series (that is, in terms of $\{c_k\}$ only) may exist. This criterion results in two ways for obtaining new and known conditions for the trigonometric series to be a Fourier series. One of them is to search for specific conditions for the function ℓ_c to be in W_0. Such an approach leaves a chance for discoveries but a general feeling is that in this case any calculations will be equivalent to those with the sequence $\{c_k\}$, just because of the structure of ℓ_c. More promising seems to be the application of general conditions for belonging to W_0 to the concrete function ℓ_c. In any case, in [29] completely new results not related to bv immediately appeared as consequences of earlier obtained results for functions to belong to W_0. For example, one of them comes as a consequence of the main result in [23].

Theorem 8 *If there is* $p \in (0, 2]$ *such that* $\{c_k\} \in l^p$, *then (3) is the Fourier series of an* L^2 *function. If* $\{c_k\} \in l^p$ *only for some* $p \in (2, +\infty)$, *then assuming* $\{c_k - c_{k+1}\} \in l^q$, *with* $q \in \left(0, 1 + \frac{1}{p-1}\right)$, *we get that (3) is a Fourier series.*

Recently, an important step towards filling the gap of integrability spaces between the discrete Hardy space and l^1 has been undertaken in [30].

6 Multidimensional Case

Naturally, people tried to extend the main results in the topic to the multivariate setting. On the one hand, certain of earlier attempts are surveyed in [22, Section 9]. More recent results can be found in [21] and in more detail in [26, Part II]. We see no reason to reproduce them here. Instead, we concentrate ourselves on the recent extension of the results of Sect. 5, or more precisely, of those in [29], due to Trigub in [43]. Again, as in [29], in [43] more spaces of trigonometric series and corresponding Banach algebras are studied; we consider here only those within the scope in question.

As in dimension one, we consider trigonometric series of the form

$$\sum_{k\in\mathbb{Z}^n} c_k e^{ikx}, \tag{29}$$

where $kx = k_1x_1 + \ldots + k_nx_n$ and we are interested in conditions under which this series is the Fourier series of an integrable function. On the basis of the coefficients $\{c_k\}$, a multivariate analog of the piece-wise linear function ℓ_c is constructed as follows. Keeping the same notation, we have

$$\ell_c = \sum_{k\in\mathbb{Z}^n} c_k \prod_{j=1}^{n} (1 - |x_j - k_j|)_+,$$

where the subscript + means that the function takes the value zero each time when the value in parentheses becomes negative. Finally, the Wiener algebra is defined as

$$W_0 = W_0(\mathbb{R}^n) = \left\{ f : f(x) = \int_{\mathbb{R}^n} g(y)e^{-ixy}\,dy, \quad g \in L^1(\mathbb{R}^n) \right\}.$$

Theorem 9 *For the series (29) to be the Fourier series of an integrable function it is necessary and sufficient that $\ell_c \in W_0(\mathbb{R}^n)$.*

Since most of the results given in the comprehensive survey paper [28] are of the multidimensional form, being applied to the piece-wise linear function they would lead to known and new assertions for multivariate trigonometric series. Hopefully, such applications will appear in the very near future.

References

1. G. Alexits, *Convergence problems of orthogonal series*, Akadémiai Kiadó, Budapest, 1961.
2. B. Aubertin and J. J. F. Fournier, *Integrability theorems for trigonometric series*, Studia Math. **107** (1993), 33–59.
3. N. K. Bary, *A Treatise on Trigonometric Series, I and II*, MacMillan, New York, 1964.
4. E. S. Belinsky, *On asymptotic behavior of integral norms of trigonometric polynomials*, Metric Questions of the Theory of Functions and Mappings, Nauk. Dumka, Kiev **6** (1975), 15–24 (Russian).
5. E. Belinsky, E. Liflyand and R. Trigub, *The Banach algebra A* and its properties*, J. Fourier Anal. Appl. **3** (1997), 103–129.
6. A. Beurling, *On the spectral synthesis of bounded functions*, Acta Math. **81** (1949), 225–238.
7. R. P. Boas, *Absolute convergence and integrability of trigonometric series*, J. Rational Mech. Anal. **5** (1956), 621–632.
8. D. Borwein, *Linear functionals connected with strong Cesáro summability*, J. London Math. Soc. **40** (1965), 628–634.
9. M. Buntinas and N. Tanović-Miller, *New integrability and L^1-convergence classes for even trigonometric series II*, Approximation Theory (J. Szabados and K Tandori, eds.), Colloq. Math. Soc. János Bolyai, North-Holland, Amsterdam **58** (1991), 103–125.
10. R. E. Edwards, *Criteria for Fourier transforms*, J. Australian Math. Soc. **7** (1967), 239–246.
11. G. A. Fomin, *A Class of Trigonometric Series*, Matem. Zametki **23** (1978), 213–222 (Russian). - English transl. in Math. Notes **23** (1978), 117–123.
12. S. Fridli, *Integrability and L^1-convergence of trigonometric and Walsh series*, Annales Univ. Sci. Budapest, Sect. Comp. **16** (1996), 149–172.
13. S. Fridli, *Hardy spaces generated by an integrability condition*, J. Approx. Theory **113** (2001), 91–109.
14. J. Garcia-Cuerva and J.L. Rubio de Francia, *Weighted Norm Inequalities and Related Topics*, North-Holland, 1985.
15. R.R. Goldberg, *Restrictions of Fourier Transforms and Extension of Fourier Sequences*, J. Approx. Theory **3** (1970), 149–155.

16. A. Iosevich and E. Liflyand, *Decay of the Fourier transform: analytic and geometric aspects*, Birkhäuser, 2014.
17. R.L. Johnson and C.R. Warner, *The convolution algebra* $H^1(R)$, J. Function Spaces Appl. **8** (2010), 167–179.
18. F.W. King, *Hilbert transforms, Vol.1 and 2*, Enc. Math. Appl., Cambridge Univ. Press, Cambridge, 2009.
19. A. N. Kolmogorov, *Sur l'ordre de grandeur des coefficients de la série de Fourier-Lebesgue*, Bulletin de l'Academie Polonaise, (1923), 83–86.
20. A. N. Kolmogorov, *Une série de Fourier-Lebesgue divergente partout*, C. R. Acad. Sci. Paris **183** (1926), 1327–1328.
21. E. Liflyand, *Fourier transforms of functions from certain classes*, Anal. Math. **19** (1993), 151–168.
22. E. Liflyand, *Lebesgue constants of multiple Fourier series*, Online J. Anal. Combin. **1** (2006), 112 p.
23. E. Liflyand, *On absolute convergence of Fourier integrals*, Real Anal. Exchange **36** (2010/11), 353–360.
24. E. Liflyand, *Fourier transforms on an amalgam type space*, Monatsh. Math. **172** (2013), 345–355.
25. E. Liflyand, *The Fourier transform of a function of bounded variation: symmetry and asymmetry*, J. Fourier Anal. Appl. **24** (2018), 525–544.
26. E. Liflyand, *Functions of Bounded Variation and their Fourier Transforms*, Birkhäuser, 2019.
27. E. Liflyand, *Harmonic Analysis on the Real Line: a Path in the Theory*, Birkhäuser, 2021.
28. E. Liflyand, S. Samko and R. Trigub, *The Wiener algebra of absolutely convergent Fourier integrals: an overview*, Anal. Math. Physics **2** (2012), 1–68.
29. E. Liflyand, R. Trigub, *Wiener algebras and trigonometric series in a coordinated fashion*, Constructive Approx. **54** (2021).
30. E. Liflyand, *A scale of spaces of functions with integrable Fourier transform*, Anal. Math. Phys. **15**(3), Paper No.63, 13 p. (2025).
31. B.M. Makarov, M.G. Goluzina, A.A. Lodkin, and A.N. Podkorytov, *Problèmes d'Analyse Reéelle*, Cassini, Paris, 2010. – French transl. of 2nd Russian edition, Nevskii Dialekt & BHV-Peterburg, St. Petersburg, 2004.
32. J. Pfleger, *Über Reihen mit r-fach monoton abnehmenden Gliedern*, Monatsh. Math. Phys. **41** (1934), 191–200.
33. P. G. Rooney, *On the representation of sequences as Fourier coefficients*, Proc. Amer. Math. Soc. **11** (1960), 762–768.
34. R. Ryan, *Fourier transforms of certain classes of integrable functions*, Trans. Amer. Math. Soc. **105** (1962), 102–111.
35. R. Salem, *Sur les transformations des séries de Fourier*, Fund. Math. **33** (1939), 108–114.
36. R. Salem, *Essais sur les séries trigonometriques*, Actual. Sci. et Industr., Vol 862, 1940, Paris.
37. S. Sidon, *Reihentheoretische Sätze und ihre Anwendungen in der Theorie der Fourierischen Reihen*, Math. Zeitschrift **10** (1921), 121–127.
38. S. Sidon, *Hinreichende Bedingungen für den Fourier-Charakter einer trigonometrischen Reihe*, J. London Math. Soc. **14** (1939), 158–160.
39. E. M. Stein and G. Weiss, *Introduction to Fourier analysis on Euclidean spaces*, Princeton Univ. Press, Princeton, N. J., 1971.
40. S.A. Telyakovskiĭ, *Integrability conditions for trigonometrical series and their application to the study of linear summation methods of Fourier series*, Izv. Akad. Nauk SSSR Ser. Mat. **28** (1964), 1209–1236 (Russian).
41. S. A. Telyakovskii, *An estimate, useful in problems of approximation theory, of the norm of a function by means of its Fourier coefficients*, Trudy Mat. Inst. Steklova **109** (1971), 65–97 (Russian). - English transl. in Proc. Steklov Math. Inst. **109** (1971), 73–109.
42. S. A. Telyakovskii, *Concerning a sufficient condition of Sidon for the integrability of trigonometric series*, Mat. Zametki **14** (1973), 317–328 (Russian). - English transl. in Math. Notes **14** (1973), 742–748.

43. R.M. Trigub, *On Fourier Series on the Torus and Fourier Transforms*, Mat. Zametki **110** (2021), 766–772 (Russian). – English transl. in Math. Notes **110** (2021), 767–772.
44. R.M.Trigub, E.S. Belinsky, *Fourier Analysis and Appoximation of Functions*, Kluwer, 2004.
45. W. H. Young, *On the Fourier series of bounded functions*, Proc. London Math. Soc.(2) **12** (1913), 41–70.
46. A. Zygmund, *Trigonometric series, Vol. I, II*, Cambridge Univ. Press, Cambridge, U.K., 1968.

Variational Conditions on Extremal Orthonormal Polynomials for Restricted Measures

Doron S. Lubinsky

Dedicated to the 80th birthday of Ed Saff.

1 Results

Let μ be a finite positive Borel measure on the real line with infinitely many points in its support, and all finite moments

$$\int t^j d\mu(t), \; j = 0, 1, 2, \ldots .$$

Then we may define orthonormal polynomials

$$p_n(\mu, x) = \gamma_n(\mu) x^n + \ldots, \gamma_n(\mu) > 0,$$

$n = 0, 1, 2, \ldots$ satisfying the orthonormality conditions

$$\int p_n(\mu, x) p_m(\mu, x) d\mu(x) = \delta_{mn}.$$

The nth reproducing kernel is

$$K_n(\mu, x, t) = \sum_{j=0}^{n-1} p_j(\mu, x) p_j(\mu, t).$$

D. S. Lubinsky (✉)
School of Mathematics, Georgia Institute of Technology, Atlanta, GA, USA
e-mail: lubinsky@math.gatech.edu

D. Bilyk et al. (eds.), *Recent Advances in Approximation and Potential Theory*, Applied and Numerical Harmonic Analysis,
https://doi.org/10.1007/978-3-031-95551-8_15

A central problem in the theory of orthonormal polynomials is to establish bounds on $p_n(\mu, x)$. Steklov conjectured in 1920 that if μ has support $[-1, 1]$, and is absolutely continuous there, while μ' is bounded below by a positive constant, then $\{p_n(\mu, x)\}_{n\geq 0}$ is uniformly bounded in compact subsets of $(-1, 1)$. Rakhmanov showed in 1979, 1981 [15] and [16] that Steklov's conjecture is false, and in fact one can find a measure satisfying its hypotheses such that for infinitely many n,

$$|p_n(\mu, 0)| \geq Cn^{1/2}/(\log n)^{3/2+\varepsilon}, \tag{1}$$

any $\varepsilon > 0$. Rakhmanov's weight μ' was unbounded. Ambroladze [2] subsequently constructed continuous weights on the unit circle that satisfy a weak Dini condition, for which the corresponding sequence of orthonormal polynomials can be unbounded. There have been further major developments in recent years, due to Aptekarev et al. [1, 3, 4, 7–11]. The relationship between spacing of zeros of successive orthogonal polynomials and bounds has been explored in [13].

There are positive results that establish boundedness, at least away from the endpoints of the support, for fairly general weights. Probably the most general result for measures supported on $[-1, 1]$ is due to Badkov [5]. He proved that if μ satisfies Szegő's condition and μ' satisfies a local Dini-Lipschitz condition, then the corresponding orthonormal polynomials are uniformly bounded in compact subsets of (a, b). This was a consequence of deeper pointwise asymptotics.

One might hope that when a measure is suitably restricted in some subinterval, one can say something about bounds in that subinterval. A recent result of the author [14] deals with this question. It involves $p_n(S^2\nu, y)$, the orthonormal polynomial for the measure $S^2\nu$, where S is a real polynomial.

Theorem 1 *Let ν be a positive measure on $[-1, 1]$, with infinitely many points in its support. Let $\mathcal{K}$ be a closed subset of $\mathbb{R}$ containing $(-1, 1)$. Let $y \in \mathbb{R}$ and $n \geq 1$. Then*

$$\sup\left\{p_n^2(\mu, y) : \mu_{|(-1,1)} = \nu \text{ and } \mu \geq \nu \text{ and } supp[\mu] \subseteq \mathcal{K}\right\}$$
$$= \sup S_J^2(y)\, p_{n-J}^2\left(S_J^2\nu, y\right),$$

where the supremum is taken over all $0 \leq J \leq n$ and monic polynomials S_J of degree J with distinct zeros in $\mathcal{K}\backslash(-1, 1)$.

In the case where $\mathcal{K}$ is an interval containing $[-1, 1]$ in its interior, it was shown in [14] that the sup can grow like a power of n. We note that it was only implied in the proofs there that $\mu \geq \nu$, not explicitly stated. In the case where ν has no masspoints at $-1, 1$, this follows from $\mu_{|(-1,1)} = \nu$. However, if μ has masspoints at ± 1, then it does not follow. It should have been stated explicitly there.

In this paper, we consider the case where the total mass of μ is bounded and in addition, supp$[\mu]$ is contained in an interval containing $[-1, 1]$. Thus let $\Lambda > 0$ and $I = [c, d]$ be an interval containing $[-1, 1]$. Let

$$\mathcal{M}(I,\Lambda)=\left\{\mu:\mu_{|(-1,1)}=\nu;\text{supp}[\mu]\subset I;\mu(I\backslash[-1,1])\leq\Lambda\right\}.$$

For a given real x, let

$$\phi_n(I,\Lambda,x)=\sup\{|p_n(\mu,x)|:\mu\in\mathcal{M}(I,\Lambda)\}.$$

We shall derive necessary conditions for measures μ that attain the sup, via variational tools. Any such measure will be called an extremal measure.

Theorem 2 *Let ν be a measure with support in $[-1,1]$ with infinitely many points in its support, and such that ± 1 are not masspoints. Let $d>1$, $c<-1$, $I=[c,d]$, $\Lambda>0$. Fix $x\in\mathbb{R}$.*

(a) There exists a measure $\mu_I\in\mathcal{M}(I,\Lambda)$ such that

$$|p_n(\mu_I,x)|=\phi_n(I,\Lambda,x). \tag{2}$$

(b) Write this measure in the form $\mu_I=\nu+\rho_I$, and let

$$R_n(\mu_I,t)=\frac{p_n(\mu_I,t)}{p_n(\mu_I,x)}\left[K_{n+1}(\mu_I,x,t)+K_n(\mu_I,x,t)\right]. \tag{3}$$

Then

$$\int R_n(\mu_I,t)\,d(\omega-\rho_I)(t)\geq 0 \tag{4}$$

for every nonnegative measure ω that has support in $I\backslash(-1,1)$ and total mass $\leq\Lambda$.

(c) Let

$$\eta=\inf\{R_n(\mu_I,t):t\in I\backslash(-1,1)\}.$$

For each t_0 in the support of ρ_I,

$$R_n(\mu_I,t_0)=\eta\leq 0. \tag{5}$$

(d) The support of ρ_I is purely discrete and consists of at most $n+1$ mass points.
(e) If ρ_I has total mass less than Λ, then supp$[\rho_I]\subset\{-1,1\}$ and $R_n(\mu_I,\cdot)=0$ in supp$[\rho_I]$.

It is not clear if μ_I is unique. We next investigate how $\phi_n(I,\Lambda,x)$ varies as the right endpoint y of $I=[c,y]$ increases. Let us set

$$\psi_n(y)=\ln\frac{1}{\phi_n([c,y],\Lambda,x)^2},\; y>1, \tag{6}$$

for fixed c, Λ and x.

Theorem 3 *Let ν be a measure with support in $[-1, 1]$ with infinitely many points in its support, and such that ± 1 are not masspoints. Let $c < -1$, $\Lambda > 0$, and fix $x \in \mathbb{R}$. Then ψ_n is absolutely continuous in $(1, \infty)$, ψ_n' exists for a.e. $y > 1$, and at every point where the derivative exists,*

$$\psi_n'(y) = \mu_{[c,y]}(\{y\}) R_n'(\mu_{[c,y]}, y). \tag{7}$$

Here $\mu_{[c,y]}$ is any extremal measure for $[c, y]$.

We next investigate how $\phi_n(I, \Lambda, x)$ varies as the total mass Λ increases. Let

$$\chi_n(\Lambda) = \ln \frac{1}{\phi_n([c, d], \Lambda, x)^2}, \ \Lambda > 0,$$

for fixed c, d and x, and study how this changes as Λ increases.

Theorem 4 *Let ν be a measure with support in $[-1, 1]$ with infinitely many points in its support, and such that ± 1 are not masspoints. Let $c < -1$, $d > 1$, $I = [c, d]$, and fix $x \in \mathbb{R}$. Then $\chi_n'(\Lambda)$ exists for a.e. $\Lambda > 0$ and at every point where the derivative exists,*

$$\chi_n'(\Lambda) = \inf\left\{R_n(\mu_{I,\Lambda}, t) : t \in I \backslash (-1, 1)\right\}. \tag{8}$$

Here $\mu_{I,\Lambda}$ is any extremal measure for the given Λ and I. Moreover, the inf is continuous in Λ.

Note that it follows that the inf is independent of the particular extremal measure.

This paper is organized as follows: in Sect. 2, we present two variational results. In Sect. 3, we prove Theorem 2. We prove Theorem 3 in Sect. 4, and Theorem 4 in Sect. 5. In the sequel $C, C_1, C_2, \ldots$ denote constants independent of n, x, t. The same symbol does not necessarily denote the same constant in different occurrences.

Dedication
It is a privilege to dedicate this paper to the 80th birthday of Ed Saff. Ed has contributed greatly to many areas of mathematics, including orthogonal polynomials. His work on exponential weights and his development of potential theory for external fields [17] dramatically advanced the area. His service to the research community is unparalleled.

2 Variational Formulae

In this section, we shall consider how $|p_n(\mu, x)|$ changes when we fix n and x, and perturb the measure μ. We shall consider two types of perturbations—the first involving adding a possibly signed measure, and the second involving shifting mass points. As far as we know, the formulae below are new, and we are believe are of independent interest.

Theorem 5 *Let μ be a positive measure with support on the real line and infinitely many points in its support. Let ω be a possibly signed measure of finite total mass, with compact support. For $\varepsilon \geq 0$, let*

$$\mu^{\varepsilon} = \mu + \varepsilon\omega \tag{9}$$

and assume that for small enough $\varepsilon \geq 0$, μ^{ε} is a nonnegative measure. Fix $n \geq 1$ and $x \in \mathbb{R}$ such that $p_n(\mu, x) \neq 0$. Then

$$\frac{\partial}{\partial\varepsilon}\left\{\ln\frac{1}{p_n(\mu^{\varepsilon}, x)^2}\right\}_{|\varepsilon=0} = \int R_n(\mu, t)\, d\omega(t), \tag{10}$$

where

$$R_n(\mu, t) = \frac{p_n(\mu, t)}{p_n(\mu, x)}\{K_n(\mu, x, t) + K_{n+1}(\mu, x, t)\}. \tag{11}$$

In (10), the derivative is a right derivative as μ_{ε} is only defined for $\varepsilon \geq 0$. In our second type of perturbation, which is more complicated, involving shifting point masses, we allow signed ε.

Theorem 6 *Let μ be a positive measure with support on the real line and infinitely many points in its support. Let $L \geq 1$, $\{\beta_j\}_{j=1}^{L}$ be real numbers, and $\{\xi_j\}_{j=1}^{L}$ be distinct numbers. For each $1 \leq j \leq L$, let $\sigma(\xi_j) \in \mathbb{R}$. For real ε, let*

$$\omega_{\varepsilon} = \sum_{j=1}^{L}\beta_j\delta_{\xi_j+\sigma(\xi_j)\varepsilon} \tag{12}$$

and

$$\mu^{\varepsilon} = \mu - \omega_0 + \omega_{\varepsilon}. \tag{13}$$

Assume that for small enough $|\varepsilon|$, μ^{ε} is a nonnegative measure. Fix $n \geq 1$, and $x \in \mathbb{R}$ such that $p_n(\mu, x) \neq 0$. Then with R_n as above,

$$\frac{\partial}{\partial\varepsilon}\left\{\ln\frac{1}{p_n(\mu^{\varepsilon}, x)^2}\right\}_{|\varepsilon=0} = \int R_n'(\mu, t)\,\sigma(t)\, d\omega_0(t). \tag{14}$$

We begin with the elementary assertion that orthogonal polynomials of fixed degree n, are smooth functions of the underlying moments:

Lemma 1

(a) *Let μ^0 be a positive measure with infinitely points in its support and all finite moments. Assume that for small enough $|h|$, we are give a measure μ^h with the following property: for each fixed $j \geq 0$, there is a continuous function g_j such that for small enough $|h_1|$ and $|h_2|$, the jth moment satisfies*

$$\int t^j d\mu^{h_1}(t) - \int t^j d\mu^{h_2}(t) = g_j(h_2)(h_1 - h_2) + o(h_1 - h_2).$$

Then for each fixed x, and n, $\frac{d}{dh}\left(p_n\left(\mu^h, x\right)\right)$ and $\frac{d}{dh}\gamma_n\left(\mu^h\right)$ exist and are continuous for small enough $|h|$.

(b) *If we know only that the moments are continuous functions of h, then for fixed x, $\gamma_n\left(\mu^h\right)$ and $p_n\left(\mu^h, x\right)$ are continuous in h.*

Proof

(a) This follows from the determinantal representation for orthogonal polynomials in terms of the moments: for a positive measure μ, write for $j \geq 0$,

$$c_j(\mu) = \int t^j d\mu$$

and for $m \geq 0$, let

$$D_m(\mu) = \det\left[c_{j+k}(\mu)\right]_{0 \leq j,k \leq m}. \tag{15}$$

It is known [12, 19, Vol. 1, p. 15] and [20, p. 27] that

$$p_n(\mu, x) = \frac{1}{\sqrt{D_{n-1}(\mu) D_n(\mu)}} \det \begin{bmatrix} c_0(\mu) & c_1(\mu) & c_2(\mu) & \cdots & c_n(\mu) \\ c_1(\mu) & c_2(\mu) & c_3(\mu) & \cdots & c_{n+1}(\mu) \\ \vdots & \vdots & \vdots & \ddots & \vdots \\ c_{n-1}(\mu) & c_n(\mu) & c_{n+1}(\mu) & \cdots & c_{2n-1}(\mu) \\ 1 & x & x^2 & \cdots & x^n \end{bmatrix}. \tag{16}$$

It follows from this representation and our hypothesis, that $p_n\left(\mu^h, x\right)$ is a differentiable function of h for small $|h|$; indeed we can expand each of the determinants as a product of finitely many terms involving the moments and powers of x. The one possible issue is that the determinants in the denominator approach 0, but this is impossible as $D_m\left(\mu^h\right) \to D_m\left(\mu^0\right) \neq 0$ as $h \to 0$. Since

$$\gamma_n(\mu) = \sqrt{\frac{D_{n-1}(\mu)}{D_n(\mu)}}, \tag{17}$$

it also follows that $\gamma_n\left(\mu^h\right)$ is differentiable with respect to h.
(b) is simpler. ■

Proof of Theorem 5 First note from Lemma 1 that $p_n\left(\mu^\varepsilon, x\right)$ is a continuously differentiable function of ε as the moments are linear functions of ε. Using the reproducing kernel property, the definition (9) of μ^ε, and then orthogonality,

$$\begin{aligned} p_n(\mu, x) &= \int p_n(\mu, t) K_{n+1}\left(\mu^\varepsilon, x, t\right) d\mu^\varepsilon(t) \\ &= p_n\left(\mu^\varepsilon, x\right) \int p_n(\mu, t) p_n\left(\mu^\varepsilon, t\right) d\mu(t) + 0 \\ &+ \varepsilon \int p_n(\mu, t) K_{n+1}\left(\mu^\varepsilon, x, t\right) d\omega(t). \end{aligned}$$

Differentiating with respect to ε gives

$$\begin{aligned} 0 = &\frac{\partial p_n\left(\mu^\varepsilon, x\right)}{\partial \varepsilon} \int p_n(\mu, t) p_n\left(\mu^\varepsilon, t\right) d\mu(t) + p_n\left(\mu^\varepsilon, x\right) \int p_n(\mu, t) \frac{\partial p_n\left(\mu^\varepsilon, t\right)}{\partial \varepsilon} d\mu(t) \\ &+ \int p_n(\mu, t) K_{n+1}\left(\mu^\varepsilon, x, t\right) d\omega(t) + \varepsilon \int p_n(\mu, t) \frac{\partial K_{n+1}\left(\mu^\varepsilon, x, t\right)}{\partial \varepsilon} d\omega(t). \end{aligned}$$

The previous lemma shows that all the derivatives exist and are continuous, so the interchanges are permissible. Letting $\varepsilon \to 0+$, we obtain

$$\begin{aligned} 0 = &\frac{\partial p_n\left(\mu^\varepsilon, x\right)}{\partial \varepsilon}_{|\varepsilon=0} \int p_n^2(\mu, t) d\mu(t) + p_n(\mu, x) \int p_n(\mu, t) \frac{\partial p_n\left(\mu^\varepsilon, t\right)}{\partial \varepsilon}_{|\varepsilon=0} d\mu(t) \\ &+ \int p_n(\mu, t) K_{n+1}(\mu, x, t) d\omega(t), \end{aligned}$$

so

$$\begin{aligned} \frac{\partial p_n\left(\mu^\varepsilon, x\right)}{\partial \varepsilon}_{|\varepsilon=0} &= -p_n(\mu, x) \int p_n(\mu, t) \frac{\partial p_n\left(\mu^\varepsilon, t\right)}{\partial \varepsilon}_{|\varepsilon=0} d\mu(t) \\ &- \int p_n(\mu, t) K_{n+1}(\mu, x, t) d\omega(t). \end{aligned} \tag{18}$$

Next, using orthogonality,

$$p_n\left(\mu^\varepsilon, x\right) = \int p_n\left(\mu^\varepsilon, t\right) K_{n+1}(\mu, x, t) d\mu(t)$$

$$= p_n(\mu, x) \int p_n\left(\mu^\varepsilon, t\right) p_n(\mu, t)\, d\mu_\varepsilon(t)$$

$$- \varepsilon \int p_n\left(\mu^\varepsilon, t\right) K_{n+1}(\mu, x, t)\, d\omega(t).$$

Differentiating with respect to ε gives

$$\frac{\partial p_n(\mu^\varepsilon, x)}{\partial \varepsilon} = p_n(\mu, x) \int \frac{\partial p_n(\mu^\varepsilon, t)}{\partial \varepsilon} p_n(\mu, t)\, d\mu_\varepsilon(t)$$

$$+ p_n(\mu, x) \int p_n\left(\mu^\varepsilon, t\right) p_n(\mu, t)\, d\omega(t)$$

$$- \int p_n\left(\mu^\varepsilon, t\right) K_{n+1}(\mu, x, t)\, d\omega(t)$$

$$- \varepsilon \int \frac{\partial p_n(\mu^\varepsilon, t)}{\partial \varepsilon} K_{n+1}(\mu, x, t)\, d\omega(t).$$

Letting $\varepsilon \to 0+$ gives

$$\frac{\partial p_n(\mu^\varepsilon, x)}{\partial \varepsilon}_{|\varepsilon=0} = p_n(\mu, x) \int \frac{\partial p_n(\mu^\varepsilon, t)}{\partial \varepsilon}_{|\varepsilon=0} p_n(\mu, t)\, d\mu(t)$$

$$+ p_n(\mu, x) \int p_n^2(\mu, t)\, d\omega(t) - \int p_n(\mu, t) K_{n+1}(\mu, x, t)\, d\omega(t).$$

Adding this and (18) gives

$$2 \frac{\partial p_n(\mu^\varepsilon, x)}{\partial \varepsilon}_{|\varepsilon=0} = p_n(\mu, x) \int p_n^2(\mu, t)\, d\omega(t)$$

$$- 2 \int p_n(\mu, t) K_{n+1}(\mu, x, t)\, d\omega(t).$$

Then

$$\frac{\partial}{\partial \varepsilon} \left\{ \ln \frac{1}{p_n(\mu^\varepsilon, x)^2} \right\}_{|\varepsilon=0} = -\frac{2}{p_n(\mu, x)} \frac{\partial p_n(\mu^\varepsilon, x)}{\partial \varepsilon}_{|\varepsilon=0}$$

$$= \int \left[-p_n^2(\mu, t) + \frac{2 K_{n+1}(\mu, x, t)}{p_n(\mu, x)} p_n(\mu, t) \right] d\omega(t)$$

$$= \int \frac{p_n(\mu, t)}{p_n(\mu, x)} \left[K_{n+1}(\mu, x, t) + K_n(x, t) \right] d\omega(t).$$

■

Proof of Theorem 6 First note from Lemma 1 that $p_n(\mu^\varepsilon, x)$ is a continuously differentiable function of ε as the moments are differentiable functions of ε. Using the reproducing kernel property, the definition (13) of μ^ε, and then orthogonality,

$$\begin{aligned} p_n(\mu, x) &= \int p_n(\mu, t) K_{n+1}\left(\mu^\varepsilon, x, t\right) d\mu^\varepsilon(t) \\ &= p_n\left(\mu^\varepsilon, x\right) \int p_n(\mu, t)\, p_n\left(\mu^\varepsilon, t\right) d\mu(t) \\ &\quad - \int p_n(\mu, t) K_{n+1}\left(\mu^\varepsilon, x, t\right) d\omega_0(t) \\ &\quad + \int p_n(\mu, t) K_{n+1}\left(\mu^\varepsilon, x, t\right) d\omega_\varepsilon(t). \end{aligned}$$

Differentiating with respect to ε and recalling the definition (12) of ω_ε gives

$$\begin{aligned} 0 = {} & \frac{\partial p_n(\mu^\varepsilon, x)}{\partial \varepsilon} \int p_n(\mu, t)\, p_n\left(\mu^\varepsilon, t\right) d\mu(t) \\ & + p_n\left(\mu^\varepsilon, x\right) \int p_n(\mu, t) \frac{\partial p_n(\mu^\varepsilon, t)}{\partial \varepsilon} d\mu(t) \\ & - \int p_n(\mu, t) \frac{\partial K_{n+1}(\mu^\varepsilon, x, t)}{\partial \varepsilon} d\omega_0(t) \\ & + \int p_n(\mu, t) \frac{\partial K_{n+1}(\mu^\varepsilon, x, t)}{\partial \varepsilon} d\omega_\varepsilon(t) \\ & + \sum_{j=1}^{L} \beta_j \frac{\partial}{\partial t} \left\{ p_n(\mu, t) K_{n+1}\left(\mu^\varepsilon, x, t\right) \right\}_{|t=\xi_j+\sigma(\xi_j)\varepsilon} \sigma\left(\xi_j\right) \end{aligned}$$

so letting $\varepsilon \to 0+$, and using continuity of the derivatives,

$$\begin{aligned} \frac{\partial p_n(\mu^\varepsilon, x)}{\partial \varepsilon}_{|\varepsilon=0} &= -p_n(\mu, x) \int p_n(\mu, t) \frac{\partial p_n(\mu^\varepsilon, t)}{\partial \varepsilon}_{|\varepsilon=0} d\mu(t) \\ &\quad - \sum_{j=1}^{L} \beta_j \frac{\partial}{\partial t} \left\{ p_n(\mu, t) K_{n+1}(\mu, x, t) \right\}_{|t=\xi_j} \sigma\left(\xi_j\right). \end{aligned} \tag{19}$$

Next,

$$\begin{aligned} p_n\left(\mu^\varepsilon, x\right) &= \int p_n\left(\mu^\varepsilon, t\right) K_{n+1}(\mu, x, t)\, d\mu(t) \\ &= p_n(\mu, x) \int p_n\left(\mu^\varepsilon, t\right) p_n(\mu, t)\, d\mu_\varepsilon(t) \end{aligned}$$

$$+ \int p_n\left(\mu^{\varepsilon}, t\right) K_{n+1}(\mu, x, t)\, d\omega_0(t)$$
$$- \int p_n\left(\mu^{\varepsilon}, t\right) K_{n+1}(\mu, x, t)\, d\omega_{\varepsilon}(t).$$

Differentiating with respect to ε and recalling the definition (12) of ω_{ε} gives

$$\begin{aligned}
&\frac{\partial p_n(\mu^{\varepsilon}, x)}{\partial \varepsilon} \\
&= p_n(\mu, x) \int \frac{\partial p_n(\mu^{\varepsilon}, t)}{\partial \varepsilon} p_n(\mu, t)\, d\mu_{\varepsilon}(t) \\
&+ p_n(\mu, x) \sum_{j=1}^{L} \beta_j \frac{\partial}{\partial t} \left\{p_n\left(\mu^{\varepsilon}, t\right) p_n(\mu, t)\right\}_{|t=\xi_j+\sigma(\xi_j)\varepsilon} \sigma\left(\xi_j\right) \\
&+ \int \frac{\partial p_n(\mu^{\varepsilon}, t)}{\partial \varepsilon} K_{n+1}(\mu, x, t)\, d\omega_0(t) \\
&- \int \frac{\partial p_n(\mu^{\varepsilon}, t)}{\partial \varepsilon} K_{n+1}(\mu, x, t)\, d\omega_{\varepsilon}(t) \\
&- \sum_{j=1}^{L} \beta_j \frac{\partial}{\partial t} \left\{p_n\left(\mu^{\varepsilon}, t\right) K_{n+1}(\mu, x, t)\right\}_{|t=\xi_j+\sigma(\xi_j)\varepsilon} \sigma\left(\xi_j\right)
\end{aligned}$$

so letting $\varepsilon \to 0+$, and using the continuity of the derivatives in ε,

$$\begin{aligned}
\frac{\partial p_n(\mu^{\varepsilon}, x)}{\partial \varepsilon}_{|\varepsilon=0} &= p_n(\mu, x) \int \frac{\partial p_n(\mu^{\varepsilon}, t)}{\partial \varepsilon}_{|\varepsilon=0} p_n(\mu, t)\, d\mu(t) \\
&+ p_n(\mu, x) \sum_{j=1}^{L} \beta_j \frac{\partial}{\partial t} \left\{p_n^2(\mu, t)\right\}_{|t=\xi_j} \sigma\left(\xi_j\right) \\
&- \sum_{j=1}^{L} \beta_j \frac{\partial}{\partial t} \left\{p_n(\mu, t) K_{n+1}(\mu, x, t)\right\}_{|t=\xi_j} \sigma\left(\xi_j\right).
\end{aligned}$$

Adding this and (19) gives

$$\begin{aligned}
&2 \frac{\partial p_n(\mu^{\varepsilon}, x)}{\partial \varepsilon}_{|\varepsilon=0} \\
&= p_n(\mu, x) \sum_{j=1}^{L} \beta_j \frac{\partial}{\partial t} \left\{p_n^2(\mu, t)\right\}_{|t=\xi_j} \sigma\left(\xi_j\right)
\end{aligned}$$

$$-2\sum_{j=1}^{L}\beta_j\frac{\partial}{\partial t}\left\{p_n(\mu,t)\,K_{n+1}(\mu,x,t)\right\}_{|t=\xi_j}\sigma\left(\xi_j\right)$$

$$=\int\frac{\partial}{\partial t}\left\{p_n(\mu,x)\,p_n^2(\mu,t)-2p_n(\mu,t)\,K_{n+1}(\mu,x,t)\right\}\sigma(t)\,d\omega_0(t)$$

so

$$\frac{\partial}{\partial\varepsilon}\left\{\ln\frac{1}{p_n(\mu^{\varepsilon},x)^2}\right\}$$

$$=\int\frac{\partial}{\partial t}\left[-p_n^2(\mu,t)+\frac{2K_{n+1}(\mu,x,t)}{p_n(\mu,x)}p_n(\mu,t)\right]\sigma(t)\,d\omega_0(t)$$

$$=\int R_n'(\mu,t)\,\sigma(t)\,d\omega_0(t)\,.$$

■

3 Proof of Theorem 2

Proof of Theorem 2(a) We note that since $\mu\geq\nu$,

$$p_n\left(\mu^2,x\right)\leq K_{n+1}(\mu,x,x)\leq K_{n+1}(\nu,x,x)\,.$$

Thus

$$\phi_n(I,\Lambda,x)=\sup\left\{\left|p_n(\mu,x)\right|:\mu\in\mathcal{M}(I,\Lambda)\right\}\leq K_{n+1}(\nu,x,x)\,.$$

Next, choose a sequence $\left\{\mu^{(m)}\right\}_{m\geq1}$ of measures in $\mathcal{M}(I,\Lambda)$ such that

$$\lim_{m\to\infty}\left|p_n\left(\mu^{(m)},x\right)\right|=\phi_n(I,\Lambda,x)\,.$$

Since $\left\{\mu^{(m)}\right\}$ are measures supported on I and all have total mass $\leq\Lambda+\nu([-1,1])$, we can choose a subsequence that converges weakly to some measure μ_I supported on I and of total mass $\leq\Lambda+\nu([-1,1])$. Since each $\mu^{(m)}=\nu$ in $(-1,1)$, the same is true of μ_I. So $\mu_I\in\mathcal{M}(I,\Lambda)$. For notational simplicity, we assume that the full sequence $\left\{\mu^{(m)}\right\}$ converges weakly to μ_I. Then for each fixed $j\geq0$,

$$\lim_{m\to\infty}\int t^j d\mu^{(m)}(t)=\int t^j d\mu_I(t)\,.$$

From Lemma 1(b),

$$|p_n(\mu_I, x)| = \lim_{m\to\infty} \left| p_n\left(\mu^{(m)}, x\right)\right| = \phi_n(I, \Lambda, x).$$

■

Proof of Theorem 2(b) Given $\varepsilon \in [0, 1]$ and a nonnegative measure ω supported on $I \setminus (-1, 1)$ with total mass $\leq \Lambda$, and the extremal measure $\mu = \mu_I = \nu + \rho_I$ above, we shall consider

$$\begin{aligned} \mu_I^\varepsilon &= \mu_I + \varepsilon(\omega - \rho_I) \\ &= \nu + (1-\varepsilon)\rho_I + \varepsilon\omega. \end{aligned} \tag{20}$$

This is a positive measure, and belongs to $\mathcal{M}(I, \Lambda)$. Observe that the function

$$f(\varepsilon) = \ln \frac{1}{p_n\left(\mu_I^\varepsilon, x\right)^2}, \varepsilon \in [0, 1],$$

has its minimum in $[0, 1]$ at $\varepsilon = 0$. It is also a differentiable function of ε, at least for sufficiently small ε, since the moments

$$\int t^j d\mu^\varepsilon(t) = \int t^j d\nu(t) + \varepsilon \int t^j d(\omega - \rho_I)(t)$$

are linear functions of ε, so we may apply Lemma 1. Then $f'(0+) \geq 0$. By Theorem 5, with ω replaced there by $\omega - \rho_I$,

$$0 \leq f'(0+) = \frac{\partial}{\partial \varepsilon}\left(\ln \frac{1}{p_n\left(\mu_I^\varepsilon, x\right)^2}\right)_{|\varepsilon=0} = \int R_n(\mu_I, t)\, d(\omega - \rho_I)(t).$$

■

Proof of Theorem 2(c) Suppose t_0 is in the support of ρ_I and $\delta > 0$. Then we set $\omega = \rho_I$ outside $[t_0 - \delta, t_0 + \delta]$ and $\omega = 0$ in $[t_0 - \delta, t_0 + \delta]$. This is still valid even if $[t_0 - \delta, t_0 + \delta]$ is not contained in $I \setminus (-1, 1)$. From (4),

$$-\int_{t_0-\delta}^{t_0+\delta} R_n(\mu_I, t)\, d\rho_I(t) \geq 0 \Rightarrow \inf_{[t_0-\delta, t_0+\delta]} R_n(\mu_I, t) \leq 0,$$

as $\rho_I([t_0 - \delta, t_0 + \delta]) > 0$. Letting $\delta \to 0+$ gives

$$R_n(\mu_I, t_0) \leq 0.$$

Next, let $\delta > 0$, $t_1 \in I \backslash ((-1,1) \cup [t_0 - \delta, t_0 + \delta])$, and $\Delta = \rho_I([t_0 - \delta, t_0 + \delta])$, and

$$\omega = \rho_I - \rho_{I|[t_0-\delta, t_0+\delta]} + \Delta \delta_{t_1}.$$

Then ω is a nonnegative measure supported in $I \backslash (-1,1)$, and still $\omega(I) \leq \Lambda$. Also then from (4),

$$\int_{t_0-\delta}^{t_0+\delta} R_n(\mu_I, t)\, d(-\rho_I)(t) + \Delta R_n(\mu_I, t_1) \geq 0$$

$$\Rightarrow -\Delta \inf_{[t_0-\delta, t_0+\delta]} R_n(\mu_I, t) + \Delta R_n(\mu_I, t_1) \geq 0$$

$$\Rightarrow R_n(\mu_I, t_1) \geq \inf_{[t_0-\delta, t_0+\delta]} R_n(\mu_I, t).$$

Letting $\delta \to 0+$, we obtain for any t_0 in the support of ρ_I, and any $t_1 \in I \backslash (-1,1)$,

$$R_n(\mu_I, t_1) \geq R_n(\mu_I, t_0),$$

which implies (5). ■

In the following lemma, the zeros of $p_n(\mu_I, \cdot)$ are denoted by

$$(c <)\, x_{nn} < x_{n-1,n} < x_{n-2,n} < \ldots < x_{1n}\, (< d).$$

These are of course not in general the zeros of $p_n(\nu, x)$. For the given x, let

$$U_n(t) = (t - x)\left[K_{n+1}(\mu_I, x, t) + K_n(\mu_I, x, t)\right].$$

Lemma 2

(a) *U_n has simple zeros $\{y_{jn}\}_{0 \leq j \leq n}$, where $y_{jn} \in (x_{j+1,n}, x_{jn})$ if $1 \leq j \leq n-1$, while $y_{0n} > x_{1n}$ and $y_{nn} < x_{nn}$.*

(b) *R_n defined by (3) has $2n$ simple zeros and at most 2 lie outside I^o. There are n zeros at those of $p_n(\mu_I, \cdot)$ and another n at the zeros of U_n, other than x. At most 1 zero can lie in $[d, \infty)$ and at most one in $(-\infty, c]$.*

(c) *R_n' has $2n-1$ simple zeros. For $1 \leq j \leq n-1$, it has one zero in $(x_{j+1,n}, y_{jn})$ and another in (y_{jn}, x_{jn}) as long as $x \notin [x_{j+1,n}, x_{jn}]$. If $x \in [x_{j+1,n}, x_{jn}]$, then this interval contains one zero of R_n'. In addition there is one zero in (x_{1n}, y_{0n}), provided $x \neq y_{0n}$ and one zero in (y_{nn}, x_{nn}), provided $x \neq y_{nn}$.*

(d) *If $x \neq y_{0n}$, then R_n is positive and increasing in (y_{0n}, ∞), so there is a local minimum in (x_{1n}, y_{0n}). If $x = y_{0n}$, then R_n is positive and increasing in (x_{1n}, ∞).*

Proof

(a) First note that from (2), $p_n(\mu_I, x) \neq 0$. We use a result of Shohat [18, p. 472, Theorem VII], which is more clearly stated as Theorem 4 in [6, p. 161, Theorem 4]. To apply this, note that

$$\begin{aligned}
&U_n(t) / \left(\frac{\gamma_n}{\gamma_{n+1}} p_n(\mu_I, x) \right) \\
&= p_{n+1}(\mu_I, t) + \left(\frac{-\frac{\gamma_n}{\gamma_{n+1}} p_{n+1}(\mu_I, x) + \frac{\gamma_{n-1}}{\gamma_n} p_{n-1}(\mu_I, x)}{\frac{\gamma_n}{\gamma_{n+1}} p_n(\mu_I, x)} \right) p_n(\mu_I, t) \\
&\quad - \left(\frac{\frac{\gamma_{n-1}}{\gamma_n}}{\frac{\gamma_n}{\gamma_{n+1}}} \right) p_{n-1}(\mu_I, t) \\
&= p_{n+1}(\mu_I, t) + C_n p_n(\mu_I, t) + D_n p_{n-1}(\mu_I, t), \text{ say.}
\end{aligned}$$

As $D_n < 0$, $C_n \in \mathbb{R}$, then the result of Shohat shows that $p_{n+1}(\mu_I, t) + C_n p_n(\mu_I, t) + D_n p_{n-1}(\mu_I, t)$ has all real simple zeros and at most two lie outside (c, d).

Next, Theorem 5 of [6, p. 161], shows that the interval $(x_{nn}(\mu_I), x_{1n}(\mu_I)) \subset (c, d)$ contains $n - 1$ zeros of U_n and there is one zero to the left of $x_{nn}(\mu_I)$ and one to the right of $x_{1n}(\mu_I)$.

(b) From (a), $K_{n+1}(\mu_I, x, t) + K_n(\mu_I, x, t)$ has simple zeros. If we can show that $p_n(\mu_I, t)$ and $K_{n+1}(\mu_I, x, t) + K_n(\mu_I, x, t)$ don't have common zeros, then we have proved that R_n has simple zeros, since $p_n(\mu_I, \cdot)$ has n simple zeros in I^o.

So suppose that $p_n(\mu_I, t)$ and $K_{n+1}(\mu_I, x, t) + K_n(\mu_I, x, t)$ have a common zero y. Firstly this cannot be x, since $K_{n+1}(\mu_I, x, x) + K_n(\mu_I, x, x) > 0$. So $y \neq x$. Then

$$\begin{aligned}
0 &= K_{n+1}(\mu_I, x, y) + K_n(\mu_I, x, y) \\
&= 2K_n(\mu_I, x, y) = 2\frac{\gamma_{n-1}}{\gamma_n} \frac{p_n(\mu_I, x)\, p_{n-1}(\mu_I, y)}{x - y}.
\end{aligned}$$

Since $p_n(\mu_I, x) \neq 0$, so $p_{n-1}(\mu_I, y) = p_n(\mu_I, y) = 0$, which is impossible. Thus R_n has $2n$ simple zeros.

It remains to show that R_n can have at most one zero in $[c, \infty)$. We know that any such zero must be a zero of $K_{n+1}(\mu_I, x, t) + K_n(\mu_I, x, t)$ as $p_n(\mu_I, \cdot) > 0$ in $[c, \infty)$. Theorem 5 of [6, p. 161], shows that the interval $(x_{nn}(\mu_I), x_{1n}(\mu_I)) \subset (c, d)$ contains $n - 1$ zeros of $U_n(t)$ and there is one zero to the left of $x_{nn}(\mu_I)$ and one to the right of $x_{1n}(\mu_I)$. Since $x_{1n}(\mu) < c$, it follows that there is at most one zero of $K_{n+1}(\mu_I, x, t) + K_n(\mu_I, x, t)$ in $[c, \infty)$.

(c) As R_n has $2n$ simple zeros, so R_n' has $2n-1$ simple zeros, that interlace the zeros of R_n. The remaining statements follow from (b).

(d) We have as $t \to \infty$,

$$R_n(\mu_I, t) = p_n^2(\mu_I, t)(1 + o(1)),$$

so $R_n(\mu_I, t) > 0$ for $t > y_{0n}$. Then if $x \neq y_{0n}$, the zero of R_n' in (x_{1n}, y_{0n}) must be a local minimum of R_n. If $x = y_{0n}$, then the largest zero of $R_n(\mu_I, \cdot)$ is x_{1n} and R_n is increasing and positive in (x_{1n}, ∞). ■

Proof of Theorem 2(d) We begin with three observations: (I) $R_n(\mu_I, \cdot)$ has exactly n local minima in $(-\infty, \infty)$. ■

Proof This follows from the fact that R_n has $2n$ simple zeros and $R_n(\mu_I, t)$ is positive for large enough $|t|$.

(II) If at least one of $-1, 1$ is a support point of ρ_I, then $R_n(\mu_I, t)$ has a local minimum in $[-1, 1]$ with the derivative vanishing at that point.

Proof We know that $R_n(\pm 1) \geq \eta = \inf\{R_n(\mu_I, t) : t \in I \backslash (-1, 1)\}$. If for example 1 is a support point, then we know $R_n(\mu_I, 1) = \eta$. If $R_n'(\mu_I, 1) = 0$, then 1 is a local minimum. Now suppose the derivative is not zero. Now necessarily $R_n(\mu_I, \cdot) > \eta$ in $(1, 1+\varepsilon)$ for some $\varepsilon > 0$ and hence also $R_n(\mu_I, \cdot) < \eta$ in $(1-\varepsilon, 1)$ for some $\varepsilon > 0$. As $R_n(\mu_I, -1) \geq \eta$, there is then a local minimum in $(-1, 1)$.

(III) If d is a support point of ρ_I, then R_n has a local minimum in $[d, \infty)$. Similarly for c.

Proof Now $R_n(\mu_I, d) = \eta$. If $R_n'(\mu_I, d) = 0$, then d must be a local minimum. Suppose the derivative is not 0. Then $R_n(\mu_I, \cdot) > \eta$ in $(d-\varepsilon, d)$ for some $\varepsilon > 0$. Then also $R_n(\mu_I, \cdot) < \eta$ in $(d, d+\varepsilon)$ for some $\varepsilon > 0$. But $R_n(\mu_I, t) \to \infty$ as $t \to \infty$, so there must be a local minimum in (d, ∞).

We can now proceed to the proof proper. Let ℓ be the number of support points amongst $\{c, d\}$. Let $k = 1$ if at least one of $-1, 1$ is a support point and 0 otherwise. Then from our observations above, there are $\leq \ell + k$ local minima of $R_n(\mu_I, \cdot)$ in $[-1, 1] \cup (-\infty, c] \cup [d, \infty)$. Then there are at most $(n - \ell - k)$ local minima of $R_n(\mu_I)$ in $(c, d) \backslash (-1, 1)$ and hence at most $n - \ell - k$ support points there. This gives altogether at most

$$(n - \ell - k) + \ell + k + 1 = n + 1$$

support points. The extra 1 arises if both $-1, 1$ are support points. ■

Proof of Theorem 2(e) Suppose ρ_I has total mass $T < \Lambda$. Then pick any $t_0 \in I \backslash (-1, 1)$ and let

$$\omega = \rho_I + (\Lambda - T)\,\delta_{t_0}.$$

Then from (4),

$$0 \leq \int R_n(\mu_I, t)\, d(\omega - \rho_I)(t) = (\Lambda - T)\, R_n(\mu_I, t_0).$$

So $R_n \geq 0$ in $I \backslash (-1, 1)$. In particular, for $t_0 \in \operatorname{supp}[\rho_I]$, we have

$$R_n(\mu_I, t_0) \geq 0.$$

But we already know that $R_n(\mu_I, t_0) \leq 0$. So $R_n = 0$ in $\operatorname{supp}[\rho_I]$. But then from (c),

$$\inf\{R_n(\mu_I, s) : s \in I\} = 0$$

so that $R_n \geq 0$ in $I \backslash (-1, 1)$. As R_n changes sign at its zeros, it follows that R_n cannot have zeros in $I^o \backslash [-1, 1]$ and $R_n > 0$ there. Hence the support of ρ_I does not intersect $I^o \backslash [-1, 1]$. So

$$\operatorname{supp}[\rho_I] \subset \{-1, 1, c, d\}.$$

Next suppose that d lies in the support. Then $R_n(\mu_I, d) = 0$ and by Lemma 2(b), there are no more zeros in (d, ∞), so that R_n is of one sign there. But since $R_n \geq 0$ to the left of d and R_n changes sign at d, we have $R_n < 0$ in (d, ∞). However, as $t \to \infty$, it follows that

$$R_n(\mu_I, t) = p_n^2(\mu_I, t)(1 + o(1))$$

so we have a contradiction. Thus d is not in the support and similarly, c is not in the support. ■

4 Proof of Theorem 3

Recall that

$$\psi_n(y) = \ln \frac{1}{\phi_n([c, y], \Lambda, x)^2}, \; y > 1.$$

Let

$$\phi_n(y) = \phi_n([c, y], \Lambda, x).$$

We begin by showing that ϕ_n is continuous:

Lemma 3

(a) Let $y > 1$, $I = [c, y]$, and $\{h_m\}_{m\geq 1}$ be a sequence of numbers with limit 0. For $m \geq 1$, let $I_m = [c, y + h_m]$ and μ_{I_m} be an extremal measure for I_m, and assume that μ_{I_m} converges weakly to $\hat{\mu}_I$ as $m \to \infty$. Then $\hat{\mu}_I$ is an extremal measure for $I = [c, y]$ and

$$\phi_n(y) = \lim_{m\to\infty} \phi_n(y + h_m). \tag{21}$$

(b) ϕ_n is continuous in $[1, \infty)$.

Proof

(a) It suffices to consider sequences in which all h_m are of one sign.
Case I: All $h_m \geq 0$.
Recall that ϕ_n is an increasing function. As $\hat{\mu}_I \in \mathcal{M}(I, \Lambda)$,

$$\begin{aligned}\left|p_n\left(\hat{\mu}_I, x\right)\right| &\leq \phi_n(y) \leq \liminf_{m\to\infty} \phi_n(y + h_m) \\ &\leq \limsup_{m\to\infty} \phi_n(y + h_m) = \lim_{m\to\infty} \left|p_n\left(\mu_{I_m}, x\right)\right| = \left|p_n\left(\hat{\mu}_I, x\right)\right|\end{aligned}$$

as in Lemma 1(b), since μ_{I_m} converges weakly to$\hat{\mu}_I$. Then we have (21) and that $\hat{\mu}_I$ is extremal for I.
Case II: All $h_m < 0$
Let μ_I be an extremal measure for $[c, y]$, so that it has the form $\nu + \rho_I$, where for some $L \leq n + 1$,

$$\rho_I = \sum_{j=1}^{L} \mu_I\left(\{\xi_j\}\right)\delta_{\xi_j},$$

where all $\xi_j \in I\backslash(-1, 1)$. We assume that ξ_1 is smallest and ξ_L is largest. Then for large enough m, and $1 \leq j \leq L - 1$, $\xi_j \in [c, y + h_m]$. Define

$$\rho_{I,m} = \rho_I + \rho_I(\{\xi_L\})\left[\delta_{\xi_L + h_m} - \delta_{\xi_L}\right]$$

so that (recall $h_m < 0$)

$$\mu^{(m)} = \nu + \rho_{I,m} \in \mathcal{M}(I_m, \Lambda).$$

Also as $m \to \infty$, $\mu^{(m)} \to \mu_I$ weakly, so

$$\lim_{m\to\infty} \left|p_n\left(\mu^{(m)}, x\right)\right| = |p_n(\mu_I, x)| = \phi_n(y)$$

so

$$\liminf_{n\to\infty} \phi_n(d+h_m) \geq \liminf_{n\to\infty} \left|p_n\left(\mu^{(m)}, x\right)\right|$$
$$= \phi_n(y) \geq \limsup_{n\to\infty} \phi_n(y+h_m).$$

Then we have (21). Also, it then follows that for the given extremal measures $\{\mu_{I_m}\}$ converging weakly to $\hat{\mu}_I$, necessarily,

$$\left|p_n\left(\hat{\mu}_I, x\right)\right| = \lim_{m\to\infty} \left|p_n\left(\mu_{I_m}, x\right)\right| = \lim_{m\to\infty} \phi_n(y+h_m) = \phi_n(y)$$

so $\hat{\mu}_I$ is extremal for I.

(b) The continuity of ϕ_n in $(1,\infty)$ follows directly from (a), while Case I there shows that ϕ_n is right continuous at 1. ■

Our next step is to show that ψ_n' is given a.e. by (7):

Lemma 4 *ψ_n' exists a.e. in $(1,\infty)$. At every point where the derivative exists, it is given by*

$$\psi_n'(y) = \mu_{[c,y]}(\{y\})\, R_n'\left(\mu_{[c,y]}, y\right),$$

where $\mu_{[c,y]}$ is any extremal measure. In particular, the right-hand side is independent of the particular $\mu_{[c,y]}$.

Proof Let μ_I be an extremal measure for the given y, so that it has a representation

$$\mu_{[c,y]} = \nu + \sum_{j=1}^{L} \mu_{[c,y]}\left(\{\xi_j\}\right)\delta_{\xi_j} \tag{22}$$

where all $\{\xi_j\}$ lie in $I\setminus(-1,1)$, while $L \leq n+1$, and the sum of the $\mu_I\left(\{\xi_j\}\right)$ is at most Λ. Let

$$\omega_h = \mu_{[c,y]}(\{y\})\,\delta_{y+h}, \tag{23}$$

so that ω_h consists of the point mass of $\mu_{[c,y]}$ at y (if any), shifted to $y+h$. Then for $h>0$,

$$\mu^{(h)} = \mu_{[c,y]} + \omega_h - \omega_0 \in \mathcal{M}([c, y+h], \Lambda). \tag{24}$$

Then for small $h>0$, as $\mu_{[c,y]}$ is extremal for $[c,y]$,

$$\psi_n(y+h) - \psi_n(y)$$

$$\leq \left[\ln \frac{1}{p_n\left(\mu^{(h)}, x\right)^2} - \ln \frac{1}{p_n\left(\mu_I, x\right)^2} \right]$$

$$= \left(\int R_n'\left(\mu_{[c,y]}, t\right) sign\,(t)\, d\omega_0\,(t) \right) h + o\,(h)$$

$$= \left(\mu_{[c,y]}\,(\{y\})\, R_n'\left(\mu_{[c,y]}, y\right)\right) h + o\,(h)\,, \tag{25}$$

by Theorem 6 and our choice of measure. So

$$\limsup\nolimits_{h \to 0+} \left[\psi_n\,(y+h) - \psi_n\,(y)\right]/h$$
$$\leq \quad \mu_{[c,y]}\,(\{y\})\, R_n'\left(\mu_{[c,y]}, y\right). \tag{26}$$

Next, let ω_h be defined by (23) and for small negative h, define $\mu^{(h)}$ by (24). As before, we have (25), but since $h < 0$, so

$$\liminf_{h \to 0-} \left[\psi_n\,(y+h) - \psi_n\,(y)\right]/h$$
$$\geq \int R_n'\left(\mu_{[c,y]}, t\right) sign\,(t)\, d\omega_0\,(t)$$
$$= \mu_{[c,y]}\,(\{y\})\, R_n'\left(\mu_{[c,y]}, y\right).$$

In summary, we have shown that for $y > 1$, and any extremal measure $\mu_{[c,y]}$,

$$\liminf_{h \to 0-} \left[\psi_n\,(y+h) - \psi_n\,(y)\right]/h$$
$$\geq \mu_{[c,y]}\,(\{y\})\, R_n'\left(\mu_{[c,y]}, y\right)$$
$$\geq \limsup_{h \to 0+} \left[\psi_n\,(y+h) - \psi_n\,(y)\right]/h. \tag{27}$$

Since ψ_n is decreasing, it is differentiable a.e. and at such points y, this last relation shows that

$$\psi_n'\,(y) = \mu_{[c,y]}\,(\{y\})\, R_n'\left(\mu_{[c,y]}, y\right),$$

for any extremal measure $\mu_{[c,y]}$. Of course the last right-hand side is independent of the particular extremal measure. ■

Next we prove uniform boundedness of R_n. In the sequel, we let

$$K_n^{(1,1)}\,(\nu, t, t) = \sum_{j=0}^{n-1} p_j'\,(\nu, t)^2 .$$

Lemma 5 *Fix $n \geq 1$, $x \in \mathbb{R}$, and $\mu \in \mathcal{M}(I, \Lambda)$. Assume that*

$$|p_n(\mu, x)| \geq r > 0.$$

Let

$$R_n(\mu, t) = \frac{p_n(\mu, t)}{p_n(\mu, x)} [K_{n+1}(\mu, x, t) + K_{n+1}(\mu, x, t)].$$

Then

(a)

$$|R_n(\mu, t)| \leq \frac{2}{r} K_{n+1}(\nu, x, x)^{1/2} K_{n+1}(\nu, t, t). \tag{28}$$

(b)

$$\left|R_n'(\mu, t)\right| \leq \frac{4}{r} \left[K_{n+1}(\nu, x, x) K_{n+1}(\nu, t, t) K_{n+1}^{(1,1)}(\nu, t, t)\right]^{1/2}. \tag{29}$$

In particular these bounds do not depend on μ or Λ.

Proof

(a) Firstly, as $\mu \geq \nu$,

$$p_n^2(\mu, t) \leq K_{n+1}(\mu, t, t) \leq K_{n+1}(\nu, t, t),$$

while by Cauchy-Schwarz,

$$(K_{n+1}(\mu, x, t))^2 \leq K_{n+1}(\nu, t, t) K_{n+1}(\nu, x, x).$$

Then

$$|R_n(\mu, t)| \leq \frac{K_{n+1}(\nu, t, t)^{1/2}}{r} 2 (K_{n+1}(\nu, t, t) K_{n+1}(\nu, x, x))^{1/2}.$$

(b) We use the well known extremal property of reproducing kernels and easy consequence of Cauchy-Schwarz: for any polynomial P of degree $\leq n-1$, and any real t

$$P'(t)^2 \leq K_n^{(1,1)}(\nu, t, t) \int P^2 \, d\nu.$$

Then

$$\left(\frac{\partial}{\partial t}K_n(\mu, x, t)\right)^2 \leq K_n^{(1,1)}(\nu, t, t)\int K_n(\mu, x, t)^2 \, d\nu(t)$$
$$\leq K_n^{(1,1)}(\nu, t, t)\int K_n(\mu, x, t)^2 \, d\mu(t)$$
$$= K_n^{(1,1)}(\nu, t, t)\, K_n(\mu, x, x)$$
$$\leq K_n^{(1,1)}(\nu, t, t)\, K_n(\nu, x, x)$$

as $\mu \geq \nu$ and the reproducing kernel K_n is decreasing in the measure. Then

$$\left|R_n'(\mu, t)\right| \leq \left|\frac{p_n'(\mu, t)}{p_n(\mu, x)}\right| \left|K_{n+1}(\mu, x, t) + K_{n+1}(\mu, x, t)\right|$$
$$+\left|\frac{p_n(\mu, t)}{p_n(\mu, x)}\right| \left|\frac{\partial}{\partial t}\left(K_{n+1}(\mu, x, t) + K_{n+1}(\mu, x, t)\right)\right|$$
$$\leq \frac{1}{r}K_{n+1}^{(1,1)}(\nu, t, t)^{1/2}\, 2\left(K_{n+1}(\nu, t, t)\, K_{n+1}(\nu, x, x)\right)^{1/2}$$
$$+\frac{1}{r}K_{n+1}(\nu, t, t)^{1/2}\, 2\left(K_{n+1}^{(1,1)}(\nu, t, t)\, K_{n+1}(\nu, x, x)\right)^{1/2}.$$

Then (29) follows. ■

To prove that ψ_n is absolutely continuous, is more complicated. We first need

Lemma 6 *Let $y > 1$, $I = [c, y]$, and for $|h| \leq 1$,*

$$\omega_h = \sum_{j=1}^{N} \beta_j \delta_{\xi_j + h\sigma(\xi_j)} \tag{30}$$

where $N \geq 1$, all $\{\xi_j\}_{j=1}^{N}$ are distinct and lie in $[c, y] \setminus (-1, 1)$, all $\sigma(\sigma_j) = \pm 1$, and all β_j are real with

$$\sum_{j=1}^{N} \left|\beta_j\right| \leq \Lambda. \tag{31}$$

Fix $n \geq 1$, $x \in (-1, 1)$, and $\mu \in \mathcal{M}(I, \Lambda)$. For $|h| \leq 1$ let

$$\mu^{(h)} = \mu - \omega_0 + \omega_h. \tag{32}$$

(a) Then there exist numbers $F, C > 0$, $h_0 > 0$ such that for $|h| \leq h_0$, subject to $\mu^{(h)}$ being a nonnegative measure,

$$\left| p_n\left(\mu^{(h)}, x\right) - p_n(\mu, x) - Fh \right| \le Ch^2. \tag{33}$$

Here the number F is continuous in the parameters $\{\beta_j\}, \{\xi_j\}$. The constant C depends only on y, Λ, ν, n and not on the particular $\mu, \{\beta_j\}, \{\xi_j\}, N, \sigma$.

(b) *Fix $r > 0$, and assume that*

$$|p_n(\mu, x)| \ge r. \tag{34}$$

Then there exists h_1 such that for $|h| \le h_1$ for which $\mu^{(h)}$ is a nonnegative measure,

$$\left| \ln \frac{1}{p_n\left(\mu^{(h)}, x\right)^2} - \ln \frac{1}{p_n(\mu, x)^2} - \left(\int R_n'(\mu, t)\,\sigma(t)\,d\omega_0(t) \right) h \right| \le C_1 h^2. \tag{35}$$

The constants h_1 and C_1 depend on y, ν, n, Λ and the lower bound r but not on the particular $\mu, \{\beta_j\}, \{\xi_j\}, N, \sigma$.

Proof

(a) Observe that for each fixed k,

$$\int t^k d\mu^{(h)}(t) - \int t^k d\mu(t)$$
$$= \left(\sum_{j=1}^{N} \beta_j k \xi_j^{k-1} \sigma(\xi_j) \right) h + O\left(h^2\right),$$

where the sum in the last line is uniformly continuous in $\{\beta_j\}$ and $\{\xi_j\}$ when the $\beta_j \ge 0$ and satisfy (31), while the $\{\xi_j\}$ lie in $[c, d] \setminus (-1, 1)$. Moreover the constant in the order term depends only on Λ, y, L, k. Then if D_m is the determinant defined by (15), we see that also

$$D_m\left(\mu^{(h)}\right) - D_m(\mu) = E_m h + O\left(h^2\right),$$

where E_m is uniformly continuous in $\{\beta_j\}$ and $\{\xi_j\}$ when the $\beta_j \ge 0$ and satisfy (31), while the $\{\xi_j\}$ lie in $[c, y] \setminus (-1, 1)$. Moreover the constant in the order term depends only on Λ, c, y, m. Then a similar assertion is true for the numerator in the determinant representation (16) of $p_n(\mu, x)$. Finally, we know that as each $\mu^{(h)} \ge \nu$, each $D_m\left(\mu^{(h)}\right)$ is bounded below by a constant depending only on $D_m(\nu)$, at least for small enough $|h|$. Then the result follows from the representation (16).

(b) From (a) and the lower bound for $p_n(\mu, x)$,

$$\left| \ln \frac{1}{p_n\left(\mu^{(h)}, x\right)^2} - \ln \frac{1}{p_n\left(\mu, x\right)^2} - Gh \right| \leq C_1 h^2,$$

where C_1 and h_1 satisfy the uniformity conclusions in (b). Also then from Theorem 6,

$$G = \frac{\partial}{\partial h}\left\{ \ln \frac{1}{p_n\left(\mu^{(h)}, x\right)^2} \right\}_{|h=0} = \int R_n'\left(\mu, t\right) \sigma\left(t\right) d\omega_0\left(t\right).$$

■

We next show that the left and right-hand derivatives exist at each $y > 1$.

Lemma 7

(a) For $y \in (1, \infty)$ there exists

$$D_+ \psi_n\left(y\right) = \lim_{h \to 0+} \frac{\psi_n\left(y+h\right) - \psi_n\left(y\right)}{h} = \inf_{\mu_{[c,y]}} \mu_{[c,y]}\left(\{y\}\right) R_n'\left(\mu_{[c,y]}, y\right).$$

(b) For $y \in (1, \infty)$ there exists

$$D_- \psi_n\left(y\right) = \lim_{h \to 0-} \frac{\psi_n\left(y+h\right) - \psi_n\left(y\right)}{h} = \sup_{\mu[c,y]} \mu_{[c,y]}\left(\{y\}\right) R_n'\left(\mu_{[c,y]}, y\right).$$

Proof

(a) We proved in Lemma 4 that

$$\begin{aligned} &\limsup_{h \to 0+} \left[\psi_n\left(y+h\right) - \psi_n\left(y\right)\right] / h \\ &\leq \inf_{\mu_{[c,y]}} \mu_{[c,y]}\left(\{y\}\right) R_n'\left(\mu_{[c,y]}, y\right). \end{aligned} \tag{36}$$

Indeed in (27), $\mu_{[c,y]}$ was any extremal measure. Next choose a sequence $\{h_m\}$ of positive numbers with limit 0 such that

$$\begin{aligned} &\lim_{m \to \infty} \left[\psi_n\left(y+h_m\right) - \psi_n\left(y\right)\right] / h_m \\ &= \liminf_{h \to 0+} \left[\psi_n\left(y+h\right) - \psi_n\left(y\right)\right] / h. \end{aligned} \tag{37}$$

By Theorem 2, each extremal measure μ_{I_m} for $I_m = [c, y + h_m]$ has a representation of the form

$$\mu_{I_m} = \nu + \sum_{j=1}^{L} \alpha_{jm}\delta_{\xi_{jm}} \tag{38}$$

We may assume that $L \leq n+1$ is the same for all m by passing to a subsequence. By passing to another subsequence, we can assume that μ_{I_m} converges weakly as $m \to \infty$ to some measure $\hat{\mu}_I$. Then $\hat{\mu}_I$ is an extremal measure for $I = [c, y]$ by Lemma 3. Next, fix a small $\delta > 0$ such that $y - \delta$ is not a mass point of $\hat{\mu}_I$, nor of μ_{I_m}, $m \geq 1$. For $m \geq 1$, $|h| \leq c$, let

$$\omega_{m,h} = \sum_{j:\xi_{jm}\in[y-\delta,y+h_m]} \alpha_{jm}\delta_{\xi_{jm}-h} \tag{39}$$

and

$$\mu^{(m)} = \mu_{[c,y+h_m]} - \omega_{m,0} + \omega_{m,h_m}. \tag{40}$$

Thus we are shifting all possible mass points of μ_{c+h_m} in the interval $[y-\delta, y+h_m]$ to the left by h_m. Then $\mu^{(m)} \in \mathcal{M}(I, \Lambda)$. By Lemma 6(b), for large enough m (recall $\left|p_n\left(\mu_{I_m}, x\right)\right|$ is bounded below for large enough m as it is maximal over the class $\mathcal{M}(\nu, S, I_m)$ and approaches $\left|p_n\left(\hat{\mu}_I, x\right)\right| = \phi_n(y)$)

$$\left| \ln \frac{1}{p_n\left(\mu^{(m)}, x\right)^2} - \ln \frac{1}{p_n\left(\mu_{I_m}, x\right)^2} - \int R_n'\left(\mu_{I_m}, t\right)(-sign\,(t))\,d\omega_{m,0}(t)\,h_m \right| \leq C_1 h_m^2 \tag{41}$$

where C_1 is independent of m. Here we are using the unformity in the measure in Lemma 6, which allows us to apply that lemma to μ_{I_m} uniformly in m. Then

$$\begin{aligned}
&\left(\ln \frac{1}{p_n\left(\mu^{(m)}, x\right)^2} - \ln \frac{1}{p_n\left(\mu_{I_m}, x\right)^2} \right) / h_m \\
&= -\int R_n'\left(\mu_{I_m}, t\right) sign\,(t)\,d\omega_{m,0}(t) + O\left(h_m\right) \\
&= -\int_{y-\delta}^{y+h_m} R_n'\left(\mu_{I_m}, t\right) d\mu_{I_m}(t) + O\left(h_m\right).
\end{aligned}$$

Of course the possible mass points of μ_{I_m} at the endpoint $y + h_m$ are included in the integrals. As μ_{I_m} converges weakly to $\hat{\mu}_I$ and $y - \delta$ is not a mass point of $\hat{\mu}_I$, and by the continuity established in Lemma 6,

$$\lim_{m\to\infty}\left(\ln\frac{1}{p_n\left(\mu^{(m)},x\right)^2}-\ln\frac{1}{p_n\left(\mu_{I_m},x\right)^2}\right)/h_m$$
$$=-\int_{y-\delta}^{y}R_n'\left(\hat{\mu}_I,t\right)d\hat{\mu}_I\left(t\right). \tag{42}$$

Then from (37), using $\phi_n\left(y\right)^2\geq p_n\left(\mu^{(m)},x\right)^2$,

$$\begin{aligned}
&\liminf_{h\to 0+}\left[\psi_n\left(y+h\right)-\psi_n\left(y\right)\right]/h\\
&=\lim_{m\to\infty}\left[\ln\frac{1}{\phi_n\left(y+h_m\right)^2}-\ln\frac{1}{\phi_n\left(y\right)^2}\right]/h_m\\
&\geq\lim_{m\to\infty}\left[\ln\frac{1}{p_n\left(\mu_{I_m},x\right)^2}-\ln\frac{1}{p_n\left(\mu^{(m)},x\right)^2}\right]/h_m\\
&=\int_{y-\delta}^{y}R_n'\left(\hat{\mu}_I,t\right)d\hat{\mu}_I\left(t\right),
\end{aligned}$$

by (42). Here $\delta>0$ can be taken arbitrarily small, subject only to $y-\delta$ not being a mass point of $\hat{\mu}_I$ and μ_{I_m} for all m, so we obtain

$$\begin{aligned}
&\liminf_{h\to 0+}\left[\psi_n\left(y+h\right)-\psi_n\left(y\right)\right]/h\\
&\geq R_n'\left(\hat{\mu}_I,y\right)\hat{\mu}_I\left(\{y\}\right)\\
&\geq\inf_{\mu_{[c,y]}}R_n'\left(\mu_{[c,y]},y\right)\mu_{\mu_{[c,y]}}\left(\{y\}\right).
\end{aligned}$$

Together with (36), this gives the result.

(b) Again let $\mu_{[c,y]}$ be an extremal measure for the given y. As before, we have (27), so that

$$\begin{aligned}
&\liminf_{h\to 0-}\left[\psi_n\left(y+h\right)-\psi_n\left(y\right)\right]/h\\
&\geq\sup_{\mu_{[c,y]}}\mu_{[c,y]}\left(\{y\}\right)R_n'\left(\mu_{[c,y]},y\right).
\end{aligned}\tag{43}$$

Next choose a sequence $\{h_m\}$ of negative numbers with limit 0 such that

$$\begin{aligned}
&\lim_{m\to\infty}\left[\psi_n\left(y+h_m\right)-\psi_n\left(y\right)\right]/h_m\\
&=\limsup_{h\to 0-}\left[\psi_n\left(y+h\right)-\psi_n\left(y\right)\right]/h.
\end{aligned}$$

By Theorem 2, each extremal measure μ_{I_m} for $I_m = [c, y+h_m]$ has a representation of the form (38). As above, by passing to a subsequence, we can assume that μ_{I_m} converges weakly as $m \to \infty$ to some measure $\hat{\mu}_I$. Then $\hat{\mu}_I$ is an extremal measure for $I = [c, y]$ by Lemma 3. Next, fix a small $\delta > 0$ such that $y - \delta$ is not a mass point of $\hat{\mu}_I$ nor any μ_{I_m}. For $m \geq 1$, define $\omega_{m,h}$ and $\mu^{(m)}$ by (39) and (40). Thus we are shifting all possible mass points of μ_{I_m} in the interval $[y-\delta, y+h_m]$ to the right by $|h_m|$. Then $\mu^{(m)} \in \mathcal{M}(I, \Lambda)$. As above, and recalling that $h_m < 0$,

$$\begin{aligned}
&\limsup_{h \to 0-} [\psi_n(y+h) - \psi_n(y)]/h \\
&= \lim_{m \to \infty} \left[\ln \frac{1}{\phi_n(y+h_m)^2} - \ln \frac{1}{\phi_n(y)^2} \right] / h_m \\
&\leq \lim_{m \to \infty} \left[\ln \frac{1}{p_n\left(\mu_{I_m}, x\right)^2} - \ln \frac{1}{p_n\left(\mu^{(m)}, x\right)^2} \right] / h_m \\
&= \int_{y-\delta}^{y} R_n'\left(\hat{\mu}_I, t\right) d\hat{\mu}_I(t).
\end{aligned}$$

As above we can let $\delta \to 0+$ through a suitable subsequence and hence

$$\begin{aligned}
&\limsup_{h \to 0-} [\psi_n(y+h) - \psi_n(y)]/h \\
&\leq R_n'\left(\hat{\mu}_I, y\right) \hat{\mu}_I(\{y\}) \\
&\leq \sup_{\mu_{[c,y]}} R_n'\left(\mu_{[c,y]}, y\right) \mu_{[c,y]}(\{y\}).
\end{aligned}$$

Together with (43), this gives the result. ■

Proof of Theorem 3 We have already proved in Lemma 4 that $\psi_n'(y)$ is given by (7) at every point where the derivative exists and in particular a.e. Now we prove the absolute continuity of ψ in $[S, T]$ where $1 < S < T < \infty$. This follows from the uniform boundedness of the left and right derivatives of ψ in compact intervals. Recall that from Lemma 5,

$$\left|R_n'(\mu, y)\right| \leq \frac{4}{\left|p_n\left(\mu_{[c,y]}, x\right)\right|} \left[K_{n+1}(\nu, x, x) K_{n+1}(\nu, y, y) K_{n+1}^{(1,1)}(\nu, y, y)\right]^{1/2},$$

so is uniformly bounded for $y \in [S, T]$. Also

$$\mu_{[c,y]}(\{y\}) \leq \Lambda.$$

Then the previous lemma shows that for each $y \in [S, T]$, there exists $\varepsilon_y > 0$ such that

$$0 \geq \frac{\psi_n(y+h) - \psi_n(y)}{h} \geq -C, h \in [0, \varepsilon_y]$$

and

$$0 \leq \frac{\psi_n(y-h) - \psi_n(y)}{-h} \leq C, h \in [0, \varepsilon_y]$$

Here C is independent of y. Thus

$$|\psi_n(y+h) - \psi_n(y)| \leq C|h| \text{ for all } |h| \leq \varepsilon_y.$$

Let $S \leq s < t \leq T$. Since the intervals $\left\{\left(y - \frac{1}{2}\varepsilon_y, y + \frac{1}{2}\varepsilon_y\right) : y \in [S, T]\right\}$ cover $[s, t]$, we can find finitely many that cover $[s, t]$. Then for some $\{y_j\}$,

$$s = y_0 < y_1 < \ldots < y_m = t$$

with $y_{j+1} - y_j \leq \varepsilon_{y_j}$ for all j. Then

$$0 \leq \psi_n(s) - \psi_n(t) = \sum_{j=0}^{m-1} \left(\psi_n(y_j) - \psi_n(y_{j+1})\right)$$

$$\leq C \sum_{j=0}^{m-1} \left(y_{j+1} - y_j\right) = C(t - s).$$

So ψ_n satisfies a Lipschitz condition of order 1 in $[S, T]$ with Lipschitz constant C. By standard results, ψ_n is absolutely continuous in $[S, T]$. ■

5 Proof of Theorem 4

Recall that we fix $I = [c, d]$ and x, and let

$$\chi_n(\Lambda) = \ln \frac{1}{\phi_n(I, \Lambda, x)^2}, \Lambda > 0.$$

Proof of Theorem 4 Let $\mu_{I,\Lambda} = \nu + \rho_{I,\Lambda}$ denote an extremal measure for the given $\Lambda > 0$ and for the given interval I. Let $\tau = \rho_{I,\Lambda}(I)$. For $\varepsilon > 0$, let

$$\mu^{\varepsilon} = \mu_{I,\Lambda} + \frac{\varepsilon}{\tau}\rho_{I,\Lambda} = \nu + \left(1 + \frac{\varepsilon}{\tau}\right)\rho_{I,\Lambda}$$

Then $\mu^{\varepsilon} \in M(\nu, \Lambda+\varepsilon, I)$, and so by Theorem 5, with $\omega = \frac{1}{\tau}\rho_I$, as $\varepsilon \to 0+$,

$$\begin{aligned}
&\frac{\chi_n(\Lambda+\varepsilon)-\chi_n(\Lambda)}{\varepsilon} \\
&\leq \frac{1}{\varepsilon}\left[\ln\frac{1}{|p_n(\mu^{\varepsilon},x)|^2}-\ln\frac{1}{|p_n(\mu_I,x)|^2}\right] \\
&= \frac{1}{\tau}\left(\int R_n\left(\mu_{I,\Lambda},t\right)d\rho_{I,\Lambda}(t)\right)+o(1) \\
&= \inf\left\{R_n\left(\mu_{I,\Lambda},t\right): t\in I\right\},
\end{aligned}$$

by Theorem 2(c). In the other direction, for small $\varepsilon > 0$, let

$$\mu^{\varepsilon} = \mu_{I,\Lambda} + \frac{\varepsilon}{\tau}(-\rho_{I,\Lambda}) = \nu + \left(1-\frac{\varepsilon}{\tau}\right)\rho_{I,\Lambda}.$$

By Theorem 5, with $\omega = -\frac{1}{\tau}\rho_{I,\Lambda}$, as $\varepsilon \to 0+$,

$$\begin{aligned}
&\frac{\chi_n(\Lambda-\varepsilon)-\chi_n(\Lambda)}{-\varepsilon} \\
&\geq \frac{1}{-\varepsilon}\left[\ln\frac{1}{|p_n(\mu^{\varepsilon},x)|^2}-\ln\frac{1}{|p_n(\mu_I,x)|^2}\right] \\
&= -\frac{1}{\tau}\left(\int R_n\left(\mu_{I,\Lambda},t\right)d(-\rho_{I,\Lambda})(t)\right)+o(1) \\
&= \inf\left\{R_n\left(\mu_{I,\Lambda},t\right): t\in I\right\},
\end{aligned}$$

by Theorem 2(c) again. Thus at every $\Lambda > 0$,

$$\begin{aligned}
\limsup_{h\to 0+}\frac{\chi_n(\Lambda+h)-\chi_n(\Lambda)}{h} &\leq \inf\left\{R_n\left(\mu_{I,\Lambda},t\right): t\in I\right\} \\
&\leq \liminf_{h\to 0-}\frac{\chi_n(\Lambda+h)-\chi_n(\Lambda)}{h}. \qquad (44)
\end{aligned}$$

Then if $\chi_n'(\Lambda)$ exists, we obtain,

$$\chi_n'(\Lambda) = \inf\left\{R_n\left(\mu_{I,\Lambda},t\right): t\in I\right\}.$$

As χ_n is decreasing in Λ, this holds a.e. The continuity of the inf in Λ follows easily from Lemma 3. ■

References

1. M. Alexis, A. Aptekarev, S.Denisov, *Continuity of weighted operators, Muckenhoupt $A_p$weights, and Steklov problem for orthogonal polynomials,* Int. Math. Res. Not. IMRN 2022, no. 8, 5935–5972.
2. M.U. Ambroladze, *On the possible rate of growth of polynomials that are orthogonal with a continuous positive weight.* Math. USSR-Sb. 72 (1992), 311–331.
3. A. Aptekarev, S. Denisov, D. Tulyakov, *Steklov's problem of estimating the growth of orthogonal polynomials*, Proc. Steklov Inst. Math. 289 (2015), 72–95.
4. A. Aptekarev, S. Denisov, D. Tulyakov, *On a problem by Steklov*, J. Amer. Math. Soc. 29 (2016), 1117–1165.
5. V.M. Badkov, *The asymptotic behavior of orthogonal polynomials.* (Russian) Mat. Sb. (N.S.) 109(151) (1979), no. 1, 46–59.
6. C. Brezinski, K.A. Driver, and M. Redivo-Zaglia, *Quasi-orthogonality with applications to some families of classical orthogonal polynomials*, Applied Numerical Mathematics, 48(2004), 157-168.
7. S. Denisov, *Remark on the formula by Rakhmanov and Steklov's conjecture,* J. Approx. Theory 205 (2016), 102–113.
8. S. Denisov, *On the size of the polynomials orthonormal on the unit circle with respect to a measure which is a sum of the Lebesgue measure and ppoint masses*, Proc. Amer. Math. Soc. 144 (2016), 1029–1039.
9. S. Denisov, *On the growth of polynomials orthogonal on the unit circle with a weight w that satisfies $w, w^{-1} \in L_\infty(T)$*, Math. Sbornik, 209 (2018), 985–1018.
10. S. Denisov, K. Rush, *Orthogonal polynomials on the circle for the weight w satisfying conditions $w, w^{-1} \in BMO$*, Constr. Approx. 46 (2017), no. 2, 285–303.
11. S. Denisov, K. Rush, *On Schur parameters in Steklov's problem.* J. Approx. Theory, 215 (2017), 68–91.
12. G. Freud, *Orthogonal Polynomials*, Pergamon Press/ Akademiai Kiado, Budapest, 1971.
13. Eli Levin, D.S. Lubinsky, *Bounds on orthogonal polynomials and separation of their zeros,* J. Spectr. Theory, 12 (2022), 497–513.
14. D.S. Lubinsky, *Bounds on Orthonormal Polynomials for Restricted Measures*, Constr. Approx., https://link.springer.com/article/10.1007/s00365-023-09671-z
15. E.A. Rahmanov, *Steklov's conjecture in the theory of orthogonal polynomials,* (Russian) Mat. Sb. (N.S.) 108(150) (1979), 581–608.
16. E.A. Rakhmanov, *Estimates of the growth of orthogonal polynomials whose weight is bounded away from zero,* (Russian) Mat. Sb. (N.S.) 114(156) (1981), 269–298.
17. E.B. Saff and V. Totik, *Logarithmic Potentials with External Fields*, Springer, Berlin, 1997.
18. J.A. Shohat, *On mechanical quadratures, in particular, with positive coefficients,* Trans. Amer. Math. Soc., 42(1937), 461-496.
19. B. Simon, *Orthogonal Polynomials on the Unit Circle*, Parts 1 and 2, American Mathematical Society, Providence, 2005.
20. G. Szegő, *Orthogonal Polynomials*, Amer. Math. Soc. Colloq. Publns., Vol. 23, 4th edn, 1975, Providence.

Behavior of Bounded Solutions, $u(x, t)$, to the Heat Equation in $\mathbb{R}^n \times (0, \infty)$ as $t \to \infty$

Wolodymyr R. Madych

1 Introduction

For $x = (x_1, \ldots, x_n) \in \mathbb{R}^n$ and $t > 0$, consider the solution, $u(x, t)$, to the heat equation that is bounded on $\mathbb{R}^n \times (0, \infty)$. Namely, $u(x, t)$ satisfies

$$\frac{\partial u}{\partial t} = \Delta u, \quad \text{on } \mathbb{R}^n \times (0, \infty),$$

where Δ is the Laplacian in $\mathbb{R}^n$, and

$$|u(x, t)| \le C < \infty \text{for all } x \in \mathbb{R}^n \text{and } t > 0.$$

In what follows, the class of all such functions u is denoted as BH.

This article is concerned with the behavior as t tends to ∞ of functions $u(x, t)$ in BH.

If u is in BH then as a consequence of weak compactness, see [3, Chapter VII, Theorems A and 1], there is a function $f(x)$ in $L^\infty(\mathbb{R}^n)$ such that

$$u(x, t) = G_t * f(x) = \int_{\mathbb{R}^n} G_t(x - y) f(y) dy, \tag{1}$$

W. R. Madych (✉)
Department of Mathematics, University of Connecticut, Mansfield, UK
e-mail: wolodymyr.madych@uconn.edu

D. Bilyk et al. (eds.), *Recent Advances in Approximation and Potential Theory*, Applied and Numerical Harmonic Analysis,
https://doi.org/10.1007/978-3-031-95551-8_16

where $G(x)$ is the Gaussian

$$G(x) = \frac{e^{-|x|^2/4}}{(\sqrt{4\pi})^n} \quad \text{and} \quad G_t(x) = t^{-n/2}G(x/\sqrt{t}) = \frac{e^{-|x|^2/4t}}{(\sqrt{4\pi t})^n}.$$

Note that $G_1(x) = G(x)$.

As a consequence of classical results concerning convolution type approximate identities, for example see [2, p. 62, Theorem 2],

$$\lim_{t\to 0} u(x,t) = f(x) \quad \text{for almost all } x \in \mathbb{R}^n. \tag{2}$$

The results and methods of [1, pp. 216 and 217] imply that the solution $u(x,t)$ defined by (1) is the unique solution that is bounded on $\mathbb{R}^n \times (0,\infty)$ and satisfies (2); see also [3, Chapter 2, Theorem 6].

If f is continuous on $\mathbb{R}^n$ then the function $u(x,t)$ defined by (1) is continuous on $\mathbb{R}^n \times [0,\infty)$ and hence is a classical solution to the initial value problem

$$\frac{\partial u}{\partial t} = \Delta u \quad \text{for } (x,t) \in \mathbb{R}^n \times (0,\infty), \textit{and} \quad u(x,0) = f(x).$$

For more details on this issue, see [1, Theorem on p, 209].

In view of (2), the essentially bounded function f in (1) may be regarded as initial data, even in cases when it fails to be continuous. We shall do so in what follows.

Notation and Conventions We use standard mathematical notation. For example, the notation $|x|$, where $|x|^2 = x_1^2 + \cdots + x_n^2$, was made use of earlier. The notation $w(x)$ is used to indicate the dependence of the distribution w on the variable x even when w may not be pointwise defined and when no confusion is possible. As customary, C is used to denote generic constants whose value depends on the context.

2 The Behavior as $t \to \infty$ of Functions $u(x,t)$ in BH

Proposition 1 *If u is in BH then, as t tends to ∞, $u(x,t)$ either (i) converges to a constant c independent of x, uniformly on compact subsets of $\mathbb{R}^n$, or (ii) fails to converge for every x in $\mathbb{R}^n$.*

This proposition is a consequence of the following three relatively simple lemmas.

Lemma 1 *Suppose $u \in BH$ and f is the $L^\infty(\mathbb{R}^n)$ function in representation* (1) *of u. If x and $\widetilde{x}$ are any pair of points in $\mathbb{R}^n$, then*

$$|u(x,t) - u(\widetilde{x},t)| \leq \omega(|x - \widetilde{x}|/\sqrt{t})\|f\|_{L^\infty} \tag{3}$$

where

$$\omega(\delta) = \sup_{|y|\leq\delta} \int_{\mathbb{R}^n} |G(x+y) - G(x)|dx.$$

Proof Write

$$\begin{aligned} u(x,t) - u(\widetilde{x},t) &= G_t * f(x) - G_t * f(\widetilde{x}) \\ &= \int_{\mathbb{R}^n} \big(G_t(x-y) - G_t(\widetilde{x}-y)\big) f(y)dy \\ &= \int_{\mathbb{R}^n} \big(G_t(y + x - \widetilde{x}) - G_t(y) f(\widetilde{x} - y)dy \\ &= \int_{\mathbb{R}^n} \left(G\left(y + \frac{x-\widetilde{x}}{\sqrt{t}}\right) - G(y)\right) f(\widetilde{x} - \sqrt{t}y)dy. \end{aligned}$$

Hence

$$\begin{aligned} |u(x,t) - u(\widetilde{x},t)| &\leq \int_{\mathbb{R}^n} \left|G\left(y + \frac{x-\widetilde{x}}{\sqrt{t}}\right) - G(y)\right| |f(\widetilde{x} - \sqrt{t}y)|dy \\ &\leq \omega(|x - \widetilde{x}|/\sqrt{t})\|f\|_{L^\infty}, \end{aligned}$$

which is the desired result. □

Lemma 2 *Suppose $u \in BH$. If x and $\widetilde{x}$ are any pair of points in $\mathbb{R}^n$, then*

$$\lim_{t\to\infty} |u(x,t) - u(\widetilde{x},t)| = 0.$$

This convergence is uniform on compact subsets of $\mathbb{R}^n$.

Proof Suppose A is a compact subset of $\mathbb{R}^n$ and $r < \infty$ is the radius of the ball that contains A. Then, for any pair of points x and $\widetilde{x}$ in A, $|x - \widetilde{x}| \leq 2r$ and, by virtue of Lemma 1 it follows that $|u(x,t) - u(\widetilde{x},t)| \leq \omega(2r/\sqrt{t})$. Hence the conclusions of Lemma 2 are consequences of the fact that $\lim_{\delta\to 0} \omega(\delta) = 0$. □

Lemma 3 *Suppose $u \in BH$. If for some point $\widetilde{x}$ in $\mathbb{R}^n$, $\lim_{t\to\infty} u(\widetilde{x},t) = c$, then $\lim_{t\to\infty} u(x,t) =$ the same constant c for all x in $\mathbb{R}^n$ uniformly on compact subsets.*

Proof Suppose A is a compact subset of $\mathbb{R}^n$ then there is an $r < \infty$ such that $|x - \widetilde{x}| \leq r$ for all x in A. Let $\epsilon > 0$, write

$$|u(x,t) - c| \leq |u(x,t) - u(\widetilde{x},t)| + |u(\widetilde{x},t) - c|,$$

and note that for t sufficiently large, $|u(x,t) - u(\widetilde{x},t)| \le \omega(r/\sqrt{t}) < \epsilon/2$ and $|u(\widetilde{x},t) - c| < \epsilon/2$. Hence, for sufficiently large t, $|u(x,t) - c| < \epsilon$ for all $x \in A$ and the conclusion of the lemma follows. □

In view of Lemma 3, if $u(x,t) \in BH$ its behavior as $t \to \infty$ at all $x \in \mathbb{R}^n$ is completely reflected by its behavior at a single value of x. For convenience and without loss of generality, we will use the origin, namely the value $x = 0$, as the point of interest in determining the convergence or divergence of u as $t \to \infty$.

We conclude this section with a corollary of Proposition 1.

Corollary 1 *Suppose* $\mathfrak{C} = \{u \in BH$ *that enjoy property (i) of Proposition 1* $\}$ *and* $\mathfrak{D} = \{u \in BH$ *that enjoy property (ii) of Proposition 1* $\}$. *Then*

$$\mathfrak{C} \neq \emptyset, \quad \mathfrak{D} \neq \emptyset, \text{ and } \quad \mathfrak{C} \cup \mathfrak{D} = BH.$$

The fact that $\mathfrak{C} \cup \mathfrak{D} = BH$ is an immediate consequence of Proposition 1.

The fact that $\mathfrak{C} \neq \emptyset$ is evident from examples such as $u(x,t) =$ constant and $u(x,t) = e^{-t} \sin\langle \xi, x \rangle$ where $\xi = (\xi_1, \ldots, \xi_n)$ is a unit vector in $\mathbb{R}^n$ and $\langle \xi, x \rangle$ is the scalar product $\xi_1 x_1 + \cdots + \xi_n x_n$.

That $\mathfrak{D} \neq \emptyset$ is not so evident; it is established in Sect. 4 via Proposition 2 by exhibiting a class of bounded functions f that, when used as initial data, result in $u \in \mathfrak{D}$.

3 A Handy Class of Initial Data

The main objective of this section is to show that $\mathfrak{D}$, the set of those functions $u(x,t)$ in BH with the property that $u(x,t)$ diverges as $t \to \infty$, is not empty. To this end, we define and establish certain useful characteristics of a class of functions $f(x)$ that, when used as initial data, result in solutions $u(x,t)$ with the desired property.

Definition 1 Given a positive constant λ, the space $\mathfrak{N}_\lambda$ consists of all complex valued functions f that satisfy

$$f(\lambda x) = -f(x) \quad \text{for all } x \text{in } \mathbb{R}^n \setminus \{0\}. \tag{4}$$

Furthermore, define $\mathfrak{N}_\lambda^\infty = \mathfrak{N}_\lambda \cap L^\infty(\mathbb{R}^n)$.

In view of the fact that if $f \in \mathfrak{N}_\lambda$ then $f(x) = f(\lambda \lambda^{-1} x) = -f(x/\lambda)$, the function f also satisfies

$$f(x/\lambda) = -f(x). \tag{5}$$

Hence, without loss of generality, in what follows we assume that $\lambda > 1$.

Identities (4) and (5) imply that if $f \in \mathfrak{N}_\lambda$ then for any integer n, $n = 0, \pm 1, \pm 2, \dots$,

$$f(\lambda^n x) = (-1)^n f(x). \tag{6}$$

Identity (6) implies that, given any positive σ, if f is in $\mathfrak{N}_\lambda$ then its values on $\mathbb{R} \setminus \{0\}$ are completely determined by its values on $\Omega_0 = \{x : \sigma \le |x| < \lambda\sigma\}$. To see this, let $\Omega_n = \{x : \lambda^n \sigma \le |x| < \lambda^{n+1}\sigma\}$ and note that for any x in Ω_n we may write

$$f(x) = (-1)^n f(x/\lambda^n)$$

where x/λ^n is in Ω_0. Since

$$\Omega_m \cap \Omega_n = \emptyset \text{ if } m \ne n \quad \text{and} \quad \mathbb{R} \setminus \{0\} = \bigcup_{n=-\infty}^{\infty} \Omega_n,$$

the desired result follows.

In view of this property, examples of functions that satisfy (4) can be readily concocted. A typical example that can be expressed in terms of elementary functions is

$$f(x) = \sin\left(\frac{\pi \log|x|}{\log \lambda}\right). \tag{7}$$

4 Members of $\mathfrak{N}_\lambda^\infty$ as Initial Data

Suppose that the initial data, the function f in formula (1), is in $\mathfrak{N}_\lambda^\infty$. Then

$$u(0, \lambda^2 t) = G_{\lambda^2 t} * f(0) = \int_{-\infty}^{\infty} G(y) f(\lambda\sqrt{t}\, y) dy = -u(0, t).$$

Hence if $u(0, 1) = c \ne 0$, then $u(0, \lambda^{2n}) = (-1)^n c$ and, since $\lambda > 1$, it follows that $u(0, t)$ fails to converge as $t \to \infty$.

There are many such functions f that will do the job. Namely, functions f that are in $\mathfrak{N}_\lambda^\infty$ and $g * f(0) = u(0, 1) = c > 0$.

For example, let $0 < \sigma < \rho < \infty$ be such that $\int_{\sigma \le |y| < \rho} G(y) dy \ge 0.9$. Let $\lambda = \rho/\sigma$ and define

$$f(x) = (-1)^n \quad \text{if } \sigma \le \frac{|x|}{\lambda^n} < \rho = \lambda\sigma \text{ for } n = 0, \pm 1, \pm 2, \dots.$$

Then $f(x)$ is well defined for all x in $\mathbb{R}^n \setminus \{0\}$ and is in $\mathfrak{N}_\lambda^\infty$. Furthermore

$$u(0, 1) = G * f(0) = c \quad \text{where } c \geq 0.8.$$

Of course, this example can be modified so that f is continuous, even in C^∞, on $\mathbb{R}^n \setminus \{0\}$.

The material in this section can be summarized as follows:

Proposition 2 *There are functions $f(x)$ in $\mathfrak{N}_\lambda^\infty$ that, when used as initial data in formula* (1), *result in functions $u(x,t)$ in BH with the property that $u(x,t)$ fails to converge for every $x \in \mathbb{R}^n$ as $t \to \infty$.*

Corollary 2 *$\mathfrak{D}$ is not empty.*

Furthermore such initial data functions f can be modified on compact subsets of $\mathbb{R}^n$ without altering the divergent behavior of the resulting $u(x,t)$ as $t \to \infty$. More precisely, if f_1 is initial data such that $u_1(x,t) = G_t * f_1(x)$ fails to converge as $t \to \infty$ and f_2 is a distribution with compact support, then $u(x,t) = G_t * (f_1 + f_2)(x)$ is also a solution to the heat equation on $\mathbb{R}^n \times (0, \infty)$ that fails to converge as $t \to \infty$. This follows from the fact that $\lim_{t\to\infty} G_t * f_2 = 0$, which will be verified in the next section.

Before closing this section, we note the following:
If u is in $\mathfrak{D}$, in view of Lemma 2, the behavior of $u(x,t)$ as $t \to \infty$ is analogous to the behavior of functions in $\mathfrak{C}$, in the sense that for fixed t that is sufficiently large $u(x,t)$ is uniformly close to $u(0,t)$, which is a constant independent of x, on compact subsets of $\mathbb{R}^n$.

5 Examples of Convergence

In addition to the examples mentioned earlier more examples of cases where $u(x,t)$ converges as $t \to \infty$ are documented in this section in terms of their initial data. For the sake of generality, classes of initial data wider than $L^\infty(\mathbb{R}^n)$, that lead to convergence of $u(x,t)$ as $t \to \infty$, are also considered.

In what follows, the relationship between f and u is given by Eq. (1), namely, $u(x,t) = G_t * f(x)$.

Recall that, in order to verify that $\lim_{t\to\infty} u(x,t) = c$ uniformly on compact subsets of $x \in \mathbb{R}^n$ it suffices to show that $\lim_{t\to\infty} u(0,t) = c$.

Example 1 If f is an odd function in $L^\infty(\mathbb{R}^n)$, namely, $f(-x) = f(x)$, then $\lim_{t\to\infty} u(x,t) = 0$ uniformly on compact subsets of $x \in \mathbb{R}^n$.

This is a consequence of the fact that

$$u(0,t) = G_t * f(0) = \int_{\mathbb{R}^n} G_t(x) f(x) dx = 0 \quad \text{for all } t > 0,$$

which follows from the fact that the integrand $G_t(x)f(x)$ is an odd function.

Example 2 Suppose the function f in $L^\infty(\mathbb{R}^n)$ satisfies

$$\lim_{r\to\infty} f(r\sigma) = 0 \quad \text{for almost all unit vectors } \sigma \in S^{n-1} = \{x : |x| = 1\}.$$

Then $\lim_{t\to\infty} u(x,t) = 0$ for all x in $\mathbb{R}^n$ uniformly on compact subsets.

Since

$$u(0,t) = \int_{\mathbb{R}^n} G_t(x)f(x)dx = \int_{\mathbb{R}^n} G(x)f(\sqrt{t}x)dx$$

where

$$|G(x)f(\sqrt{t}x)| \leq |G(x)|\|f\|_{L^\infty}$$

while the hypothesis implies that

$$\lim_{t\to\infty} G(x)f(\sqrt{t}x) = 0$$

almost everywhere, the dominated convergence theorem implies the desired result.

Note that if the function f in $L^\infty(R^n)$ satisfies the stronger condition that $\lim_{|x|\to\infty} f(x) = 0$, then $\lim_{t\to\infty} u(x,t) = 0$ uniformly for all x in $\mathbb{R}^n$.

In addition to providing an example of a wide subclass in $\mathfrak{C}$, this example implies that the behavior of the initial data f on compact subsets of R^n does not influence the behavior of $u(x,t)$ as $t \to \infty$. Indeed:

Example 3 If f is a distribution with compact support then $u(x,t) = G_t * f(x)$ is a solution to the heat equation in $\mathbb{R}^n \times (0,\infty)$ that satisfies $\lim_{t\to\infty} u(x,t) = 0$ uniformly in x on $\mathbb{R}^n$.

The fact that $u(x,t) = G_t * f(x)$ is well defined and a solution to the heat equation in $\mathbb{R}^n \times (0,\infty)$ follows from the properties of G_t, $t > 0$. In particular, for fixed $t > 0$, G_t is a function in the Schwartz space $\mathscr{S}$ of smooth, rapidly decreasing functions and

$$|G_t * f(x)| \leq C \sum_{|\alpha|\leq N} \sup_{|y|<r} |D^\alpha G_t(x-y)| \to 0 \quad \text{as } |x| \to \infty, \tag{8}$$

where N is a constant that depends on f and the set $\{y : |y| \leq r\}$ contains the support of f.

That $\lim_{t\to\infty} u(x,t) = 0$ follows from

$$u(x,t) = G_{t-\epsilon} * G_\epsilon * f(x) \quad \text{where } 0 < \epsilon < t,$$

Example 2, and relation (8) which also implies that $G_\epsilon * f$ is in $L^\infty(\mathbb{R}^n)$.

Example 4 If f in $L^\infty(\mathbb{R}^n)$ and homogeneous of degree 0, namely, $f(x) = f(\lambda x)$ for all positive λ and almost all $x \in \mathbb{R}^n$, then $\lim_{t\to\infty} u(x,t) = G * f(0)$ uniformly on compact subsets of $x \in \mathbb{R}^n$.

This is a consequence of the fact that if f is homogeneous of degree 0 then, for all $t > 0$, $u(0,t) = G * f(0)$. Namely,

$$u(0,t) = G_t * f(0) = \int_{\mathbb{R}^n} G(x) f(\sqrt{t}x)dx = \int_{\mathbb{R}^n} G(x) f(x)dx = G * f(0).$$

Example 5 Suppose the function f in $L^\infty(\mathbb{R}^n)$ is such that $\lim_{r\to\infty} f(r\sigma)$ exists for almost all unit vectors $\sigma \in S^{n-1}$. In other words, f satisfies

$$\lim_{r\to\infty} f(r\sigma) = c(\sigma) \quad \text{for almost all unit vectors } \sigma \in S^{n-1}.$$

If h is defined by $h(x) = c(x/|x|)$, then h is a well defined function in $L^\infty(\mathbb{R}^n)$ that is homogeneous of degree 0 and $\lim_{t\to\infty} u(x,t) = G * h(0)$ for all x in $\mathbb{R}^n$ uniformly on compact subsets.

In the univariate case, $n = 1$, this reduces to the following:

Example 6 Suppose f in $L^\infty(\mathbb{R})$ satisfies

$$\lim_{x\to\infty} f(x) = c_1 \quad \text{and} \quad \lim_{x\to-\infty} f(x) = c_2.$$

Then $\lim_{t\to\infty} u(x,t) = (c_1 + c_2)/2$ for all x in $\mathbb{R}$ uniformly on compact subsets.

Example 7 If f is in $L^p(\mathbb{R}^n)$ for some p that satisfies $1 \le p < \infty$ then $\lim_{t\to\infty} u(x,t) = 0$ uniformly in x on $\mathbb{R}^n$.

References

1. F. John, *Partial differential equations* . Fourth edition. Applied Mathematical Sciences, 1. Springer-Verlag, New York, 1982. x+249 pp.
2. E. M. Stein, *Singular integrals and differentiability properties of functions.* Princeton Mathematical Series, No. 30 Princeton University Press, Princeton, N.J. 1970. xiv+290 pp.
3. D. V. Widder, *The heat equation. Pure and Applied Mathematics*, Vol. 67. Academic Press [Harcourt Brace Jovanovich, Publishers], New York-London, 1975. xiv+267 pp.

Weighted Equilibrium and the Flow of Derivatives of Polynomials

Andrei Martínez-Finkelshtein and Evgenii A. Rakhmanov

Dedicated to Ed Saff on the occasion of his 80th birthday, in recognition of his profound contributions to approximation theory and his enduring inspiration as a mentor, colleague, and friend.

1 Introduction

Given a sequence of (monic) polynomials Q_n, $\deg Q_n = n$, $n \in \mathbb{N}$, with zeros on the interval $[-1, 1]$, we consider the triangular table of their iterated derivatives

$$Q_{n,k}(x) \stackrel{\text{def}}{=\!=} \frac{(n-k)!}{n!} \frac{d^k}{dx^k} Q_n(x) = x^{n-k} + \text{lower degree terms},$$
$$k = 0, 1, \dots, n-1. \tag{1}$$

The main result of this paper is the following theorem:

Theorem 1 *Let Q_n be a sequence of polynomials,* $\deg Q_n = n$*, whose zeros belong to the interval* $[-1, 1] \subset \mathbb{R}$*, and such that the sequence of probability zero-counting measures*[1]

[1] In all zero-counting measures the zeros are considered with account of their multiplicity.

A. Martínez-Finkelshtein (✉)
Department of Mathematics, Baylor University, Waco, TX, USA

Department of Mathematics, University of Almería, Almería, Spain
e-mail: A_Martinez-Finkelshtein@baylor.edu

E. A. Rakhmanov
Department of Mathematics, University of South Florida, Tampa, FL, USA
e-mail: rakhmano@usf.edu

D. Bilyk et al. (eds.), *Recent Advances in Approximation and Potential Theory*, Applied and Numerical Harmonic Analysis,
https://doi.org/10.1007/978-3-031-95551-8_17

$$\sigma_{n,0} \stackrel{\text{def}}{=\!=} \frac{1}{n}\chi\left(Q_n\right), \quad \chi(Q_n) \stackrel{\text{def}}{=\!=} \sum_{Q_n(x)=0} \delta_x,$$

has a weak- limit μ_0, given by the arc-sine distribution, namely,*

$$d\mu_0(x) = \frac{dx}{\pi\sqrt{1-x^2}}, \quad x \in (-1,1). \tag{2}$$

Let

$$\sigma_{n,k} \stackrel{\text{def}}{=\!=} \frac{1}{n}\chi\left(Q_{n,k}\right) = \frac{1}{n}\sum_{Q_{n,k}(x)=0} \delta_x. \tag{3}$$

Then for any $t \in [0,1)$ and any sequence k_n of natural numbers with $k_n/n \to t$, the sequence σ_{n,k_n} has a weak- limit μ_t, where μ_t is absolutely continuous and*

$$d\mu_t(x) = \frac{1}{\pi}\frac{\sqrt{1-t^2-x^2}}{1-x^2}\,dx, \qquad x \in [-\sqrt{1-t^2}, \sqrt{1-t^2}]. \tag{4}$$

A simple illustration of this result is given by the sequence $Q_n = P_n^{(\alpha,\beta)}$, $\alpha, \beta > -1$, of the (monic) Jacobi polynomials, see [2, §18.3]. The well-known property [2, §18.9]

$$Q_n'(x) = nP_{n-1}^{(\alpha+1,\beta+1)}(x)$$

allows for the explicit computation of the weak-* limit μ_t of the sequence (3), as $k/n \to t$, in this case. This yields the expression in (4). Theorem 1 shows that this behavior is not intrinsic to Jacobi polynomials, but it is generic for any sequence of polynomials whose initial distribution is given by the arc-sine law.

Theorem 1 can be derived, in particular, from the results contained in the recent paper [3], where the description of the Cauchy transform of the limiting measure μ_t (under very general assumptions) is done in terms of solutions of the inviscid Burgers equation. The main contribution of this paper is the potential-theoretic approach, which characterizes the flow in terms of the equilibrium measure in an external field generated by two mass points. Since one of the main steps in our analysis is an estimate of the Cauchy integral expressing the polynomials $Q_{n,k}$, it is worth mentioning a recent work of [1] on zeros of polynomials generated by a Rodrigues-type formula.

2 Proof of Theorem 1

Let us start by setting up the following notation: for $0 < \tau \leq 1$, denote $\mathcal{M}_\tau(F)$ the set of Borel measures of total mass τ supported on $F \subset \mathbb{R}$.

Given a finite compactly supported Borel measure μ on $\mathbb{R}$, we define its *logarithmic potential* as

$$U^{\mu}(z) \stackrel{\text{def}}{=} \int \log \frac{1}{|z-y|} \, d\mu(y). \tag{5}$$

For our purposes, it will be sufficient to consider measures on an interval of the real line, without loss of generality, $\Delta \stackrel{\text{def}}{=} [-1, 1] \subset \mathbb{R}$. Recall the following weighted equilibrium problem that we state in its simplified form, see e.g. [4]: given $\tau \in (0, 1]$ and a continuous function $\varphi : (-1, 1) \to \mathbb{R}$ (the *external field*), there is the unique equilibrium measure $\lambda = \lambda(\tau, \varphi, \Delta) \in \mathcal{M}_{\tau}(\Delta)$ minimizing the weighted energy

$$\int U^{\mu} \, d\mu + 2 \int \varphi \, d\mu$$

in the class $\mathcal{M}_{\tau}(\Delta)$.[2] It is known that there is a constant $m_{\tau,\varphi}$ such that

$$U^{\lambda}(x) + \varphi(x) \begin{cases} = m_{\tau,\varphi}, & x \in \operatorname{supp}(\lambda), \\ \geq m_{\tau,\varphi}, & x \in \Delta. \end{cases} \tag{6}$$

The equilibrium measure λ is also characterized by the following property:

$$m_{\tau,\varphi} = \min_{x \in \Delta} \left(U^{\lambda}(x) + \varphi(x) \right) = \max_{\mu \in \mathcal{M}_{\tau}(\Delta)} \min_{x \in \Delta} \left(U^{\mu}(x) + \varphi(x) \right). \tag{7}$$

In other words, for any $\mu \in \mathcal{M}_{\tau}(\Delta)$,

$$\min_{x \in \Delta} \left(U^{\mu}(x) + \varphi(x) \right) \geq m_{\tau,\varphi} \text{ quasi-everywhere} \quad \Rightarrow \quad \mu = \lambda(\tau, \varphi, \Delta). \tag{8}$$

This will be the key property we will use to prove Theorem 1, taking into account the following well-known fact:

Lemma 1 *For $t \in [0, 1)$, measure μ_t in (4) is the equilibrium measure $\lambda(1 - t, \varphi, \Delta) \in \mathcal{M}_{1-t}(\Delta)$ in the external field*

$$\varphi(x) = \varphi_t(x) \stackrel{\text{def}}{=} \frac{t}{2} \log \frac{1}{|x^2 - 1|}, \quad x \in (-1, 1). \tag{9}$$

For the proof, see e.g. [4, Examples IV.1.18 and IV.6.2].

[2] We could clearly reformulate the problem in the class $\mathcal{M}_1(\Delta)$; the parameter τ is introduced here for convenience.

Lemma 2 *Let φ be a continuous external field on Δ, $\tau \in (0,1]$, $\lambda = \lambda(\tau, \varphi, \Delta) \in \mathcal{M}_\tau(\Delta)$ the equilibrium measure. If the support $S \stackrel{\text{def}}{=} \operatorname{supp}\left(\lambda_{\tau,\varphi}\right) \subset \Delta$ is an interval then*

$$m_{\tau,\varphi} = u_\varphi(\infty) + \tau \log \frac{1}{\operatorname{cap}(S)}, \tag{10}$$

where u_φ is the solution of the Dirichlet problem in $\overline{\mathbb{C}} \setminus S$ with the boundary values φ, and $\operatorname{cap}(S) = \operatorname{length}(S)/4$ is the logarithmic capacity of S.

In particular, with $\varphi = \varphi_t$ given in (9),

$$m_{1-t,\varphi} = \log(2) - \frac{1+t}{2}\log(1+t) - \frac{1-t}{2}\log(1-t). \tag{11}$$

Proof Let Ψ be the conformal mapping of $\overline{\mathbb{C}} \setminus S$ onto the exterior of the unit circle, such that $\Psi(\infty) = \infty$, and normalized by $\Psi'(\infty) > 0$. Then $\Psi'(\infty) = 1/\operatorname{cap}(S)$. Consider the function

$$H(z) \stackrel{\text{def}}{=} U^\lambda(z) + \tau \log|\Psi(z)| + u_\varphi(z).$$

By (6), H is harmonic in $\overline{\mathbb{C}} \setminus S$ and identically equal to $m_{\tau,\varphi}$ on S, which implies that $H(z) = m_{\tau,\varphi}$ for all $z \in \overline{\mathbb{C}} \setminus S$. In particular,

$$\begin{aligned} m_{\tau,\varphi} &= \lim_{z\to\infty} \left(U^\lambda(z) + \tau \log|\Psi(z)| + u_\varphi(z)\right) \\ &= \lim_{z\to\infty} \left(U^\lambda(z) + \tau \log|z|\right) + \lim_{z\to\infty} \tau \log\left|\frac{\Psi(z)}{z}\right| + u_\varphi(\infty) \end{aligned}$$

and (10) follows from the well-known properties of the logarithmic potential and the conformal mapping Ψ.

By Lemma 1 and formula (4), for $\varphi = \varphi_t$, we have $\tau = 1 - t$, $S = [-s, s]$ with $s = \sqrt{1-t^2}$. In this case, $\Psi(z) = \Phi(z/s)$, where $\Phi(z)$ is the inverse Zhukovski map,

$$\Phi(z) = z + (z^2 - 1)^{1/2}, \tag{12}$$

and we take the single-valued branch of the square root in $\mathbb{C} \setminus \Delta$ fixed by $(z^2 - 1)^{1/2} > 0$ for $z > 1$.

On the other hand,

$$u_\varphi(z) = \operatorname{Re}\left(\frac{\sqrt{z^2 - s^2}}{\pi} \int_{-s}^{s} \frac{\varphi_t(x)}{\sqrt{s^2 - x^2}} \frac{dx}{z - x}\right),$$

so that

$$
\begin{aligned}
u_\varphi(\infty) &= \frac{1}{\pi}\int_{-s}^{s}\frac{\varphi_t(x)}{\sqrt{s^2-x^2}}dx \\
&= \frac{t}{2}\left(\frac{1}{\pi}\int_{-s}^{s}\log\frac{1}{1-x}\frac{dx}{\sqrt{s^2-x^2}}+\frac{1}{\pi}\int_{-s}^{s}\log\frac{1}{1+x}\frac{dx}{\sqrt{s^2-x^2}}\right) \\
&= \frac{t}{\pi}\int_{-s}^{s}\log\frac{1}{1-x}\frac{dx}{\sqrt{s^2-x^2}} = t\,U^{\lambda}(1),
\end{aligned}
$$

where $\lambda = \lambda(1, 0, [-s, s])$ is the equilibrium (Robin) measure of the interval $[-s, s]$. Since

$$
U^{\lambda}(z) = -\log(\text{cap}([-s, s])) - \log|\Phi(z/s)| = \log\left(\frac{2}{s}\right) - \log|\Phi(z/s)|,
$$

we get that

$$
\begin{aligned}
u_\varphi(\infty) &= t\left(\log\left(\frac{2}{s}\right) - \log|\Phi(1/s)|\right) \\
&= t\left(\log\left(\frac{2}{s}\right) - \frac{1}{2}\log(1+t) + \frac{1}{2}\log(1-t)\right).
\end{aligned}
$$

In consequence,

$$
\begin{aligned}
m_{\tau,\varphi} &= -t\log|\Phi(1/s)| + \log\left(\frac{2}{s}\right) \\
&= \log(2) - \frac{1+t}{2}\log(1+t) - \frac{1-t}{2}\log(1-t).
\end{aligned}
$$

□

Remark 1 Clearly, formula (10) is valid under much milder assumptions on φ. We have stated the restricted version here, sufficient for our purposes, for the sake of simplicity.

Now we turn to the sequence of zero-counting measures $\{\sigma_{n,k_n}\}_{n\in\Lambda}$, defined in (3), with the assumption that

$$
\sigma_{n,0} \xrightarrow{*} \mu_0 \in \mathcal{M}_1(\Delta), \quad n \to \infty. \tag{13}
$$

The following lemma is valid even if μ_0 is not the arc-sine distribution on Δ:

Lemma 3 *Let* $\{k_n\}$, $n \in \mathbb{N}$, *be a sequence such that*

$$
\lim_n \frac{k_n}{n} = t \in [0, 1) \tag{14}
$$

exists, and let $\Lambda \subseteq \mathbb{N}$ *be such that the following weak-* limit exists:*

$$\sigma_{n,k_n} \xrightarrow{*} \nu \in \mathcal{M}_{1-t}(\Delta), \quad n \in \Lambda. \tag{15}$$

Then, for any $\zeta \in \Delta$ *where* U^ν *is continuous, and quasi-everywhere on* Δ*, for any Jordan contour* Γ *enclosing* Δ*,*

$$U^\nu(\zeta) \geq -c_t + \min_{z\in\Gamma} \left(U^{\mu_0}(z) + t \log|z-\zeta| \right), \tag{16}$$

where

$$c_t \stackrel{\text{def}}{=\!=} t \log t + (1-t)\log(1-t). \tag{17}$$

Proof Let first ζ be in the bounded complement of $\mathbb{C} \setminus (\Gamma \cup \Delta)$. We have the integral representation

$$Q_{n,k}(\zeta) = \binom{n}{k}^{-1} \frac{1}{2\pi i} \oint_\Gamma \frac{Q_n(z)}{(z-\zeta)^{k+1}} dz,$$

from which we get that

$$U^{\sigma_{n,k}}(\zeta) = -\frac{1}{n}\log\left|Q_{n,k}(\zeta)\right| = \frac{1}{n}\log\binom{n}{k} + \frac{1}{n}\log(2\pi) - \frac{1}{n}\log\left|\oint_\Gamma \frac{Q_n(z)}{(z-\zeta)^{k+1}}dz\right|.$$

By Stirling's formula,

$$\lim_n \frac{1}{n}\log\binom{n}{k_n} = -c_t.$$

On the other hand,

$$\begin{aligned}
\frac{1}{n}\log\left|\oint_\Gamma \frac{Q_n(z)}{(z-\zeta)^{k+1}}dz\right| &\leq \frac{1}{n}\log\left(\max_{z\in\Gamma}\left|\frac{Q_n(z)}{(z-\zeta)^{k+1}}\right| \cdot \text{length}(\Gamma)\right) \\
&= \max_{z\in\Gamma}\left(-U^{\sigma_{n,0}}(z) - \frac{k+1}{n}\log|z-\zeta|\right) + \frac{1}{n}\log\left(\text{length}(\Gamma)\right) \\
&= -\min_{z\in\Gamma}\left(U^{\sigma_{n,0}}(z) + \frac{k+1}{n}\log|z-\zeta|\right) + \frac{1}{n}\log\left(\text{length}(\Gamma)\right).
\end{aligned}$$

Thus, we conclude that

$$U^{\sigma_{n,k_n}}(\zeta) \geq -c_t + \min_{z\in\Gamma}\left(U^{\sigma_{n,0}}(z) + \frac{k_n+1}{n}\log|z-\zeta|\right) + o(1), \quad n \to \infty. \tag{18}$$

Taking limits in both sides and using assumptions (13) and (15) we conclude that (16) is valid for ζ in the bounded complement of $\mathbb{C} \setminus (\Gamma \cup \Delta)$.

By the Continuity Theorem [4, Theorem A.IV.4.6], we can extend this inequality to any $\zeta \in \Delta$ where the potential U^ν is continuous. Moreover, by the Lower Envelope Theorem [4, Theorem A.IV.4.8], it holds quasi-everywhere on Δ. □

We need to estimate the right-hand side of (16) in the case when μ_0 is the arcsine distribution given in (2). In this case,

$$U^{\mu_0}(z) = \log(2) - \log|\Phi(z)|, \tag{19}$$

with Φ given in (12).

Also, following [4], we use the notation

$$\text{“}\inf_{z\in\Delta}\text{”}h(z)$$

for essential infimum, that is, the largest number L such that on Δ the real function h takes values smaller than L only on a set of zero capacity.

Lemma 4 *With the assumptions of Lemma 3,*

$$\text{“}\inf_{x\in[-1,1]}\text{”}\left(U^\nu + \varphi\right)(x) \geq m_{1-t,\varphi}, \tag{20}$$

where $m_{1-t,\varphi}$ *is given in* (11).

Proof Fix $\zeta \in (-1, 1)$ where the potential U^ν is continuous; as a contour Γ in (16) we take the level curve of Φ, namely

$$\Gamma = \Gamma_A \stackrel{\text{def}}{=} \{z \in \mathbb{C} \setminus \Delta : |\Phi(z)| = \Phi(A)\}, \quad A > 1.$$

It is easy to see that Γ_A is an ellipse whose equation in Cartesian coordinates on the (x, y) plane is

$$\frac{x^2}{A^2} + \frac{y^2}{(A^2 - 1)} = 1.$$

By (19), it is also a level curve of U^{μ_0}. Thus, the problem of minimizing $\log|z - \zeta|$ with $z \in \Gamma_A$ is equivalent to the real-valued constrained optimization problem

$$\begin{aligned} &\text{minimize} \quad (x-\zeta)^2 + y^2 \\ &\text{subject to} \quad \frac{x^2}{A^2} + \frac{y^2}{A^2-1} = 1. \end{aligned}$$

Assume w.o.l.g. that $0 \leq \zeta \leq 1$. Standard methods show that the only possible points of minimum are $(x, y) = (A, 0)$ and, if $\zeta A^2 \leq 1$,

$$(x, y) = \left(\zeta A^2, \pm\sqrt{(A^2-1)(1-\zeta)}\right).$$

Hence, the global minimum is

$$\begin{cases} (A-\zeta)^2, & \text{if } \zeta A^2 \geq 1, \\ \min\left\{(A-\zeta)^2, (A^2-1)(1-\zeta^2)\right\}, & \text{if } \zeta A^2 \leq 1. \end{cases}$$

Since

$$(A-\zeta)^2 = (A^2-1)(1-\zeta^2) + (\zeta A - 1)^2 \geq (A^2-1)(1-\zeta^2),$$

for $-1 \leq \zeta \leq 1$ and $z \in \Gamma_A$,

$$\log|z-\zeta| \geq \frac{1}{2}\log(A^2-1) + \frac{1}{2}\log(1-\zeta^2),$$

so that, combining it with (19),

$$U^{\mu_0}(z) + t\log|z-\zeta| \geq \log(2) - \log|\Phi(A)| + \frac{t}{2}\log(A^2-1) + \frac{t}{2}\log(1-\zeta^2).$$

Using it in (16), we see that for $-1 \leq \zeta \leq 1$,

$$U^{\mu_t}(\zeta) \geq -c_t + \log(2) - \log|\Phi(A)| + \frac{t}{2}\log(A^2-1) + \frac{t}{2}\log(1-\zeta^2). \tag{21}$$

With φ as in (9) we can rewrite it as

$$\left(U^{\mu_t} + \varphi\right)(\zeta) \geq -c_t + \log(2) - \log|\Phi(A)| + \frac{t}{2}\log(A^2-1),$$

and we conclude that

$$\min_{x\in[-1,1]}\left(U^{\mu_0} + \varphi\right)(x) \geq -c_t + \log(2) - \log|\Phi(A)| + \frac{t}{2}\log(A^2-1). \tag{22}$$

The global maximum of the right-hand side over $A > 1$ is attained at

$$A = \frac{1}{\sqrt{1-t^2}} = \frac{1}{s}$$

and is equal to

$$\begin{aligned} &-c_t + \log(2) - \log|\Phi(1/s)| + \frac{t}{2}\log(s^{-2}-1) = \\ &-c_t + \log(2) - \log|\Phi(1/s)| + t\log(t/s). \end{aligned}$$

Simplifying, we obtain that

$$\min_{x\in[-1,1]} \left(U^{\mu_t} + \varphi\right)(x) \geq -c_t + \log(2) + (1-t)\log(1-t) + \left(\frac{t}{2} - 1\right)\log(2-t) + \frac{t}{2}\log(t)$$

and (20) follows. □

The assertion of Theorem 1 follows from Lemma 4 and the characterization (8).

Acknowledgments The first author was partially supported by Simons Foundation Collaboration Grants for Mathematicians (grant 710499). He also acknowledges the support of the project PID2021-124472NB-I00, funded by MICIU/AEI/10.13039/501100011033 and by "ERDF A way of making Europe", as well as the support of Junta de Andalucía (research group FQM-229 and Instituto Interuniversitario Carlos I de Física Teórica y Computacional).

Competing Interests The authors have no conflicts of interest to declare that are relevant to the content of this chapter.

References

1. R. Bøgvad, C. Hägg, and B. Shapiro. Rodrigues' descendants of a polynomial and Boutroux curves. *Constr. Approx.*, 59(3):737–798, 2024.
2. *NIST Digital Library of Mathematical Functions*. https://dlmf.nist.gov/, Release 1.2.3 of 2024-12-15. F. W. J. Olver, A. B. Olde Daalhuis, D. W. Lozier, B. I. Schneider, R. F. Boisvert, C. W. Clark, B. R. Miller, B. V. Saunders, H. S. Cohl, and M. A. McClain, eds.
3. A. Martínez-Finkelshtein and E. A. Rakhmanov. Flow of the zeros of polynomials under iterated differentiation. Preprint arXiv:2408.13851, 2024.
4. E. B. Saff and V. Totik. *Logarithmic Potentials with External Fields*, volume 316 of *Grundlehren der Mathematischen Wissenschaften*. Springer-Verlag, Berlin, second edition, 2024.

A Linear Programming Bound in a Sign-alternating Case

Alexander Reznikov

1 Introduction

This paper is motivated by the linear programming technique for, sometimes asymptotically, solving optimization problems. Some of the most recent examples of this technique include solutions to the Best Packing and Minimal Riesz Energy problems in $\mathbb{R}^8$ and $\mathbb{R}^{24}$ [3–5]; this technique can be also applied to the p-frame problem [1].

The latter problem was solved in some specific cases, while other cases remain unknown. One of the reasons why those cases are hard is because a classical linear programming proof relies on the fact that most of the coefficients in a certain Fourier series expansion are non-negative (see Theorem 4 below) and, therefore, can be neglected for an estimate from below. This is not the case for the p-frame potential $f(t) = |t|^p$, which creates a big obstacle.

In this note we show another example of such a situation, where we are still able to neglect most of the terms in the expansion. Besides the result itself, we think that the technique to obtain this result can be useful for other applications.

We start by introducing the minimal energy problem, which is one of the central problems addressed in Book [2]. Let $\Omega \subset \mathbb{R}^d$ be an infinite compact set, and a *potential* be a symmetric and lower semi-continuous function $K\colon \Omega \times \Omega \to \mathbb{R} \cup \{+\infty\}$. The *minimal K-energy constant on* Ω is defined by

$$W_K(\Omega) := \min_{\mu} I_K[\mu],$$

A. Reznikov (✉)
Department of Mathematics, Florida State University, Tallahassee, FL, USA
e-mail: areznikov@fsu.edu

D. Bilyk et al. (eds.), *Recent Advances in Approximation and Potential Theory*, Applied and Numerical Harmonic Analysis,
https://doi.org/10.1007/978-3-031-95551-8_18

where the minimum is taken over all probability measures μ supported on Ω. For a fixed measure μ on Ω we define

$$I_K[\mu] := \int_\Omega \int_\Omega K(x, y) dx dy.$$

We remark that under our assumptions on K the minimum is, in fact, attained (see Lemma 4.1.3 in [2]). A classical example of such a potential K is the s-Riesz potential

$$K_s(x, y) := |x - y|^{-s}, s \neq 0$$

or the logarithmic potential

$$K_{\log}(x, y) := \log \frac{1}{|x - y|}.$$

Our work is partially motivated by the well-known formula

$$r^{-s} = \frac{1}{\Gamma(s/2)} \int_0^\infty e^{-tr^2} \cdot t^{s/2-1} dt,$$

which implies that the s-Riesz potential is an infinite convex combination of *Gaussian potentials*

$$G_a(x, y) := e^{-a \cdot |x-y|^2}, a > 0.$$

These potentials are continuous and bounded, which is technically helpful, and share all the basic properties of the Riesz potentials; for example, Theorem 4.2.7 and Proposition 4.4.1 from [2] imply that the minimizing measure for $W_{G_a}(\Omega)$ is unique for every $a > 0$.

We are interested in the asymptotic behavior of $W_{G_a}(\Omega)$ as $a \to \infty$. In this paper we will focus on a very specific case $\Omega = [-1, 1]$ and show how can linear programming technique help with estimating W_{G_a} from below. We note that this is the "hard" estimate since to estimate the minimal energy from below we need to estimate $I_{G_a}[\mu]$ for every probability measure μ on Ω.

2 Main Results

Our first main result provides a very strong structure of the measure that minimizes $W_{G_a}([-1, 1])$. For simplicity, we denote

$$I_a[\mu] = I_{G_a}[\mu],$$

and

$$W_a := W_{G_a}([-1, 1]).$$

Theorem 1 *Assume $a > 0$ and μ_a is the probability measure on $[-1, 1]$ such that*

$$I_a[\mu_a] = W_a.$$

Then the measure μ_a is discrete; that is, for some natural number $n = n(a)$ and distinct points $\{x_1(a), \ldots, x_n(a)\} \subset [-1, 1]$ we have

$$\mu_a = \sum_{j=1}^{n} \alpha_j(a)\delta_{x_j(a)},$$

where each $\alpha_j(a) > 0$ and

$$\sum_{j=1}^{n} \alpha_j(a) = 1.$$

We note that, even though this is a strong result, it provides no information on the cardinality $n = n(a)$ of the support of μ_a. Thus, if this cardinality is very large (which is the case for large values of a), then the measure μ_a can macroscopically look like a continuous measure. To estimate this cardinality, it can be helpful to know the order of decay of W_a, which is our next result.

Theorem 2 *There exist two positive constants c_1 and c_2 such that for every $a > 0$ we have*

$$c_1 a^{-1/2} \leqslant W_a \leqslant c_2 a^{-1/2}.$$

The following remark shows the connection between this estimate and the cardinality $n(a)$.

Remark 1 Theorem 2 implies an estimate on the cardinality $n = n(a)$ of the support of μ_a. Indeed, denote

$$U_a^{\mu_a}(x) := \int\limits_{[-1,1]} G_a(x, y)d\mu_a(y) = \int\limits_{[-1,1]} e^{-a|x-y|^2} d\mu(y).$$

It is known (see Theorem 3 below) that for every $x = x_j \in \mathrm{supp}(\mu_a)$, we have $U_a^{\mu_a}(x_j) = W_a$. Therefore, for every j we have

$$c_2 a^{-1/2} \geqslant W_a = U_a^{\mu_a}(x_j) = \sum_{k=1}^{n} \alpha_k(a) e^{-a|x_j - x_k|^2} \geqslant \alpha_j(a).$$

Finally, the estimate

$$1 = \sum_{j=1}^{n} \alpha_j(a) \leqslant n \cdot c_2 a^{-1/2}$$

implies

$$n(a) \geqslant c_3 \sqrt{a}.$$

In particular, we see that as $a \to \infty$, we have $n(a) \to \infty$, so the measures μ_a become less and less discrete.

3 Being a Postdoc of Ed Saff

It goes without saying that Ed Saff is a brilliant mathematician, and his results, publications and monographs speak for themselves. I would like to share a few personal memories that made some strong impressions and, I think, made me a better researcher.

In 2014, after completing my PhD in Michigan State University, I moved to Nashville, TN to work with Ed Saff. I had a competing postdoc offer from another excellent university, and I was not sure I made the right choice. Three years later I could confidently say that the choice to work with Ed was one of the best in my life.

I came to Vanderbilt very energized and, as many people in that state of mind, I wanted to prove as much as possible without spending too much time on writing something down. When I brought Ed my draft of our first (future) paper, he immediately told me this is not how papers should be written. He then very carefully proofread all my subsequent drafts; if I remember correctly, we sent this paper to a journal in March, even though the results were obtained in September of the previous year. It was a huge surprise that a person of Ed's level can spend half a year proofreading his postdoc's writeup; but Ed was very serious about it, and I hope I picked some of his proofreading methods. This very detailed approach to writing made his Minimal Energy book come out a bit late; but what an excellent book it is!

What was particularly great about being Saff's postdoc is that he always treated me as equal. He invited us (me, my wife and our kid) to his house for dinner several times; we went for many lunches together, and he always had time to talk about math and life. For example, it was Ed who told me that Florida has several different retirement plans, and because he told me about it in advance, I had a lot of time to research them and make a decision. It was also Ed who discouraged me from growing mango trees in Florida, which saved me a lot of money.

Largely because of Ed, and the research group he always has at Vanderbilt, the three years spent in Nashville bring only great memories. I do hope that in 10 years I will have another opportunity to sent a paper for a 90th anniversary edition!

4 Proofs

4.1 Proof of Theorem 1

To prove this theorem we need a classical fact from potential theory. For a measure μ supported on $[-1, 1]$, denote it's Gaussian potential by

$$U_a^{\mu}(x) := \int\limits_{[-1,1]} G_a(x, y)d\mu(y) = \int\limits_{[-1,1]} e^{-a|x-y|^2} d\mu(y).$$

With this notation, we clearly have

$$I_a[\mu] = \int\limits_{[-1,1]} U_a^{\mu}(x)d\mu(x).$$

Theorem 3 *Let μ_a be the measure that attains W_a; i.e., $I_a[\mu_a] = W_a$. Then the potential of this measure satisfies the following two properties:*

$$U_a^{\mu_a}(x) \geqslant W_a, \quad \text{for every } x \in [-1, 1]; \tag{1}$$

$$U_a^{\mu_a}(x) = W_a, \quad \text{for every } x \in supp(\mu_a). \tag{2}$$

Remark 2 We remark that the in the general potential theory, the second property is satisfied K-quasieverywhere on the support of the optimal measure. In our case, since the function $K = G_a$ is continuous on $[-1, 1] \times [-1, 1]$, G_a-quasieverywhere is equivalent to everywhere.

We further remark that the theorem implies that every point in the support of μ_a attains the global minimum of $U_a^{\mu_a}(\cdot)$.

We are ready to prove our first main theorem.

Proof of Theorem 1 For every $z \in \mathbb{C}$, we define an auxiliary function

$$f(z) := \int\limits_{[-1,1]} e^{-a\cdot(z-y)^2} d\mu_a(y).$$

We note that we have just the square, without the absolute value, in the exponent; however, if $x \in [-1, 1]$, we have

$$f(x) = \int_{[-1,1]} e^{-a\cdot(x-y)^2} d\mu_a(y) = \int_{[-1,1]} e^{-a\cdot|x-y|^2} d\mu_a(y) = U_a^{\mu_a}(x).$$

We further observe that the function $f(z)$ is an entire function. Theorem 3 implies that

$$f(z) = W_a, \quad z \in \operatorname{supp}(\mu_a).$$

Assume now $\operatorname{supp}(\mu_a)$ is not finite. Since it is a compact subset of $[-1, 1]$, it must have a limit point. Then the uniqueness theorem for entire functions implies f is a constant function on $\mathbb{C}$. Since $f(z) \to 0$ when $z \in \mathbb{R}$ and $z \to \infty$, this constant must be zero. However, an obvious estimate $G_a(x, y) \geqslant e^{-4a}$ implies that $W_a > 0$, which contradicts the fact that for any $z \in \operatorname{supp}(\mu_a)$ we have

$$0 = f(z) = W_a.$$

This implies that our assumption is wrong and $\operatorname{supp}(\mu_a)$ is finite. □

4.2 A Classical Approach to Linear Programming

In this section we briefly show how a classical linear programming bound should work. We will need it to point out the main difficulty of applying this technique to our case, and thus we will skip most of the technical details and focus on the main obstacle. For more (and complete) details, we refer the reader to Chapter 6.4 in [2].

Let us, in this case, consider the *discrete s-Riesz energy problem* on the unit sphere $\mathbb{S}^d$. To do that, fix a positive number $s < d$. For an N-point configuration $\omega_N = \{x_1, \ldots, x_N\} \subset \mathbb{S}^d$, denote

$$E(\omega_N) := \sum_j \sum_{k \neq j} K_s(x_j, x_j) = \sum_j \sum_{k \neq j} |x_j - x_k|^{-s},$$

where the distance $|x - y|$ is the Euclidean distance on $\mathbb{R}^{d+1}$. The goal of the problem is to minimize this quantity over all N-point configurations:

$$\mathcal{E}(N) := \min_{\omega_N \subset \mathbb{S}^d} E(\omega_N).$$

Remark 3 We remark that if

$$\mu = \sum_{j=1}^{N} \delta_{x_j},$$

then $E(\omega_N)$ is almost equal to the double integral

$$\int\int K_s(x, y)d\mu(x)d\mu(y).$$

However, the terms $K_s(x_j, x_j)$ make this double integral equal to infinity for every discrete measure μ. This shows an advantage of continuous kernels G_a, for which those terms are just equal to 1.

When $s < d$, the quantity $\mathcal{E}(N)$ can be estimated from below by expanding the kernel K_s in a certain orthogonal series. Namely, denote

$$f(t) = t^{-s/2}, \quad t \in (0, \infty). \tag{3}$$

Then, for $x, y \in \mathbb{S}^d$ we can use that $|x| = |y| = 1$ and write

$$K_s(x, y) = |x - y|^{-s} = f(|x - y|^2) = f(2 - 2x \cdot y).$$

Now fix a number $\delta > 0$ and denote

$$g_0^\delta(t) = f(\delta + 2 - 2t). \tag{4}$$

For $x, y \in \mathbb{S}^d$ we have $2 - 2x \cdot y \geqslant 0$, which implies that $\delta + 2 - 2x \cdot y \geqslant \delta > 0$; thus, we can view g_0^δ as a mollified version of the function f. Further, denote

$$g_k^\delta(t) := (-1)^k f^{(k)}(\delta + 2 - 2t), \quad k \geqslant 1.$$

To state the main feature of the Riesz kernels K_s that allow the linear programming methods to work, we need to define the Gegenbauer polynomials.

Definition 1 Fix $d \in \mathbb{N}$. The *normalized Gegenbauer polynomials* $P_n(t) = P_n^{(d)}(t)$ are the polynomials of degree n that are orthogonal on the interval $[-1, 1]$ with respect to the measure

$$dw(t) = (1 - t^2)^{d/2-1}dt$$

and normalized by $P_n(1) = 1$.

With this notation, the following theorem holds.

Theorem 4 *For the functions $g_k^\delta(t)$ defined above, define $a_j^{k,\delta}$ to be the coefficients in the decomposition*

$$g_k^\delta(t) = \sum_{j=0}^{\infty} a_j^{k,\delta} P_j(t).$$

Then for every $j \geqslant 0$, $k \geqslant 1$ and $\delta > 0$ we have

$$a_j^{k,\delta} \geqslant 0.$$

Moreover, if $j \geqslant 1$, then

$$a_j^{0,\delta} \geqslant 0.$$

The next step of the estimate, which is the most crucial for our motivation, goes as follows. First, by the Taylor formula, for any $u > 0$ we have

$$f(u) = \sum_{m=0}^{n} \frac{\delta^m}{m!}(-1)^m f^{(m)}(\delta + u) + \frac{(-1)^{n+1}}{n!} \int_u^{\delta+u} (\theta - u)^n f^{(n+1)}(\theta) d\theta.$$

Since the signs of the derivatives of f alternate, we immediately estimate

$$f(u) \geqslant \sum_{m=0}^{n} \frac{\delta^m}{m!}(-1)^m f^{(m)}(\delta + u).$$

Therefore, for every configuration $\omega_N = \{x_1, \ldots, x_N\} \subset \mathbb{S}^d$, we have

$$E(\omega_N) = \sum_{j \neq k} f(2 - 2x_j \cdot x_k) \geqslant \sum_{j \neq k} \sum_{m=0}^{n} \frac{\delta^m}{m!}(-1)^m f^{(m)}(\delta + 2 - 2x_j \cdot x_k)$$

$$= \sum_{j \neq k} \sum_{m=0}^{n} \frac{\delta^m}{m!} g_m^\delta(x_j \cdot x_k) = \sum_{m=0}^{n} \left(\frac{\delta^m}{m!} \cdot \sum_{j \neq k} g_m^\delta(x_j \cdot x_k) \right).$$

We notice that the function g_m^δ does not have a singularity; therefore, it is not harmful to compute $g_m^\delta(x_j \cdot x_j)$. Second, we can always choose the numbers δ and n; thus, it remains to estimate the individual sums

$$\sum_{j \neq k} g_m^\delta(x_j \cdot x_k)$$

for every m. Those sums, however, must be estimated for **any** choice of $\{x_1, \ldots, x_N\}$, independently of this choice. To do that, we write

$$\sum_{j\neq k} g_m^\delta(x_j \cdot x_k) = \sum_{j,k} g_m^\delta(x_j \cdot x_k) - N g_m^\delta(1) = \sum_{j,k}\sum_{r\geqslant 0} a_r^{m,\delta} P_r(x_j \cdot x_k) - N g_m^\delta(1) =$$
$$\sum_{r\geqslant 0} a_r^{m,\delta} \sum_{j,k} P_r(x_j \cdot x_k) - N g_m^\delta(1).$$

Since $P_0(t) = 1$ for every t, we further get

$$\sum_{j\neq k} g_m^\delta(x_j \cdot x_k) = \sum_{r\geqslant 1} a_r^{m,\delta} \sum_{j,k} P_r(x_j \cdot x_k) + N^2 a_0^{m,\delta} - N g_m^\delta(1).$$

Remark 4 It is very important to note that, up to this moment, the choice of the orthogonal system $\{P_r\}_{r=0}^\infty$ was not significant. However, the next formula would not be true for a general orthogonal system of functions (polynomials).

We now proceed with the famous "addition formula" for Gegenbauer polynomials. Namely, for every r and every pair of points $x, y \in \mathbb{S}^d$, we have for some integer $Z(r, d)$

$$P_r(x \cdot y) = Z(r, d)^{-1} \cdot \sum_{\nu=1}^{Z(r,d)} Y_{r,\nu}(x) Y_{r,\nu}(y). \tag{5}$$

The functions $Y_{r,\nu}(\cdot)$ are called *spherical harmonics*, but for us it is not important what they are. What is incredibly important, however, is that the function $P_r(x \cdot y)$, which depends on two variables x and y, can be sliced into functions that depend on x and y separately. In particular, we deduce that for a fixed r,

$$\sum_{j,k} P_r(x_j \cdot x_k) = Z(r, d)^{-1} \cdot \sum_{\nu=1}^{Z(r,d)} \sum_{j,k} Y_{r,\nu}(x_j) Y_{r,\nu}(x_k) =$$
$$Z(r, d)^{-1} \cdot \sum_{\nu=1}^{Z(r,d)} \left(\sum_j Y_{r,\nu}(x_j) \right)^2 \geqslant 0, \tag{6}$$

which is a very powerful estimate because it holds for any choice of the configuration $\{x_1, \ldots, x_N\}$. Together with Theorem 4, this implies one of the most crucial step of linear programming:

$$\sum_{j\neq k} g_m^\delta(x_j\cdot x_k)=\sum_{r\geqslant 1} a_r^{m,\delta}\sum_{j,k} P_r(x_j\cdot x_k)+a_0^{m,\delta}-Ng_m^\delta(1)\geqslant N^2a_0^{m,\delta}-Ng_m^\delta(1). \tag{7}$$

We are not going to show the rest of the estimate for the following reason. In this paper we address the case when Theorem 4 does not hold; and this is the specific place where it is used. We remark, however, that for the last estimate to hold, it is important for Theorem 4 to hold for a specific family of orthogonal functions that satisfy a formula like (5) to guarantee that the term $\sum_{j,k} P_r(x_j\cdot x_k)$ is non-negative.

4.3 *Proof of Theorem 2*

We first prove the upper estimate. Indeed,

$$W_a\leqslant I_a(dx/2)=\frac{1}{4}\int_{-1}^{1}\int_{-1}^{1} e^{-a(x-y)^2}dxdy.$$

Further, for a fixed y we have by making a substitution $x=y+u/\sqrt{a}$

$$\int_{-1}^{1} e^{-a(x-y)^2}dx=\frac{1}{\sqrt{a}}\int_{\sqrt{a}(-1-y)}^{\sqrt{a}(1-y)} e^{-u^2}du\leqslant\frac{1}{\sqrt{a}}\int_{\mathbb{R}} e^{-u^2}du=\frac{c_1}{\sqrt{a}},$$

and therefore

$$W_a\leqslant\frac{c_1}{2}\cdot\frac{1}{\sqrt{a}}.$$

□

We will now use the idea described in the previous section to prove the estimate from below. Since we want to follow the general linear programming path, we first note that for $x, y\in\mathbb{S}^d$ and the Riesz potential, the meaningful real variable was $t=x\cdot y\in[-1,1]$. Since now we have $x, y\in[-1,1]$ and we can not replace $|x-y|^2=(x-y)^2$ with something related to the product of x and y, we need to introduce $t=x-y\in[-2,2]$. Similarly to (3), we then introduce the function

$$f(t)=e^{-at^2},\quad t\in[-2,2],$$

and this is the function we want to expand in some orthogonal basis. We remind the reader that in the Riesz potential case, we needed to introduce an extra variable δ (see (4)) to mollify the singularity of $(2-2t)^{-s/2}$ at $t=1$. We do not need this step here as our Gaussian potential has no singularity, so we can work with $f(t)$

directly. Since $f(t)$ is defined on an interval, one of the logical bases is the basis of exponentials $e^{\pi int/2}$. This choice will make even more sense as we will see that an analog of the formula (5) holds. We now write the Fourier series expansion

$$f(t) = \sum_{n=-\infty}^{\infty} \lambda_n e^{\frac{i\pi nt}{2}}, \quad t \in [-2, 2].$$

In this expansion,

$$\lambda_0 = \frac{1}{4}\int_{-2}^{2} f(t)dt = \frac{1}{4}\int_{-2}^{2} e^{-at^2}dt \geqslant \frac{c}{\sqrt{a}}.$$

Furthermore, for any probability measure μ we have

$$I_a[\mu] = \int_{[-1,1]}\int_{[-1,1]} e^{-a(x-y)^2} d\mu(x)d\mu(y) =$$

$$\int_{[-1,1]}\int_{[-1,1]} \sum_{n=-\infty}^{\infty} \lambda_n e^{\frac{i\pi n(x-y)}{2}} d\mu(x)d\mu(y).$$

We now want to interchange summation and integration. To do that, we recall that by Theorem 1, the constant W_a is attained by a discrete measure μ_a. Since we only need to estimate $I_a[\mu]$ for $\mu = \mu_a$, we can assume that the integrals against $d\mu$ are just finite sums. This implies that

$$I_a[\mu_a] = \sum_{n=-\infty}^{\infty} \lambda_n \int_{[-1,1]}\int_{[-1,1]} e^{\frac{i\pi n(x-y)}{2}} d\mu_a(x)d\mu_a(y) =$$

$$\sum_{n=-\infty}^{\infty} \lambda_n \cdot \left| \int_{[-1,1]} e^{\frac{i\pi nx}{2}} d\mu_a(x) \right|^2.$$

We call the readers attention to the fact that this equation is the direct analog of Eq. (6), which supports our chose of the orthogonal basis. If, similarly to Theorem 4, we knew that $\lambda_n \geqslant 0$ for $n \geqslant 1$, we would be able to write an estimate analogous to (7) and get

$$I_a[\mu_a] \geqslant \lambda_0 \geqslant \frac{c}{\sqrt{a}}. \tag{8}$$

This is, however, not true; indeed, one can numerically observe that for n large enough, the signs of λ_n alternate (we will never need this fact, but we will rigorously explain it at the end of the paper). For the rest of the proof, our goal is to show why we can still neglect the terms with negative λ_n.

We first prove that initially many coefficients λ_n are, indeed, positive.

Lemma 1 *There exists a positive c such that if $|n| \leqslant c \cdot a$, then $\lambda_n > 0$.*

Proof Up to a normalizing constant β that does not depend on n or on a,

$$\lambda_n = \beta \int_{-2}^{2} e^{-at^2} e^{\frac{i\pi nt}{2}} dt = 2\beta \int_{0}^{2} e^{-at^2} \cos\left(\frac{\pi nt}{2}\right) dt.$$

Thus, without loss of generality we can assume $n > 0$. Furthermore,

$$\lambda_n = \beta \int_{\mathbb{R}} e^{-at^2} \cos\left(\frac{\pi nt}{2}\right) dt - 2\beta \int_{t>2} e^{-at^2} \cos\left(\frac{\pi nt}{2}\right) dt = \beta(I_1 - 2I_2). \tag{9}$$

The first integral is equal to the Fourier transform of the Gaussian evaluated (up to another normalizing constant) at $\xi = n$, thus

$$I_1 = c_2 \cdot \frac{1}{\sqrt{a}} e^{-c_3 n^2/a}.$$

To estimate the second integral, we write

$$|I_2| \leqslant \int_{t>2} e^{-at^2} dt \leqslant \int_{t>2} e^{-2at} dt = \frac{c_4}{a} e^{-4a}.$$

Therefore,

$$\lambda_n \geqslant \frac{\beta}{\sqrt{a}} \cdot \left(c_2 e^{-c_3 n^2/a} - \frac{c_4}{\sqrt{a}} e^{-4a} \right), \tag{10}$$

which is obviously positive when a is large enough and $n^2/a < c_5 a$. □

This lemma is crucial for our estimate, as we can now write

$$W_a = \sum_{n=-\infty}^{\infty} \lambda_n \cdot \left| \int_{[-1,1]} e^{\frac{i\pi nx}{2}} d\mu_a(x) \right|^2 \geqslant \lambda_0 + \sum_{|n|>ca} \lambda_n \cdot \left| \int_{[-1,1]} e^{\frac{i\pi nx}{2}} d\mu_a(x) \right|^2 \geqslant \lambda_0 - \sum_{|n|>ca} |\lambda_n|. \tag{11}$$

Therefore, it remains to proof that the second sum (which is the tail of a series) is negligible compared to $1/\sqrt{a}$. Examining the formula for λ_n we got in Eq. (9), we see that for I_1 from we got the precise value, and estimated only I_2. To estimate I_2, we threw away the factor that depended on n. The resulted estimate (10) does not work anymore since for $n \to \infty$ it gives

$$\lambda_n \geqslant -\frac{c_4}{\sqrt{a}} e^{-4a},$$

which diverges even when summed over $|n| > ca$. This suggests that we need a better estimate for I_2, the one that utilizes the oscillation of $\cos(\pi nt/2)$. Lemma 1 implies that we need to do it only for $n > ca$. The following estimate (12) is the last key ingredient in our proof.

Lemma 2 *Denote*

$$\lambda_n := \beta \cdot \int_{-2}^{2} e^{-at^2} e^{-\frac{i\pi nt}{2}} dt,$$

where β is the normalizing constant such that

$$e^{-at^2} = \sum_{n=-\infty}^{\infty} \lambda_n e^{\frac{i\pi nt}{2}}, \quad t \in [-2, 2].$$

Let c be the constant from Lemma 1. If a is sufficiently large, then for any $|n| \geqslant c \cdot a$ we have

$$|\lambda_n| \leqslant A \frac{e^{-c_2 n}}{\sqrt{a}} + B \cdot \frac{a e^{-4a}}{n^2}, \tag{12}$$

where the positive constants A, B, and c_2 do not depend on n or a.

Assuming this lemma holds, we can now show how to finish the proof. Indeed, estimates (8) and (11) imply that

$$W_a \geqslant \lambda_0 - \sum_{|n|>ca} |\lambda_n| \geqslant \frac{c_3}{\sqrt{a}} - 2 \sum_{n>ca} \left(A \frac{e^{-c_2 n}}{\sqrt{a}} + B \cdot \frac{a e^{-4a}}{n^2} \right) \geqslant$$
$$\frac{c_3}{\sqrt{a}} - D \frac{e^{-c_4 a}}{\sqrt{a}} - E a e^{-4a}.$$

We conclude that for sufficiently large values of a,

$$W_a \geqslant \frac{c_5}{\sqrt{a}},$$

which is the estimate from below that we need. □

We now carefully prove Lemma 2.

Proof of Lemma 2 As before, we can assume $n > 0$. We recall that

$$\lambda_n = \beta \int_{\mathbb{R}} e^{-at^2} \cos\left(\frac{\pi n t}{2}\right) dt - 2\beta \int_{|t|>2} e^{-at^2} \cos\left(\frac{\pi n t}{2}\right) dt = \beta(I_1 - 2I_2),$$

where since $n > ca$, we have

$$I_1 = c_2 \cdot \frac{1}{\sqrt{a}} e^{-c_3 n^2/a} \leqslant c_2 \cdot \frac{1}{\sqrt{a}} e^{-c_4 n}.$$

This estimate is one of the terms from (12), so we focus on estimating I_2. We integrate by parts twice to get

$$I_2 = \int_2^\infty e^{-at^2} \cos\left(\frac{\pi n t}{2}\right) dt = \frac{4a}{\pi n} \int_2^\infty t e^{-at^2} \sin\left(\frac{\pi n t}{2}\right) dt =$$

$$\frac{4a}{\pi n}\left[-\frac{2}{\pi n} t e^{-at^2} \cos\left(\frac{\pi n t}{2}\right)\Big|_2^\infty + \frac{2}{\pi n} \int_2^\infty \left(e^{-at^2} - 2at^2 e^{-at^2}\right) \cos\left(\frac{\pi n t}{2}\right) dt \right] =$$

$$\frac{16a}{\pi^2 n^2} e^{-4a} (-1)^n + \frac{8a}{\pi^2 n^2} I_2 - \frac{16a^2}{\pi^2 n^2} \int_2^\infty t^2 e^{-at^2} \cos\left(\frac{\pi n t}{2}\right) dt. \quad (13)$$

If $n > ca$, we have for sufficiently large a that $8a/(\pi^2 n^2) < c_5/a < 1/2$, and after sending the term $8a/(\pi^2 n^2) I_2$ to the left-hand side, we obtain

$$|I_2| \leqslant c_6 \frac{a e^{-4a}}{n^2} + c_7 \frac{a^2}{n^2} \left| \int_2^\infty t^2 e^{-at^2} \cos\left(\frac{\pi n t}{2}\right) dt \right|. \quad (14)$$

We remark that we gained a factor of $1/n^2$ in front of the integral from the $\cos(\pi n t/2)$, and now we do not need to utilize its oscillation anymore; we can get rid of it. Namely,

$$\left| \int_2^\infty t^2 e^{-at^2} \cos\left(\frac{\pi n t}{2}\right) dt \right| \leqslant \int_2^\infty t^2 e^{-at^2} dt \leqslant \int_2^\infty t^2 e^{-2at} dt.$$

The latter integral can be easily computed, but to estimate it we can just use that for $t \geqslant 2$, we have $t^2 < e^t$, which implies

$$\int_2^\infty t^2 e^{-2at}\,dt \leqslant \int_2^\infty e^{(-2a+1)t}\,dt = \frac{1}{2a-1}e^{-4a+2} \leqslant c_8 \frac{e^{-4a}}{a}.$$

Plugging this into (14), we deduce

$$|I_2| \leqslant c_9 \frac{ae^{-4a}}{n^2},$$

and since $\lambda_n = \beta(I_1 - 2I_2)$, our estimate is complete. □

Remark 5 Although we do not need it for the estimate, we remark on a bit more specific behavior of the Fourier coefficients λ_n for large values of n. Equation (13) implies that

$$\left(1 - \frac{16a}{\pi^2 n^2}\right) I_2 = \frac{16a}{\pi^2 n^2} e^{-4a}(-1)^n - \frac{16a^2}{\pi^2 n^2} \int_2^\infty t^2 e^{-at^2} \cos\left(\frac{\pi n t}{2}\right).$$

If we estimated the second term more carefully (perhaps by integrating by parts four times instead of two), we would see that the first term dominates it. I.e., informally speaking, for $n > c \cdot a$,

$$I_2 \sim (-1)^n \cdot \frac{16a}{\pi^2 n^2} e^{-4a}.$$

Moreover, by choosing c large enough, we can see that $I_1 \ll I_2$, which (informally) implies that for very large values of n (how large depends on a), the coefficients λ_n become sign-alternating.

Competing Interests The author have no conflicts of interest to declare that are relevant to the content of this chapter.

References

1. Bilyk, D., Glazyrin, A., Matzke, R., Park, J., and Vlasiuk, O.: Optimal measures for p-frame energies on spheres. Rev. Mat. Iberoam. 38, 4 (2022), 1129–1160.
2. Borodachov, S. V., Hardin, D. P., and Saff, E. B.: Discrete energy on rectifiable sets. In: Springer Monographs in Mathematics. Springer, New York, [2019] ©2019.
3. Cohn, H., Kumar, A., Miller, S. D., Radchenko, D., and Viazovska, M.: The sphere packing problem in dimension 24. Ann. of Math. (2) 185, 3 (2017), 1017–1033.
4. Cohn, H., Kumar, A., Miller, S. D., Radchenko, D., and Viazovska, M.: Universal optimality of the E_8 and Leech lattices and interpolation formulas. Ann. of Math. (2) 196, 3 (2022), 983–1082.
5. Viazovska, M. S.: The sphere packing problem in dimension 8. Ann. of Math. (2) 185, 3 (2017), 991–1015.

Approximation by Fourier Sums on the Classes of Generalized Poisson Integrals

Anatoly S. Serdyuk and Tetiana A. Stepanyuk

Dedicated to Edward Barry Saff on the occasion of his eightieth birthday.

In the present paper, we give a survey of current results about the best possible estimates of the norm of deviations of partial Fourier sums on the sets of generalized Poisson integrals in the uniform and integral metrics.

The partial Fourier sums

$$S_n(f) = S_n(f;x) = \frac{a_0}{2} + \sum_{k=1}^{n} a_k \cos kx + b_k \sin kx,$$

$$a_k = a_k(f) = \frac{1}{\pi} \int_{-\pi}^{\pi} f(t) \cos kt dt, \quad k = 0, 1, 2, ...,$$

$$b_k = b_k(f) = \frac{1}{\pi} \int_{-\pi}^{\pi} f(t) \sin kt dt, \quad k = 1, 2, ...$$

occupy a prominent place among linear methods of polynomial trigonometric approximation, due to their simplicity and good approximative properties. Since the beginning of the twentieth century, the problem of investigating the speed of convergence of the Fourier series, depending on the structural and smoothness properties of the function, has been studied by many mathematicians.

Results related to certain approximate properties of Fourier sums can be found in fundamental monographs [1]–[4], [10, 34, 37, 44, 46].

A. S. Serdyuk (✉) · T. A. Stepanyuk
Institute of Mathematics of Ukrainian National Academy of Sciences, Kyiv, Ukraine
e-mail: serdyuk@imath.kiev.ua; tstepaniuk@imath.kiev.ua

D. Bilyk et al. (eds.), *Recent Advances in Approximation and Potential Theory*, Applied and Numerical Harmonic Analysis,
https://doi.org/10.1007/978-3-031-95551-8_19

Let L_p, $1 \le p < \infty$, be the space of 2π–periodic functions f summable to the power p on $[0, 2\pi)$, in which the norm is given by the formula $\|f\|_p = \left(\int\limits_0^{2\pi} |f(t)|^p dt\right)^{\frac{1}{p}}$; L_∞ be the space of measurable and essentially bounded 2π–periodic functions f with the norm $\|f\|_\infty = \operatorname{ess\,sup}\limits_t |f(t)|$; C be the space of continuous 2π–periodic functions f, in which the norm is specified by the equality $\|f\|_C = \max\limits_t |f(t)|$.

If we consider the estimates of the deviation of norms of the Fourier sums in the uniform metric, then the Lebesgue inequality is the fundamental result regarding the estimate of the approximate properties of the mentioned sums (see, e.g., [11, 14, 34, 37])

$$\|f(\cdot) - S_{n-1}(f;\cdot)\|_C \le (1 + L_{n-1}) E_n(f)_C, \quad f \in C, \quad n \in \mathbb{N}, \tag{1}$$

where

$$L_{n-1} = \frac{1}{\pi} \int\limits_{-\pi}^{\pi} |D_{n-1}(t)| dt,$$

are Lebesgue constants of Fourier sums,

$$D_{n-1}(t) := \frac{1}{2} + \sum_{k=1}^{\infty} \cos kt = \frac{\sin(n - \frac{1}{2})t}{2 \sin \frac{t}{2}}$$

are Dirichlet kernels, and

$$E_n(f)_C = \inf_{T_{n-1} \in \mathfrak{T}_{2n-1}} \|f - T_{n-1}\|_C$$

is the best uniform approximation of the function f by trigonometric polynomials of the order $n - 1$.

For the Lebesgue constants L_n the following asymptotic equality takes place as $n \to \infty$ (see [5])

$$L_n = \frac{4}{\pi^2} \ln n + O(1), \quad n \to \infty. \tag{2}$$

More exact estimates for the differences $L_n - \frac{4}{\pi^2} \ln(n + a)$, where $a > 0$, can be found in the works [1, 4, 6, 14, 30, 45].

Taking into account (2), the inequality (1) can be written in the form

$$\|f(\cdot) - S_{n-1}(f;\cdot)\|_C \le \left(\frac{4}{\pi^2} \ln n + O(1)\right) E_n(f)_C. \tag{3}$$

On the whole space C the inequality (3) is asymptotically exact. At the same time, there exist subsets of functions from C and for elements of these subsets, the inequality (3) is not exact even by order.

Existence of such subsets follows, in particular, from Oskolkov results [17], who proved that for any $f \in C$ the following inequality holds

$$\|f(\cdot) - S_{n-1}(f; \cdot)\|_C \le K \sum_{v=n}^{2n-1} \frac{E_v(f)_C}{v-n+1}, \ f \in C, \tag{4}$$

where K is some absolute constant and he showed that on the classes

$$C(\varepsilon) := \{f \in C : \ E_v(f)_C \le \varepsilon_v, \ v \in \mathbb{Z}_+\},$$

generated by sequences $\varepsilon = \{\varepsilon_k\}_{k=0}$ of nonnegative numbersm such that $\varepsilon_k \downarrow 0$, $k \to \infty$, the estimate (4) is exact by the order, i.e.

$$K_1 \sum_{v=n}^{2n-1} \frac{\varepsilon_v}{v-n+1}, \le \sup_{f \in C(\varepsilon)} \|f(\cdot) - S_{n-1}(f; \cdot)\|_C \le K_2 \sum_{v=n}^{2n-1} \frac{\varepsilon_v}{v-n+1}, \tag{5}$$

where K_1 and K_2 are some constants. Setting

$$\varepsilon_k = \rho^k, \quad \rho \in (0, 1),$$

from (5) it follows the following exact order estimate

$$\mathcal{E}_n(C(\varepsilon))_C = \sup_{f \in C(\varepsilon)} \|f(\cdot) - S_{n-1}(f; \cdot)\|_C \asymp \rho^n,$$

at the same time the formulas (3) allows to write only the following estimate

$$\mathcal{E}_n(C(\varepsilon))_C = o\left(\rho^n \log n\right).$$

Based on the mentioned above, the problem arose to establish asymptotic Lebesgue inequalities, which can not be improved for certain subsets of the space C arose. The works [31, 33, 35] and others are devoted to solving this problem.

Investigation of asymptotic behavior of the deviations

$$\rho_n(f; \cdot) = f(\cdot) - S_{n-1}(f; \cdot)$$

on certain classes of functions in the uniform metric was initiated in the work of Kolmogorov [9], who introduced the quantity

$$\mathcal{E}_n(\mathfrak{N})_X = \sup_{f \in \mathfrak{N}} \|f(\cdot) - S_{n-1}(f; \cdot)\|_X, \ \mathfrak{N} \subset X, \tag{6}$$

where $\mathfrak{N}$ is some class of linear normed space X. For $X = C$ and for the class of differentiable functions $\mathfrak{N} = W^r$, where W^r, $r \in \mathbb{N}$ is the set of continuous functions, which have the absolutely continuous derivatives up to and including the r-th order, such that $\|f^{(r)}\|_\infty \leq 1$, Kolmogorov [9] established the following asymptotic equality as $n \to \infty$

$$\mathcal{E}_n(W^r_\infty)_C = \frac{4}{\pi^2}\frac{\ln n}{n^r} + O\Big(\frac{1}{n^r}\Big), \tag{7}$$

where O is the quantity uniformly bounded with respect to n.

Later Nikolsky [15] established the asymptotic equality as $n \to \infty$ for the quantities $\mathcal{E}_n(W^r H^\alpha)_C$, where

$$W^r H^\alpha = \{f \in C^r : |f^{(r)}(x) - f^{(r)}(x')| \leq |x - x'|^\alpha,\ \alpha \in (0,1), r \in \mathbb{Z}_{+]}\},$$

C^r is the set of continuous functions, which have the absolutely continuous derivatives up to and including the r-th order.

Namely, he showes that as $n \to \infty$

$$\mathcal{E}_n(W^r H^\alpha)_C = \frac{2^{\alpha+1}\ln n}{\pi^2 n^{r+\alpha}} \int\limits_0^{\frac{\pi}{2}} t^\alpha \sin t dt + O\Big(\frac{1}{n^{r+\alpha}}\Big), \tag{8}$$

Nikolsky has also obtained the fundamental results regarding the research of integral norms of deviations of Fourier sums $\|\rho_n(f;\cdot)\|_{L_1}$, namely he has established the analogues of the equalities (7) and (8) in the integral metric L_1. It should be noted that the application of Lebesgue inequality (1) for the functions from the classes W^r, $r \in \mathbb{N}$ or $W^r H^\alpha$, $r \in \mathbb{Z}^+$, $\alpha \in (0,1)$ in order to obtain asymptotically accurate estimates of (7) and (8) turns out to be ineffective.

The research of Kolmogorov and Nikolsky started a whole direction in the theory of approximations. Their results were extended to wider classes of functions $\mathfrak{N}$, to other normalized spaces X and to other linear methods of polynomial approximation. The problem about finding the asymptotic equalities as $n \to \infty$ for the quantities

$$\mathcal{E}(\mathfrak{N}, U_n)_X = \sup_{f \in \mathfrak{N}} \|f(\cdot) - U_n(f;\cdot)\|_X, \tag{9}$$

where $\mathfrak{N}$ is a given class of 2π–periodic functions from the normed space X, U_n is a fixed linear polynomial method of approximation, is called the Kolmogorov-Nikolsky problem for the class $\mathfrak{N}$ and the method U_n in the space X. At the same time, the constant in the main term of the asymptotic expansion of the quantity of the form (9) is sometimes called as the Kolmogorov-Nikolsky constant. As it follows from above, the behavior of the Kolmogorov-Nikolsky constants for Fourier sums is not related to the behavior of the Lebesgue constants.

More details on the results about finding the solution of the Kolmogorov-Nikolsky problem for Fourier sums on classes of differential functions in the sense of Weyl-Nagy $W^r_{\beta,p}$ and classes (ψ,β)–differentiable functions in the sense of Stepants $L^{\psi}_{\beta,p}$ can be found in the works [8, 10, 15, 19, 20, 25, 31, 32, 36–38, 40, 41] and others.

In this work, we will limit ourselves to a review of known results for solving the Kolmogorov-Nikolsky problem for Fourier sums on the classes of generalized Poisson integrals $C^{\alpha,r}_{\beta,p}$, which are defined below.

Denote by $C^{\alpha,r}_{\beta,p}$, $\alpha > 0,\ r > 0,\ \beta \in \mathbb{R},\ 1 \le p \le \infty$, the set of all 2π–periodic functions, representable for all $x \in \mathbb{R}$ as convolutions of the form (see, e.g., [37, p. 133])

$$f(x) = \frac{a_0}{2} + \frac{1}{\pi}\int\limits_{-\pi}^{\pi} P_{\alpha,r,\beta}(x-t)\varphi(t)dt,\ a_0 \in \mathbb{R},\ \varphi \in B^0_p, \tag{10}$$

$$B^0_p = \{\varphi :\ ||\varphi||_p \le 1,\ \varphi \perp 1\},\ 1 \le p \le \infty,$$

with fixed generated kernels

$$P_{\alpha,r,\beta}(t) = \sum_{k=1}^{\infty} e^{-\alpha k^r}\cos\big(kt - \frac{\beta\pi}{2}\big),\ \alpha > 0,\ r > 0,\ \beta \in \mathbb{R}.$$

The kernels $P_{\alpha,r,\beta}(t)$ are called as generalized Poisson kernels. For $r = 1$ and $\beta = 0$ the kernels $P_{\alpha,r,\beta}(t)$ are usual Poisson kernels of harmonic functions.

For any $r > 0$ the classes $C^{\alpha,r}_{\beta,p}$ belong to set of infinitely differentiable 2π–periodic functions D^∞, i.e., $C^{\alpha,r}_{\beta,p} \subset D^\infty$ (see, e.g., [37, p. 128], [38]). For $r \ge 1$ the classes $C^{\alpha,r}_{\beta,p}$ consist of functions f, admitting a regular extension into the strip $|\text{Im } z| \le c,\ c > 0$ in the complex plane (see, e.g., [37, p. 141]), i.e., are the classes of analytic functions. For $r > 1$ the classes $C^{\alpha,r}_{\beta,p}$ consist of functions regular on the whole complex plane, i.e., of entire functions (see, e.g., [37, p. 131]). Besides, it follows from Theorem 1 in [39] that for any $r > 0$ the embedding holds $C^{\alpha,r}_{\beta,p} \subset \mathcal{J}_{1/r}$, where $\mathcal{J}_a$, $a > 0$, are known Gevrey classes

$$\mathcal{J}_a = \left\{ f \in D^\infty :\ \sup_{k\in\mathbb{N}} \Big(\frac{\|f^{(k)}\|_C}{(k!)^a}\Big)^{1/k} < \infty \right\}.$$

Approximation properties of classes of generalized Poisson integrals $C^{\alpha,r}_{\beta,p}$ in metrics of spaces L_s, $1 \le s \le \infty$, were considered in [18–20], [22–24], [26], [27], [32], [34], [35], [37], [41–43], from the viewpoint of order or asymptotic estimates for approximations by Fourier sums, best approximations, and widths.

On the classes of generalized Poisson integrals $C^{\alpha,r}_{\beta,p}$, we are interested in asymptotic equalities as $n \to \infty$ of the quantities

$$\mathcal{E}_n(C^{\alpha,r}_{\beta,p})_C = \sup_{f\in C^{\alpha,r}_{\beta,p}} \|f(\cdot) - S_{n-1}(f;\cdot)\|_C, \ r>0, \ \alpha>0, \ \beta\in\mathbb{R}, \ 1\le p\le\infty. \tag{11}$$

Nikolsky [16] investigated the approximation properties of the Fourier sums on some classes of analytic functions and obtained the results, from which it follows that in the case $r = 1$, $p = \infty$ the following asymptotic equality is true

$$\mathcal{E}_n(C^{\alpha,1}_{\beta,\infty})_C = e^{-\alpha n}\Big(\frac{8}{\pi^2}\mathbf{K}(e^{-\alpha}) + O(1)n^{-1}\Big), \tag{12}$$

where

$$\mathbf{K}(q) := \int\limits_0^{\frac{\pi}{2}} \frac{dt}{\sqrt{1-q^2\sin^2 t}}, \ q\in(0,1),$$

is a complete elliptic integral of the first kind, and O is a quantity uniformly bounded in parameters n and β.

Later, the equality (12) was clarified by Stechkin [31, p. 139], who established the asymptotic formula

$$\mathcal{E}_n(C^{\alpha,1}_{\beta,\infty})_C = e^{-\alpha n}\Big(\frac{8}{\pi^2}\mathbf{K}(e^{-\alpha}) + O(1)\frac{e^{-\alpha}}{(1-e^{-\alpha})n}\Big), \ \alpha>0, \ \beta\in\mathbb{R}, \tag{13}$$

where $O(1)$ is a quantity uniformly bounded in all analyzed parameters.

In work [18] for $r = 1$ and arbitrary values of $1 \le p \le \infty$ for quantities $\mathcal{E}_n(C^{\alpha,r}_{\beta,p})_C$, $\alpha > 0$, $\beta\in\mathbb{R}$, the following equality was established

$$\mathcal{E}_n(C^{\alpha,1}_{\beta,p})_C = e^{-\alpha n}\left(\frac{2}{\pi^{1+\frac{1}{p'}}}\|\cos t\|_{p'}K(p',e^{-\alpha}) + O(1)\frac{e^{-\alpha}}{n(1-e^{-\alpha})^{s(p)}}\right), \tag{14}$$

where $p' = \frac{p}{p-1}$,

$$s(p) := \begin{cases} 1, & p=\infty, \\ 2, & p\in[1,2)\cup(2,\infty), \\ -\infty, & p=2, \end{cases}$$

$$K(p',q) := \frac{1}{2^{1+\frac{1}{p'}}}\left\|(1-2q\cos t+q^2)^{-\frac{1}{2}}\right\|_{p'}, \ q\in(0,1),$$

and $O(1)$ is a quantity uniformly bounded in n, p, α and β. For $p = \infty$, by virtue of the known equality $K(1, q) = \mathbf{K}(q)$ (see, e.g., formula 3.674(1) of [7, p. 401]), the estimate (14) coincides with the estimate (13).

In [20] it was proved that in the case $1 \le p' < \infty$ the following equality takes place

$$K(p', q) = \frac{\pi^{\frac{1}{p'}}}{2} F^{\frac{1}{p'}}\Big(\frac{p'}{2}, \frac{p'}{2}; 1; q^2\Big), \ q \in (0, 1), \tag{15}$$

where $F(a, b; c; d)$ is Gauss hypergeometric function

$$F(a, b; c; z) = 1 + \sum_{k=1}^{\infty} \frac{(a)_k (b)_k}{(c)_k} \frac{z^k}{k!},$$

$$(x)_k := x(x + 1)(x + 2)...(x + k - 1).$$

So, taking into account the equality (15), where $q = e^{-\alpha}$, the estimate (14) can be written in the form

$$\mathcal{E}_n(C^{\alpha,1}_{\beta,p})_C = e^{-\alpha n}\Bigg(\frac{\|\cos t\|_{p'}}{\pi} F^{\frac{1}{p'}}\left(\frac{p'}{2}, (\frac{p'}{2}; 1; e^{-2\alpha}\right) + O(1)\frac{e^{-\alpha}}{n(1 - e^{-\alpha})^{s(p)}}\Bigg). \tag{16}$$

Note that for $p = 2$ and $r = 1$ formula (14), and also (16) becomes the equality

$$\mathcal{E}_n(C^{\alpha,1}_{\beta,2})_C = \frac{1}{\sqrt{\pi(1 - e^{-2\alpha})}} e^{-\alpha n}, \ \alpha > 0, \ \beta \in \mathbb{R}, \ n \in \mathbb{N}, \tag{17}$$

(see [18]). Moreover, it follows from [22] that for $p = 2$ for the quantities $\mathcal{E}_n(C^{\alpha,r}_{\beta,p})_C$ the following equalities take place

$$\mathcal{E}_n(C^{\alpha,r}_{\beta,2})_C = \frac{1}{\sqrt{\pi}}\Big(\sum_{k=n}^{\infty} e^{-2\alpha k^r}\Big)^{\frac{1}{2}}, \ \alpha > 0, \ r > 0, \ \beta \in \mathbb{R}, \ n \in \mathbb{N}. \tag{18}$$

It is clear that for $r = 1$ the formula (18) transforms into (17).

In the case of $r > 1$ and $p = \infty$ the asymptotic equalities for the quantities $\mathcal{E}_n(C^{\alpha,r}_{\beta,p})_C$, $\alpha > 0$, $\beta \in \mathbb{R}$, were obtained by Stepanets [34, Chapter 3, Section 9], who showed that for any $n \in \mathbb{N}$

$$\mathcal{E}_n(C^{\alpha,r}_{\beta,\infty})_C = \Big(\frac{4}{\pi} + \gamma_n\Big)e^{-\alpha n^r}, \tag{19}$$

where

$$|\gamma_n| < 2\Big(1 + \frac{1}{\alpha r n^{r-1}}\Big)e^{-\alpha r n^{r-1}}.$$

Later Telyakovsky [41] clarified the asymptotic equality (19) by improving the estimate for the residual term. Namely, it follows from [41] that

$$\mathcal{E}_n(C^{\alpha,r}_{\beta,\infty})_C = \frac{4}{\pi}e^{-\alpha n^r} + O(1)\Big(e^{-\alpha(2(n+1)^r-n^r)} + \Big(1+\frac{1}{\alpha r(n+2)^r}\Big)e^{-\alpha(n+2)^r}\Big), \tag{20}$$

where $O(1)$ is a quantity uniformly bounded in all analyzed parameters.

For $r > 1$ and for arbitrary values of $1 \le p \le \infty$ the asymptotic equalities for the quantities $\mathcal{E}_n(C^{\alpha,r}_{\beta,p})_C$, $\alpha > 0$, $\beta \in \mathbb{R}$, are found in [18] and have the form

$$\mathcal{E}_n(C^{\alpha,r}_{\beta,p})_C = e^{-\alpha n^r}\Big(\frac{\|\cos t\|_{p'}}{\pi} + O(1)\Big(1+\frac{1}{\alpha r n^{r-1}}\Big)e^{-\alpha n^{r-1}}\Big), \tag{21}$$

where O is a quantity uniformly bounded in all analyzed parameters. For $p = \infty$ the formula (21) follows from (19) and (20).

Concerning the case $0 < r < 1$, except the presented above case $p = 2$, asymptotic equalities for quantities $\mathcal{E}_n(C^{\alpha,r}_{\beta,p})_C$, $\alpha > 0$, $\beta \in \mathbb{R}$, were known only for $p = \infty$ due to the work of Stepanets [32], who showed that

$$\mathcal{E}_n(C^{\alpha,r}_{\beta,\infty})_C = \frac{4}{\pi^2}e^{-\alpha n^r}\ln n^{1-r} + O(1)e^{-\alpha n^r}, \ r \in (0,1), \ \beta \in \mathbb{R}, \tag{22}$$

where $O(1)$ is a quantity uniformly bounded in n and β.

In case of $0 < r < 1$ and $1 \le p < \infty$ the following order estimates for quantities $\mathcal{E}_n(C^{\alpha,r}_{\beta,p})_C$, $\alpha > 0$, $\beta \in \mathbb{R}$, hold (see, e.g., [43], [23])

$$\mathcal{E}_n(C^{\alpha,r}_{\beta,p})_C \asymp e^{-\alpha n^r} n^{\frac{1-r}{p}}. \tag{23}$$

We remark that for $0 < r < 1$ and $1 \le p < \infty$ Fourier sums provide the order of the best approximations of classes $C^{\alpha,r}_{\beta,p}$, $\alpha > 0$, $\beta \in \mathbb{R}$, in the uniform metric, i.e. (see, e.g., [23], [24])

$$\mathcal{E}_n(C^{\alpha,r}_{\beta,p})_C \asymp E_n(C^{\alpha,r}_{\beta,p})_C \asymp e^{-\alpha n^r} n^{\frac{1-r}{p}},$$

where

$$E_n(C^{\alpha,r}_{\beta,p})_C = \sup_{f \in C^{\alpha,r}_{\beta,p}} \inf_{t_{n-1} \in \mathcal{T}_{2n-1}} \|f - t_{n-1}\|_C,$$

and $\mathcal{T}_{2n-1}$ is the subspace of all trigonometric polynomials t_{n-1} of degree not higher than $n-1$.

On the one hand, this fact encourages research more deeply approximation properties of Fourier sums in given situations, and on the other hand it separates the case $1 \le p < \infty$ from the considered earlier case $p = \infty$, where order equality $\mathcal{E}_n(C^{\alpha,r}_{\beta,p})_C \asymp E_n(C^{\alpha,r}_{\beta,p})_C$, $0 < r < 1$, doesn't take place.

Besides, as follows from Temlyakov work [43] for $2 \le p < \infty$ the quantities of approximations by Fourier sums provide the order of the linear widths λ_{2n} (definition of λ_m see, e.g., [10, Chapter 1, Section 1.2]) of the classes $C^{\alpha,r}_{0,p}$, i.e.

$$\lambda_{2n}(C^{\alpha,r}_{0,p}, C) \asymp \mathcal{E}_n(C^{\alpha,r}_{0,p})_C.$$

In the paper [27] the authors established the asymptotically sharp estimates of the quantities $\mathcal{E}_n(C^{\alpha,r}_{\beta,p})_C$, $\alpha > 0$, $\beta \in \mathbb{R}$, for any $0 < r < 1$ and $1 \le p \le \infty$. In particular, it is proved that for $r \in (0, 1)$, $\alpha > 0$, $\beta \in \mathbb{R}$ and $1 \le p \le \infty$ as $n \to \infty$ the following asymptotic equalities take place

$$\mathcal{E}_n(C^{\alpha,r}_{\beta,p})_C = e^{-\alpha n^r} n^{\frac{1-r}{p}} \left(\frac{\|\cos t\|_{p'}}{\pi^{1+\frac{1}{p'}} (\alpha r)^{\frac{1}{p}}} F^{\frac{1}{p'}}\left(\frac{1}{2}, \frac{3-p'}{2}; \frac{3}{2}; 1\right) + \right.$$

$$\left. + \frac{O(1)}{n^{\min\{\frac{1-r}{p}, r\}}} \right), \quad 1 < p \le \infty, \quad \frac{1}{p} + \frac{1}{p'} = 1, \tag{24}$$

$$\mathcal{E}_n(C^{\alpha,r}_{\beta,1})_C = e^{-\alpha n^r} n^{1-r} \left(\frac{1}{\pi \alpha r} + \frac{O(1)}{n^{\min\{1-r,\, r\}}} \right), \tag{25}$$

where $O(1)$ is a quantity uniformly bounded with respect to n and β.

Formulas (24) and (25) together with formulas (12)–(14), (19)–(22) give the solution of Kolmogorov–Nikolsky problem about strong asymptotic of quantities (11) as $n \to \infty$ for all admissible values of parameters of problem.

Herewith, in the paper [27] the authors found the estimates for remainders in (24) and (25), which are expressed via absolute values and the parameters of the problem α, r, p in explicit form. These estimates can be used for practical application since they allow effectively to estimate the errors of uniform approximations of functions from the classes $C^{\alpha,r}_{\beta,p}$ by their partial Fourier sums.

The following table contains the exact values of constants (Kolmogorov–Nikolsky constants) in the main term A_n of the asymptotic expansion of quantities $\mathcal{E}_n(C^{\alpha,r}_{\beta,p})_C$ of the form

$$\mathcal{E}_n(C^{\alpha,r}_{\beta,p})_C = e^{-\alpha n^r}(A_n + o(A_n)).$$

Let us make a few remarks about the solution of the Kolmogorov-Nikolsky problem in the metrics of spaces L_p .

Currently the asymptotic equalities as $n \to \infty$ of the quantities

$$\mathcal{E}_n(C^{\alpha,r}_{\beta,1})_{L_s} = \sup_{f \in C^{\alpha,r}_{\beta,1}} \|f(\cdot) - S_{n-1}(f;\cdot)\|_s, \; r > 0, \; \alpha > 0, \; \beta \in \mathbb{R}, \; 1 \le s \le \infty, \tag{26}$$

A_n		r		
		(0, 1)	**1**	**(1, ∞)**
	∞	Stepanets [32] $\frac{4}{\pi^2}(1-r)\ln n$	Nikolsky [16] Stechkin [31] $\frac{8}{\pi^2}\mathbf{K}(e^{-\alpha})$	Stepanets [34] Telyakovskii [41] $\frac{4}{\pi}$
p	**(1, ∞)**	Serdyuk Stepanyuk [27] $\frac{\\|\cos t\\|_{p'} F^{\frac{1}{p'}}(\frac{1}{2}, \frac{3-p'}{2}; \frac{3}{2}; 1)}{\pi^{1+\frac{1}{p'}}(\alpha r)^{\frac{1}{p}}} n^{\frac{1-r}{p}}$	Serdyuk [18] $\frac{\\|\cos t\\|_{p'}}{\pi} F^{\frac{1}{p'}}(\frac{p'}{2}, \frac{p'}{2}; 1; e^{-2\alpha})$	Serdyuk [18] $\frac{\\|\cos t\\|_{p'}}{\pi}$
	1	Serdyuk Stepanyuk [27] $\frac{1}{\pi\alpha r} n^{1-r}$	Serdyuk [18] $\frac{1}{\pi(1-e^{-\alpha})}$	Serdyuk [18] $\frac{1}{\pi}$

are known for all admissible values of the problem parameters. At the same time, it turned out that the following asymptotic equality takes place

$$\lim_{n\to\infty} \frac{\mathcal{E}_n(C^{\alpha,r}_{\beta,p})_C}{\mathcal{E}_n(C^{\alpha,r}_{\beta,1})_{L_{p'}}} = 1, \quad \frac{1}{p} + \frac{1}{p'} = 1, \tag{27}$$

which holds for all $\alpha > 0$, $r > 0$ and $\beta \in \mathbb{R}$, $1 \leq p \leq \infty$. The works [19, 24, 32, 34, 37] are devoted to the establishment of asymptotic equalities for quantities of the form (26).

In the works [12, 13, 21, 28, 29, 33, 35] for the functions f from the sets $C^{\alpha,r}_{\beta} L_p$ or $C^{\alpha,r}_{\beta} C$, defined by convolutions of the form (10), in which the functions φ belong to the spaces L_p or C, respectively, the asymptotically best possible Lebesgue-type inequalities were established. In these equalities the uniform norms of the deviations of Fourier sums $\|f - S_{n-1}\|_C$ were estimated via the best approximations of the functions φ in the metrics of spaces L_p or C.

References

1. *N.I. Akhiezer*, Lectures on approximation theory. Mir, Moscow (1965).
2. *N.K. Bari*, A treatise on trigonometric series, Oxford-London-New York-Paris-Frankfurt: Pergamon Press (1964).
3. P. L. Butzer, R. J. Nessel, Fourier Analysis and Approximation, Birkhauser Verlag Basel (1971).
4. *V.K. Dzyadyk*, Introduction to the theory of uniform approximation of functions by polynomials [in Russian], Nauka, Moscow (1977).
5. *L. Fejer*, Lebesguesche konstanten und divergente Fourierreihen, J. Reine Angew Math. 138 (1910), 22–53.
6. *P.V. Galkin*, Estimate for Lebesgue constants, Proc. Steklov Inst. Math., (1971) 109, 1-4.
7. *I.S. Gradshtein, I.M. Ryzhik*, Tables of Integrals, Sums, Series, and Products, Academic Press 1966.

8. *U.Z. Hrabova, A.S. Serdyuk*, Order estimates for the best approximations and approximations by Fourier sums of the classes of (ψ, β)-differential functions, Ukr. Math. J. 65:9 (2014) 1319–1331.
9. *A. Kolmogorov*, Zur Grössennordnung des Restgliedes Fourierschen Reihen differenzierbarer Funktionen, (in German) Ann. Math.(2), 36:2 (1935) 521–526.
10. *N.P. Korneichuk*, Exact Constants in Approximation Theory, Vol. 38, Cambridge Univ. Press, Cambridge, New York (1990).
11. *H. Lebesgue*, Sur la representation trigonometrique approchee des fonctions satisfaisant a une condition de Lipschitz, Bull. Soc. Math. France 38 (1910) 182–210.
12. *A.P. Musienko, A.S. Serdyuk*, Lebesgue-type inequalities for the de la Valle-Poussin sums on sets of entire functions. Ukr. Math. J. 65:5 (2013) 709–722.
13. *A.P. Musienko, A.S. Serdyuk*, Lebesgue-type inequalities for the de la Valee-Poussin sums on sets of analytic functions. Ukr. Math. J. 65:4, (2013) 575–592.
14. *G.I. Natanson*, An estimate for Lebesgue constants of de la Vallee-Poussin sums (in Russian), in "Geometric questions in the theory of functions and sets, Kalinin State Univ., Kalinin (1986).
15. *S. M. Nikolsky*, Approximations of periodic functions by trigonometrical polynomials, [in Russian] Travaux Inst. Math. Stekloff, 15, Acad. Sci. USSR, Moscow–Leningrad, (1945) 3–76, 76p.
16. *S. M. Nikolsky*, Approximation of functions in the mean by trigonometrical polynomials, [in Russian] Izv. Akad. Nauk SSSR, Ser. Mat. 10 (1946) 207–256.
17. *K.I. Oskolkov*, Lebesgue's inequality in a uniform metric and on a set of full measure, Math. Notes, 18:4 (1975), 895–902.
18. *A.S. Serdyuk*, Approximation of classes of analytic functions by Fourier sums in uniform metric, Ukr. Math. J. 57:8 (2005) 1275–1296.
19. *A.S. Serdyuk*, Approximation of classes of analytic functions by Fourier sums in the metric of the space L_p, Ukr. Math. J. 57:10, (2005) 1395–1408.
20. *A.S. Serdyuk*, Approximation by interpolation trigonometric polynomials on classes of periodic analytic functions. Ukr. Math. J. 64:5 (2012) 797–815.
21. *A.S. Serdyuk, A.P. Musienko*, The Lebesgue type inequalities for the de la Valle Poussin sums in approximation of Poisson integrals, [n Ukrainian] Zb. Pr. Inst. Mat. NAN Ukr., 7:1, (2010) 298–316 .
22. *A.S. Serdyuk, I.V. Sokolenko*, Uniform approximation of the classes of $(\psi, \overline{\beta})$–differentiable functions by linear methods, [in Ukrainian] Zb. Pr. Inst. Mat. NAN Ukr. 8:1 (2011) 181–189.
23. *A.S. Serdyuk, T.A. Stepanyuk*, Order estimates for the best approximation and approximation by Fourier sums of classes of infinitely differentiable functions, [in Ukrainian] Zb. Pr. Inst. Mat. NAN Ukr. 10:1 (2013) 255–282.
24. *A.S. Serdyuk, T.A. Stepanyuk*, Estimates for the best approximations of the classes of infinitely differentiable functions in uniform and integral metrics, Ukr. Math. J. 66:9 (2015) 1393–1407.
25. *A.S. Serdyuk, T.A. Stepanyuk*,Order estimates for the best approximations and approximations by Fourier sums of the classes of convolutions of periodic functions of low smoothness in the integral metric, Ukr. Math. J. 65:12 (2015) 1862–1882.
26. *A.S. Serdyuk, T.A. Stepanyuk*, Approximations by Fourier sums of classes of generalized Poisson integrals in metrics of spaces L_s, Ukr. Mat. J. 69:5 (2017).
27. *A.S. Serdyuk, T.A. Stepanyuk*,Uniform approximations by Fourier sums on classes of generalized Poisson integrals, Analysis Mathematica 45, 201–236 (2019).
28. *A.S. Serdyuk, T.A. Stepanyuk*, Asymptotically best possible Lebesque-type inequalities for the Fourier sums on sets of generalized Poisson integrals, FILOMAT, 34:14, (2020) 4697–4707.
29. *A.S. Serdyuk, T.A. Stepanyuk*, About Lebesgue inequalities on the classes of generalized Poisson integrals, Jaen J. Approx. 12 (2021), 25–40.
30. *I. A. Shakirov*, Approximation of the Lebesgue constant of a Lagrange polynomial by a logarithmic function with shifted argument, J. Math. Sci. (N. Y.), 252:3 (2021), 445–452.
31. *S.B. Stechkin*, An estimate of the remainder term of Fourier series for differentiable functions, (in Russian) Tr. Mat. Inst. Steklova 145 (1980) 126–151.

32. *A.I. Stepanets*, Deviation of Fourier sums on classes of infinitely differentiable functions, Ukr. Mat. J. 36:6 (1984) 567–573.
33. *A.I. Stepanets*, On the Lebesgue inequality on classes of (ψ, β)-differentiable functions, Ukr. Math. J. 41:4 (1989), 435–443.
34. *A.I. Stepanets*, Classification and approximation of periodic functions, Kluwer Academic Publishers, Dordrecht 1995.
35. *A.I. Stepanets, A.S. Serdyuk*, Lebesgue inequalities for Poisson integrals, Ukr. Math. J., 52:6, (2000), 798–808.
36. *A.I. Stepanets*, Approximation of Convolution Classes by Fourier Sums. New Results, Ukr. Math. J., 54:3, (2002) 713–740.
37. *A.I. Stepanets*, Methods of Approximation Theory. VSP: Leiden, Boston (2005).
38. *A.I. Stepanets, A.S. Serdyuk, A.L. Shidlich*, On some new criteria for infinite differentiability of periodic functions, Ukr. Math. J. 59:10 (2007) 1569–1580 .
39. *A.I. Stepanets, A.S. Serdyuk, A.L. Shidlich*, On relationship between classes of $(\psi, \overline{\beta})$–differentiable functions and Gevrey classes, Ukr. Math. J. 61:1 (2009) 171–177 .
40. *S.A. Telyakovsky*, Approximation of differentiable functions by partial sums of their Fourier series, Mathematical Notes 4:3 (1968) 668–673.
41. *S.A. Telyakovsky*, Approximation of functions of high smoothness by Fourier sums, Ukr. Math. J. 41:4 (1989) 444–451.
42. *V.N. Temlyakov*, To the question on estimates of the diameters of classes of infinite- differentiable functions, [in Russian] Mat. Zametki 47:5 (1990) 155-157.
43. *V.N. Temlyakov*, On estimates of the diameters of classes of infinite- differentiable functions, [in Russian] Dokl. razsh. zased. semin. Inst. Prykl. Mat. im. I.N.Vekua 5:2 (1990) 111–114.
44. *A.F. Timan*, Theory of Approximation of functions of a real variable, Dover Publications (1994).
45. *V.V. Zhuk, G.I. Natanson*, Trigonometrical Fourier series and elements of approximation theory, Izdat. Leningr. Univ. (1983). [in Russian].
46. *A. Zygmund*, Trigonometric series. Cambridge Mathematical Library. Vol. I and II. Cambridge University Press (1988).

On Linear Spaces with Ideal Complements

Boris Shekhtman

Dedicated to Ed Saff on the occasion of his 80th birthday.

1 Introduction

In this article we pose and partially answer the following general

Question Let A be an algebra over the field of real or complex numbers. Let $E \subset A$ be a linear subspace. For what subspaces E does there exist an ideal (closed ideal, one-sided ideal) $J \subset A$ such that J complements E?

An equivalent formulation of this questions involves "ideal projectors".

Definition 1 A projection operator P on A is called an ideal projector (cf. [1]) if $\ker P$ is an ideal in A.

Thus the question is equivalent to the question of what linear subspaces of A are ranges of an ideal projector?

The answer often depends on particular algebra. Here is a very general result for commutative algebras of functions.

Theorem 1 *Let $A = \mathcal{F}(K)$ be an algebra of scalar valued functions on a set K. Let E be a linear subspace of A such that for some subset $K_0 \subset K$ we have*

(1) The restriction $E_{K_0} = \{f \in E$ defined as function K_0 only$\}$ coincides with A_{K_0}.
(2) If $f \in E$ is identically zero on K_0 then $f = 0$.

Then the ideal $J_{K_0} := \{f \in A$ such that $f_{K_0} = 0\}$ is an ideal in A that complements E.

B. Shekhtman (✉)
Department of Mathematics and Statistics, University of South Florida, Tampa, FL, USA
e-mail: shekhtma@usf.edu

D. Bilyk et al. (eds.), *Recent Advances in Approximation and Potential Theory*, Applied and Numerical Harmonic Analysis,
https://doi.org/10.1007/978-3-031-95551-8_20

Conversely if an ideal J_{K_0} *defined above complements a subspace* $E \subset A$ *then* E *satisfies Properties 1 and 2.*

Proof Clearly, J_{K_0} is an ideal. Condition 2 then implies that $E \cap J_{K_0} = \{0\}$. Finally take any $f \in A$ let $g \in E$ be such that $g_{K_o} = f_{K_0}$. Then $f - g$ vanishes on K_0, hence $f - g \in J_{K_0}$ hence $f = g + (f - g)$ and $A = E_{K_0} + J_{K_0}$.

Conversely, if $A = E \dotplus J_{K_0}$ then $E \cap J_{K_0} = \{0\}$ and hence 2 holds. Also if $A = E \dotplus J_{K_0}$ for every $f \in A$ there exist $g \in E$ such that $f - g \in J_{K_0}$ thus f and g coincide on K_0 and Condition 2 holds. □

In general, not every ideal in any algebra A has the form

$$\left\{f \in A \text{ such that } f_{K_0} = 0\right\}$$

for some subset $K_0 \subset K$. Hence Conditions 1 and 2 are sufficient but may not be necessary for E to be complemented by an ideal. In the next sections we will show that depending on A these conditions are sometimes necessary and sometimes not.

Before we go to these cases we will prove one other simple theorem:

Theorem 2 *Let* E *be a finite-dimensional subspace of* A*. Then* E *has an ideal complement.*

Proof Let E be an n-dimensional subspace of A spanned by linearly independent functions $f_1, \dots, f_n$ in A. Then, it is well-known that there exists points $k_1, \dots, k_n \in K$ such that

$$\det\left(f_j\left(k_m\right)\right)_{j,k=1}^{n} \neq 0.$$

Hence a set $K_0 := \{k_1, \dots, k_n\} \subset K$ satisfies the assumptions of Theorem 2. □

As a corollary we have:

Corollary 1 *Every finite-dimensional subspace of* $\mathcal{F}(K)$ *has an ideal complement.*

2 The Algebra of Continuous Functions

Theorem 3 *Let* $A := C(K)$ *be the algebra of continuous functions on a compact topological space. Then a closed subspace* $E \subset K$ *has a closed ideal compliment if and only if there exists a closed subset* $K_0 \subset K$ *such that Conditions 1 and 2 of Theorem 1 are satisfied.*

Proof It follows from Theorem 1 that if Conditions 1 and 2 of Theorem 1 for closed $K_0 \subset K$ hold then E is complemented by the ideal J_{K_0}. Since K_0 is closed, so is the ideal J_{K_0}. On the other hand, it is well-known that every closed ideal of $C(K)$ has a form $J_{K_0} := \{f \in A \text{ such that } f_{K_0} = 0\}$ for some closed subset $K_0 \subset K$. If J_{K_0} complements a subspace E, then for every $f \in A$ there exists $g \in E$ such that

$f - g \in J_{K_0}$ hence Condition 1 of Theorem 1 holds. Similarly, since $E \cap J_{K_0} = \{0\}$ Condition 2 of Theorem 1 holds. □

For instance the subspace $E \subset C[-1, 1]$ of even functions is complemented by an ideal $J \subset C[-1, 1]$. Since $E_{[0,1]} = A_{[0,1]}$ it is complemented by the ideal consisting of all functions that are identically zero on $[0, 1]$. The subspace of all odd functions is not complemented by an ideal.

3 The Algebra of Polynomials of Several Variables

In this section we will analyze the question of ideal complements in the algebra $\mathbb{C}[x_1, \ldots, x_m]$ of polynomials in m variables over the complex field. Here is a simple case:

Theorem 4 *Let $\mathbb{C}[x]$ be the algebra of polynomials of one variable. A subspace $E \subset \mathbb{C}[x]$ has an ideal complement if and only if E is finite-dimensional.*

Proof The fact that every finite-dimensional subspace has an ideal complement follows immediately from the Theorem 2. Conversely, every ideal in $\mathbb{C}[x]$ is principle ideal $J = \langle p \rangle$ hence codimension of J is $\deg p - 1$. □

General, Let $\mathbb{C}[\mathbf{x}] := \mathbb{C}[x_1, \ldots, x_m]$ be the algebra of polynomials in m variables. With every ideal $J \subset \mathbb{C}[\mathbf{x}]$ we associate a variety

$$\mathcal{V}(J) := \{\mathbf{x} \in \mathbb{C}^m : f(\mathbf{x}) = 0 \text{ for all } f \in J\}.$$

Similarly with every variety $\mathcal{V} \subset \mathbb{C}^m$ we associate an ideal

$$J(\mathcal{V}) = \{f \in \mathbb{C}[\mathbf{x}] : f(\mathbf{x}) = 0 \text{ for all } x \in \mathcal{V}\}.$$

The ideal J is called a "radical ideal" if $f^n \in J$ implies $f \in J$. Equivalently J is radical if and only if $J(\mathcal{V}(J)) = J$. Here is a simple consequence of Theorem 1:

Theorem 5 *A subspace $E \subset \mathbb{C}[\mathbf{x}]$ is complemented by a radical ideal if and only if there exists a variety $\mathcal{V} \subset \mathbb{C}^m$ such that $E_{\mathcal{V}} = \mathbb{C}[\mathbf{x}]_{\mathcal{V}}$ and $f \in E$ and $f_{\mathcal{V}} = 0$ implies that $f \equiv 0$.*

On the other hand, Condition 2 of Theorem 1 is not necessary for a space to have an ideal complement. Here is a simple example:

Let $E = \text{span}\{1, x\} \subset \mathbb{C}[x]$ and let $J = \langle x^2 \rangle$. Then clearly $E \oplus J = \mathbb{C}[x]$ and $x \in E$, $\mathcal{V}(J) = \{0\}$ and $x \in E$ vanishes on $\mathcal{V}(J)$. Notice that in this case J is not a radical ideal.

This leaves us with a general question:

0.1 What subspaces of $\mathbb{C}[\mathbf{x}]$ are complemented by an ideal?

Clearly every finite-dimensional space is complemented by an ideal and no subspace of $\mathbb{C}[\mathbf{x}]$ of finite codimension is complemented by ideal since no ideals in $\mathbb{C}[\mathbf{x}]$ has finite dimension. No proper ideal in $\mathbb{C}[\mathbf{x}]$ is complemented by an ideal. Indeed if J_1 and J_2 are two non-trivial ideals then for every $f_1 \in J_1$ and $f_2 \in J_2$ we have $0 \neq f_1 f_2 \in J_1 \cap J_2$. On the other hand an infinite-dimensional space $E = \mathbb{C}[x_1] \subset \mathbb{C}[x_1, x_2]$ is complemented by the ideal $J = \langle x_2 \rangle$. There is one more type of subspaces in $\mathbb{C}[\mathbf{x}]$ that have ideal complements. These are the $\mathcal{D}$-invariant subspaces:

Let D_j denotes the differential operator $\frac{d}{dx_j}$. A subspace $E \subset \mathbb{C}[x_1, \ldots, x_m]$ is called $\mathcal{D}$-invariant if for every $f \in E$ and every $j = 1, \ldots, m$ we have $D_j f \in E$.

Before stating the next theorem we will need a few definitions:

Define an inner product on $\mathbb{C}[x_1, \ldots, x_m]$ by

$$\langle f.g \rangle := f(D_1, \ldots D_m)(\bar{g})(0)$$

It is well-known (cf. [2], [3]) that $\{g : \langle f.g \rangle = 0 \text{ for all } f \in E\} = E^{\perp}$ is an ideal if and only if E is $\mathcal{D}$-invariant.

Theorem 6 *Let $E \subset \mathbb{C}[\mathbf{x}]$ be a closed $\mathcal{D}$-invariant subspace. Then E has an ideal complement.*

Proof If E is $\mathcal{D}$-invariant then the inner-product space $\mathbb{C}[x_1, \ldots, x_m] = E \oplus E^{\perp}$. The result follows since $E^{\perp}$ is an ideal. □

0.2 Is every $\mathcal{D}$-invariant subspace of $\mathbb{C}[\mathbf{x}]$ is closed with respect to the norm defined by the above inner product?

4 The Algebra of Matrices

In this section we will answer the question for algebra of matrices. Since there are no non-trivial ideals in this space we will characterize the subspaces that are complemented by left ideals. But we will start with a few interesting results that are true for the linear mappings between any linear space. These results provide a form of the Hilbert Nullstelensatz for linear operators.

Lemma 1 *Let X and Y and Z be linear spaces and Let $A_1, \ldots, A_n$ be linear operators from X to Y. The following are equivalent:*

$$\bigcap_{k=1}^{n} \ker A_k = \{0\}. \tag{1}$$

For any operator B from X to Z there exists a sequence $B_1, \ldots, B_n$ of operators from Y to Z such that

$$B = \sum_{k=1}^{n} B_k A_k. \tag{2}$$

Proof Assume that (1) holds. We will prove (2) by induction on n. For $n = 1$ we have $\ker A_1 = \{0\}$ hence A_1 is invertible from the left. Let A_1^{-l} be left inverse of A_1. Then $A_1^{-l} A_1 = I$ on X and hence

$$B = \left(B A_1^{-l}\right) A_1.$$

Suppose the result holds for n. If

$$(0) = \bigcap_{k=1}^{n} \ker A_k \tag{3}$$

then the result is trivial, just take $B_{n+1} = 0$. If (2) does not hold then choose

$$X_0 = \ker A_{n+1} \subset X.$$

Then $X_0 \bigcap \left(\bigcap_{k=1}^{n} \ker A_k \right) = \{0\}$ and from inductive assumption the restriction B_{X_0} of B on X_0 can be written as

$$B_{X_0} = \sum_{k=1}^{n} B_k \,(A_k)_{X_0} = \sum_{k=1}^{n+1} B_k \,(A_k)_{X_0} \tag{4}$$

since $(A_{n+1})_{X_0} = 0$, where B_k are defined on X_0. Let X_1 be a complement to X_0 in X, then $\ker A_{n+1} \cap X_1 = \{0\}$ and hence there exists B_{n+1} on X_1 such that

$$B - \sum_{k=1}^{n} B_k A_k = B_{n+1} A_{n+1},$$

hence

$$B = \sum_{k=1}^{n+1} B_k A_k$$

on X_0 and on X_1 and thus on all of X.

The converse is trivial. If $x_0 \in X$ and $0 \neq x_0 \in \bigcap_{k=1}^{n} \ker A_k$ then any B such that $B x_0 \neq 0$ does not satisfy (2). □

Corollary 2 *Let X and Y and Z be linear spaces and let $A_1, \dots, A_n$ be linear operators from X to Y. The following are equivalent:*

$$B = \sum_{k=1}^{n} B_k A_k \tag{5}$$

and

$$\bigcap_{k=1}^{n} \ker A_k \subset \ker B. \tag{6}$$

Proof If (5) holds then (6) is obvious. If (6) holds, let $X_0 = \bigcap_{k=1}^{n} \ker A_k$ and X_1 be a complement to X_0. Then (5) holds on X_0 for any (B_k) and, by Lemma 1 there exist B_k on X_1 such that (5) holds on X_1. This proves the corollary. □

As another corollary we obtain a matrix version of **The Hilbert Nullstelensatz**:

Theorem 7 *Let X be a finite-dimensional linear space, then a subspace $J \subset L(X)$ is a left ideal if and only if there exists a subspace $X_0 \subset X$ such that*

$$X_0 = \{\cap \ker A : A \in J\}. \tag{7}$$

Proof It is clear that $\{A \in L(X) : X_0 \subset \ker A\}$ is a left ideal in $L(X)$.

On the other hand if a subspace J satisfies (7) for some X_0. Since X_0 is a subspace in a finite-dimensional space there exist a finitely many operators $A_1, \dots, A_n$ in J such that $\bigcap_{k=1}^{n} \ker A_k = X_0$ and by Corollary 2 every $B \in L(X)$ can be written as $B = \sum_{k=1}^{n} B_k A_k \in J$. □

The next theorem is similar to the existence of ideal complements in $C(K)$.

Theorem 8 *Let X be a finite-dimensional space. Then a subspace $E \subset L(X)$ is complemented by a left ideal if and only if there exist a subspace $X_0 \subset X$ such that the restrictions $E_{X_0} :=$ span$\{A_{X_0} \in L(X_0, X) : A \in E\} = L(X_0, X)$ and if $A \in X_0$ and $A_{X_0} = 0$ then $A = 0$.*

Proof If $E_{X_0} = L(X_0, X)$ choose left ideal $J = \{A \in L(X) : X_0 \subset \ker A\}$. It is clear that $E_{X_0} + J = 0$ and $E_{X_0} \cap J = \{0\}$. On the other hand if $E \subset L(X)$ and $E \dot{+} J = L(X)$, Choose X_0 such that $J = \{A \in L(X) : X_0 \subset \ker A\}$. Since every operator $A = B + C$ with $B_{X_0} = 0$ then $C \in E$ span all the operators on X_0. □

One simple corollary of this theorem is that if $\dim X = n$ and $E \subset L(X)$ is complemented by a left ideal then $\dim E$ is divisible by n. Hence not every finite-dimensional subspace of $L(X)$ is complemented by a left ideal.

Using the obvious relation between left and right ideal in $L(X)$: $J \subset L(X)$ is a left ideal if and only if $J^*; = \{A^* : A \in J\}$ is the right ideal. Hence we have

Theorem 9 *Let X be a finite-dimensional space. Then a subspace $E \in L(X)$ is complemented by a right ideal if and only if the subspace E^* is complemented by the left ideal.*

Finally the theorem for continuous operators on a Banach space is true in one direction:

Theorem 10 *Let $L(X)$ be the algebra of all bounded operators on a Banach space X. Then a subspace $E \in L(X)$ for which there exist a subspace $X_0 \subset X$ such that the restrictions*

$$E_{X_0} := \text{span}\left\{A_{X_0} \in L\left(X_0, X\right) : A \in E\right\} = L\left(X_0, X\right)$$

and if $A \in X_0$ and $A_{X_0} = 0$ then $A = 0$. Then E is complemented by a left ideal.

The proof follows easy since $J \subset L(X)$ is a left ideal if for some subspace $X_0 \subset X$ we have $X_0 \subset \ker A$ for all $A \in J$.

However we do not know the general form of all left ideals in $L(X)$.

Acknowledgments I wish to thank the referee and Ivan Rothstein for corrections to the original manuscript.

References

1. Birkhoff, G. The algebra of multivariate interpolation. *Constructive Approaches To Mathematical Models (Proc. Conf. In Honor Of R. J. Duffin, Pittsburgh, Pa., 1978)*. pp. 345–363 (1979)
2. Boor, C. & Ron, A. On polynomial ideals of finite codimension with applications to box spline theory. *J. Math. Anal. Appl.*. **158**, 168–193 (1991), https://doi.org/10.1016/0022-247X(91)90275-5
3. Shekhtman, B. Ideal interpolation: translations to and from algebraic geometry. *Approximate Commutative Algebra*. pp. 163–192 (2009), https://doi.org/10.1007/978-3-211-99314-9_6

On Analytic Functions with Taylor Coefficients Taken from a Finite Set

Alexander V. Tovstolis

1 Introduction

Functions analytic in the unit disk $\mathbb{D}$ with Taylor series expansion

$$f(z) = \sum_{n=0}^{\infty} a_n z^n, \quad (1)$$

where a_n's belong to a finite set or are all integers have been considered since work of Fatou [10] and Carlson [4, 5]. In 1922, Szegö [16] proved the following striking result.

Theorem 1 *A power series (1) whose coefficients are taken from a finite set, is either a rational function, or it cannot be continued analytically beyond the unit disk.*[1]

In the case of a rational function, starting from some index, coefficients a_n form a periodic sequence, and

$$f(z) = \frac{P(z)}{1 - z^m},$$

where $P(z)$ is a polynomial, and $m \in \mathbb{Z}_+$.

[1] In the latter case, we will say that the domain of f has the unit circle as its natural boundary.

A. V. Tovstolis (✉)
Department of Mathematics, University of Central Florida, Orlando, FL, USA

D. Bilyk et al. (eds.), *Recent Advances in Approximation and Potential Theory*,
Applied and Numerical Harmonic Analysis,
https://doi.org/10.1007/978-3-031-95551-8_21

It is interesting to notice that if $f(z)$ in Theorem 1 is a rational function, then it is either a polynomial, or it has poles only at roots of unity.

A similar question is about a function with Taylor series having radius of convergence 1 and all the Taylor coefficients integer. Can we conclude that it is a rational function under some extra conditions? What are those conditions? For example, function belongs to some Hardy space in the unit disk, or even to the Nevanlinna space. There are a few extraordinary results in this direction.

R. J. Duffin and A. C. Schaeffer published the following result in 1945 (see [7, Part II, Th. I]):

Theorem 2 *Assume that the series (1) has only finitely many distinct coefficients* a_n. *If* $f(z)$ *is bounded in a sector of the unit disk, then it is a rational function.*

The following deep result is due to C. Pisot [14]:

Theorem 3 *Assume that the series (1) represents a function meromorphic in the closed unit disk* $\overline{\mathbb{D}}$ *and its coefficients* $a_n \pmod 1$ *belong to a finite set, then (1) is a rational function.*

A perfect introduction into this area can be found in L. Bieberbach's book [1, Ch. 6]. We will mention just one extremely helpful result—Kronecker's Criterion:

Theorem 4 *Function* $f(z)$ *represented by the series (1) is a rational function if and only if among determinants*

$$F_0^{(n+1)} := \begin{vmatrix} a_0 & a_1 & \dots & a_n \\ a_1 & a_2 & \dots & a_{n+1} \\ \vdots & \vdots & \ddots & \vdots \\ a_n & a_{n+1} & \dots & a_{2n} \end{vmatrix},$$

there are only finitely many non-zero ones.

Please note that $F_0^{(n+1)}$ are determinants of Hankel matrices.

Our main focus is on what can be said if $f(z)$ represented by (1) belongs to a Hardy space $H^p(\mathbb{D})$ or even to the Nevanlinna class $\mathcal{N}(\mathbb{D})$.

Let us start with a trivial case: $f \in H^p(\mathbb{D})$ with $p \geq 1$. In that case, f belongs to $H^1(\mathbb{D})$, and has non-tangential boundary values almost everywhere on the unit circle $\mathbb{T}$ (see, e.g., [12, Ch. II, § 3, Th. 3.1]). This function of boundary values (let us denote it by $F(t)$) can be considered as a 2π-periodic function of real variable t. It belongs to $L^1[0, 2\pi]$ and its Fourier coefficients are exactly a_n's from (1). Hence, by Riemann-Lebesgue Theorem, we obtain

$$\lim_{n\to\infty} a_n = 0.$$

If a_n's are taken from a finite set or are all integers, we immediately conclude that $a_n = 0$ for $n \geq K$. This means that $f(z)$ is a polynomial.

If now function f belongs to $H^p(\mathbb{D})$ for some $p \in (0, 1)$, then this trick does not work obviously.

For integer coefficients, D. G. Cantor proved even more general result in [2]:

Theorem 5 *Let $g(z)$ belong to the Nevanlinna class in the unit disk $\mathbb{D}$. If its Laurent series*

$$g(z) = \sum_{n=-\infty}^{\infty} a_n z^n$$

has all integer coefficients a_n, then g is a rational function.

Let us note that the Nevanlinna class $\mathcal{N}(\mathbb{D})$ includes all the Hardy classes $H^p(\mathbb{D})$, $p \in (0, \infty]$.

For coefficients taken from a finite set, there is a striking result by H. Helson obtained in [13, Theorem] with a very elegant proof:

Theorem 6 *Let $u(r, x) = \sum_{k=-\infty}^{\infty} a_k r^{|k|} e^{ikx}$ have only finitely many distinct coefficients, and for some arc (α, β) of the circle satisfy*

$$\int_{\alpha}^{\beta} |u(r, x)| \, dx \leq M < \infty, \qquad r < 1.$$

Then $\{a_k\}$ is ultimately periodic to the right, and ultimately periodic to the left.

Note that the statement of this theorem is equivalent to saying that $f(z) = u(|z|, \arg(z))$ is a rational function.

It is interesting that this result does not have an analogue for the case of integer coefficients. Namely, R. Salem [15] gave an example of a power series with integral coefficients, analytic in the unit disk $\mathbb{D}$ and **bounded in a sector**, which is not a rational function (see also discussion at the end of [13]).

Thus, for integral coefficients, the problem looks fully solved. For coefficients taken from a finite set, there are two problems still open:

Problem 1 Assume that function represented by (1) belongs to a Hardy space $H^p(\mathbb{D})$ for some $p \in (0, 1)$ and all a_n's belong to a finite set. Does it imply that f is a rational function? What about the Nevanlinna class?

Problem 2 If we replace the arc (α, β) in Theorem 6 by a set of positive Lebesgue measure, will it still imply that $f(z) = u(|z|, \arg(z))$ is a rational function?

The rest of the article gives some results which the author hopes would help to answer these questions. We also gave some insight for the case of $f(z)$ represented by (1) belonging to Bergman spaces $A^p(\mathbb{D})$.

2 Boundedness in a Sector Approach

Lemma 1 *Let X be a Lebesgue measurable subset of a non-trivial closed interval $[a, b]$ such that $[a, b] \setminus X$ is not dense in $[a, b]$. Then for any function $f : X \to \mathbb{C}$ Lebesgue measurable on X, there is a constant M, a nonempty open interval $(\alpha, \beta) \subset [a, b]$, and its dense subset E, such that*

$$|f(x)| < M, \quad \forall x \in E.$$

Proof Since $[a, b] \setminus X$ is not dense in $[a, b]$, there is a non-trivial interval $\left[a', b'\right] \subset [a, b]$, which does not contain points of $[a, b] \setminus X$, i.e., $\left[a', b'\right] \subset X$. Furthermore, since $f : X \to \mathbb{C}$ is Lebesgue measurable, the function $F : \left[a', b'\right] \to \mathbb{C}$ given by $F := f|_{[a',b']}$ is Lebesgue measurable.

Consider the following sets

$$E_n := \left\{x \in \left[a', b'\right] \mid |f(x)| < n\right\}, \quad n \in \mathbb{N}.$$

Since $f|_{[a',b']}$ is defined for every $x \in \left[a', b'\right]$, we conclude that

$$\left[a', b'\right] = \bigcup_{n=1}^{\infty} E_n.$$

Now, the closed interval $\left[a', b'\right]$ equipped with Euclidean metric is a complete metric space. Baire Category Theorem (see, e.g., [11, Ch. 5, § 5.3, Th. 5.9]) implies that there exists n_0 such that E_{n_0} is not nowhere dense in $\left[a', b'\right]$. Thus, there exists a nontrivial interval $[\alpha, \beta] \subset \left[a', b'\right] \subset [a, b]$, such that E_{n_0} is dense in $[\alpha, \beta]$. Thus, the statement follows with $E = E_{n_0}$, and $M = n_0$. □

Lemma 2 *Let $f \in H^p(\mathbb{D})$, for some $p \in (0, 1)$. Consider the following function*

$$F(\theta) := \sup_{r<1} \left| f\left(re^{i\theta}\right)\right|, \quad \theta \in [0, 2\pi],$$

and let

$$N := \{\theta \in [0, 2\pi] \mid F(\theta) = \infty\}.$$

If the set N is not dense in $[0, 2\pi]$, then there is a non-trivial interval $[\alpha, \beta] \subset [0, 2\pi]$, such that f is bounded in the sector $\{z \in \mathbb{D} \mid \alpha \le \arg z \le \beta\}$.

Proof According to the Hardy-Littlewood theorem (see, e.g., [8, Ch. 1, § 1.6, Th. 1.9]), $F \in L^p[0, 2\pi]$, whence it is Lebesgue measurable. Moreover, we conclude that N is a subset of $[0, 2\pi]$ of Lebesgue measure zero. Applying Lemma 1 to F with $[a, b] = [0, 2\pi]$ and $X = [0, 2\pi] \setminus N$, we conclude that there is a constant M, a non-trivial interval $[\alpha, \beta] \subset [0, 2\pi]$, and its dense subset E, such that

$$\sup_{r<1} \left| f\left(re^{i\theta}\right) \right| = F(\theta) < M, \quad \forall \theta \in E. \tag{2}$$

Note that since E is dense in $[\alpha, \beta]$, we have that for every $r \in (0, 1)$, the set

$$\left\{re^{i\theta} \mid \theta \in E\right\}$$

is dense in

$$\left\{re^{i\theta} \mid \theta \in [\alpha, \beta]\right\}.$$

Since also f is continuous in $\mathbb{D}$, (2) implies that

$$\left| f\left(re^{i\theta}\right) \right| \leq M, \quad \theta \in [\alpha, \beta].$$

Since the constant M in this inequality is independent of r, the statement follows.

□

Note that the assumption on the set N in Lemma 2 not to be dense in $[0, 2\pi]$ is essential. As an example, take any $p \in (0, 1)$, and consider the following function:

$$f(z) = \sum_{k=1}^{\infty} \frac{a_k}{z - e^{i\theta_k}}, \quad z \in \mathbb{D},$$

where $\{\theta_k\}_{k=1}^{\infty}$ is the enumeration of all rational numbers from $[0, 2\pi)$, and $\{a_k\}_{k=1}^{\infty}$ is any sequence of non-zero complex numbers, such that

$$\sum_{k=1}^{\infty} |a_k|^p < \infty.$$

It is clear that $f \in H^p(\mathbb{D})$, but there is no non-trivial interval $[\alpha, \beta] \subset [0, 2\pi]$, such that

$$\sup_{r<1} \int_{\alpha}^{\beta} \left| f\left(re^{i\theta}\right) \right| d\theta < \infty.$$

This immediately implies that f is unbounded in $\{z \in \mathbb{D} \mid \alpha \leq \arg z \leq \beta\}$.

From Lemma 2 and the Theorem 2 by R. J. Duffin and A. C. Schaeffer (see also Theorem 6 by H. Helson), we deduce the following statement:

Theorem 7 *Let $f \in H^p(\mathbb{D})$, for some $p \in (0,1)$, and assume that the Taylor series of f in $\mathbb{D}$ has only finitely many distinct coefficients. Consider the following function*

$$F(\theta) := \sup_{r<1} \left| f\left(re^{i\theta}\right) \right|, \quad \theta \in [0, 2\pi],$$

and let

$$N := \{\theta \in [0, 2\pi] \mid F(\theta) = \infty\}.$$

If the set N is not dense in $[0, 2\pi]$, then f is a rational function.

In fact, the note made after Lemma 2 makes it impossible to apply Helson's Theorem 6, which is more powerful than the one due to Duffin and Schaeffer, in case we drop the assumption on the set N not to be dense in $[0, 2\pi]$.

3 Boundedness on a Radial Set of Positive Lebesgue Measure

Theorem 8 *Let $f \in H^p(\mathbb{D})$, for some $p \in (0,1)$. Then for any $a \in (0, 2\pi)$, there exist a Lebesgue measurable set $\mathcal{E}_a \subset [0, 2\pi]$ and a constant $\mathcal{M}_{a,p}$, such that $m(\mathcal{E}_a) = a$, and*

$$\sup_{r<1, \theta \in \mathcal{E}_a} \left| f\left(re^{i\theta}\right) \right| \le \mathcal{M}_{a,p}.$$

Proof Let

$$F(\theta) := \sup_{r<1} \left| f\left(re^{i\theta}\right) \right|, \quad \theta \in [0, 2\pi],$$

According to the Hardy-Littlewood theorem (see, e.g., [8, Ch. 1, § 1.6, Th. 1.9]), $F \in L^p[0, 2\pi]$. Let us denote

$$N := \{\theta \in [0, 2\pi] \mid F(\theta) = \infty\}.$$

Consider the following function

$$\mathcal{F}(\theta) := \begin{cases} |F(\theta)|^p, & \theta \notin N, \\ -1, & \theta \in N. \end{cases}$$

Since the set N has Lebesgue measure 0, $\mathcal{F} \in L^1[0, 2\pi]$, and is defined on the whole closed interval $[0, 2\pi]$. By Chebyshev's inequality, for any $M > 0$,

$$m\left(\{\theta \in [0, 2\pi] \mid |\mathcal{F}(\theta)| \geq M\}\right) \leq \frac{1}{M} \|\mathcal{F}\|_{L^1[0,2\pi]} \cdot$$

Thus,

$$\lim_{M\to\infty} m\left(\{\theta \in [0, 2\pi] \mid |\mathcal{F}(\theta)| \geq M\}\right) = 0.$$

Therefore, for an arbitrary $a \in (0, 2\pi)$, there exists $M_a > 0$ such that

$$m\left(\{\theta \in [0, 2\pi] \mid |\mathcal{F}(\theta)| \geq M_a\}\right) < \frac{2\pi - a}{2}.$$

Hence, if we denote $A_a := \{\theta \in [0, 2\pi] \mid |\mathcal{F}(\theta)| < M_a\}$, then we get

$$m\left(A_a\right) > \frac{2\pi + a}{2}.$$

Now, let

$$g\left(t\right) := \int_{A_a} \chi_{[0,t]}\left(x\right) dx,$$

where $\chi_{[0,t]}\left(x\right)$ is the indicator function of $[0, t]$. Clearly, $g\left(t\right)$ is continuous in $[0, 2\pi]$ and

$$g\left(0\right) = 0, \qquad g\left(2\pi\right) = m\left(A_a\right) > \frac{2\pi + a}{2} > a.$$

By the Intermediate Value Theorem, there is $t_a \in (0, 2\pi)$ such that $g\left(t_a\right) = a$. Furthermore,

$$|\mathcal{F}(\theta)| < M_a, \qquad \forall \theta \in A_a \cap [0, t_a],$$

which implies

$$F\left(\theta\right) < M_a^{1/p}, \qquad \forall \theta \in (A_a \cap [0, t_a]) \setminus N.$$

Finally, since N has Lebesgue measure zero, $m\left((A_a \cap [0, t_a]) \setminus N\right) = g\left(t_a\right) = a$. Thus, the statement holds with $\mathcal{E}_a := (A_a \cap [0, t_a]) \setminus N$ and $\mathcal{M}_{a,p} := M_a^{1/p}$. □

4 Rational Functions

Proposition 1 *Let* $f(z) = \sum_{n=0}^{\infty} c_n z^n$ *belong to the Nevanlinna class* $\mathcal{N}(\mathbb{D})$, *and the Taylor coefficients* a_n *be Gaussian rationals*

$$c_n = \frac{p_n}{q_n} + i\frac{p_n'}{q_n'}, \qquad p_n, p_n' \in \mathbb{Z}, \ q_n, q_n' \in \mathbb{N}.$$

If, in addition, the denominators $q_1, \dots, q_1', \dots$ *are coming from a finite set of natural numbers, then* f *is a rational function.*

In particular, f *is a rational function if its Taylor coefficients are taken from a finite set of Gaussian rationals.*

Proof Since the denominators $q_1, \dots, q_1', \dots$ are coming from a finite set of natural numbers, $Q := lcm\left(q_1, \dots, q_1', \dots\right) \in \mathbb{N}$. Let us denote $a_n := c_n Q$, $n \in \mathbb{Z}_+$. The function

$$F(z) := Qf(z) = \sum_{n=0}^{\infty} a_n z^n \tag{3}$$

is clearly in the Nevanlinna class $\mathcal{N}(\mathbb{D})$, and has Gaussian integer Taylor coefficients.

Let us consider the following matrix

$$A_n := \left(a_{i+j}\right)_{0 \le j, r \le n} = \begin{pmatrix} a_0 & a_1 & a_2 & \dots & a_n \\ a_1 & a_2 & a_3 & \dots & a_{n+1} \\ a_2 & a_3 & a_4 & \dots & a_{n+2} \\ \vdots & \vdots & \vdots & \ddots & \vdots \\ a_n & a_{n+1} & a_{n+2} & \dots & a_{2n} \end{pmatrix}.$$

Cantor [2, Lemma 1] showed that for a function from the Nevanlinna class in the unit disk,

$$\lim_{n \to \infty} |\det(A_n)|^{1/n} = 0.$$

Therefore,

$$\lim_{n \to \infty} |\det(A_n)|^{2/n} = 0. \tag{4}$$

Since all the a_j's have integer real and imaginary parts, $|\det(A_n)|^2 \in \mathbb{Z}$, whence (4) implies there exists $N \in \mathbb{N}$, such that

$$\det(A_n) = 0, \qquad n \ge N.$$

Now, Kronecker's theorem implies that there is a recursive relation between the coefficients a_n (see, e.g., [6, Ch. VIII, Lemma 4]), i.e., $F(z)$ is a rational function. Relation (3) implies that $f(z)$ is a rational function, immediately. □

Let us note that in Bergman spaces $A^p(\mathbb{D})$, statements like the one obtained in [2], Proposition 1, etc., are not valid. One can construct a function g that belongs to every $A^p(\mathbb{D})$, $0 < p < \infty$, Taylor coefficients are all integers and are taken from a finite set, but it has no radial limits $\lim_{r\to 1-} g\left(re^{i\theta}\right)$, for any given θ, i.e., it has radial limits nowhere on the unit circumference. More precisely, the following statement holds true.

Lemma 3 *Let* $\{b_k\}$ *be a bounded sequence of complex numbers not satisfying* $\lim_{k\to\infty} b_k = 0$. *Then, there exists a subsequence of integers* $\{n_k\}$ *such that*

$$g(z) := \sum_{k=0}^{\infty} b_k z^{n_k} \tag{5}$$

belongs to every $A^p(\mathbb{D})$, $0 < p < \infty$, *but has radial limits* $\lim_{r\to 1-} g\left(re^{i\theta}\right)$ *nowhere on the unit circumference.*

Proof Consider the following function

$$\varphi(r) = 1 + |\ln(1-r)|\,, \quad 0 \le r < 1.$$

This function is continuous, strictly positive on $[0, 1)$, and $\lim_{r\to 1-} \varphi(r) = \infty$.

According to [3, Th. 2], for any sequence $\{b_k\}$ not satisfying $\lim_{k\to\infty} b_k = 0$, there exists a subsequence of integers $\{n_k\}$ such that $g(z)$ given by (5) is analytic in the unit disk, has radial limits nowhere on the unit circumference, and

$$\max_{|z|=r} |g(z)| \le \varphi(r), \quad 0 \le r < 1.$$

Since, clearly,

$$\int_0^1 |\varphi(r)|^p\, r\, dr = \int_0^1 (1 + |\ln(1-r)|)^p\, r\, dr < \infty, \quad 0 < p < \infty,$$

we obtain $g \in A^p(\mathbb{D})$, for any $0 < p < \infty$. □

For the details on constructing the subsequence $\{n_k\}$, see [3].

Speaking about Bergman spaces, the following statement holds.

Lemma 4 *Let* f *be holomorphic in the unit disk* $\mathbb{D}$ *and has the Taylor expansion* $\sum_{n=0}^{\infty} a_n z^n$, *where* $0 < m \le |a_n| \le M < \infty$, $n \in \mathbb{Z}_+$. *Then* $f \in A^p(\mathbb{D})$ *if and only if* $0 < p < 2$.

Proof Horowitz's theorem (see, e.g., [9, Ch. 3, § 3.3, Th. 5(i)]; applied with $\lambda = 1$) yields $f \in A^p(\mathbb{D})$, for any $0 < p < 2$.

From another side, if f were in $A^2(\mathbb{D})$, then we would have (see [9, Ch. 3, § 3.3, Th. 2(i)])

$$\sum_{n=0}^{\infty} \frac{|a_n|^2}{n} < \infty.$$

But this is impossible since $|a_n| \geq m > 0$. Therefore, $f \notin A^2(\mathbb{D})$, and whence $f \notin A^p(\mathbb{D})$ for $p \geq 2$. □

Note Considering relations between Hardy and Bergman spaces ($H^p(\mathbb{D}) \subset A^{2p}(\mathbb{D})$ for $p \in (0, \infty)$—see, e.g., [9, Ch. 3, § 3.2]), we conclude that a function satisfying the conditions of Lemma 4 **does not** belong to $H^p(\mathbb{D})$ for $1 \leq p \leq \infty$.

Acknowledgments The author is grateful to Professor Igor E. Pritsker (Oklahoma State University) for introduction to the area and helpful discussions, and to the anonymous referee for their useful suggestions (in particular, on simplification of the proof of Theorem 8).

References

1. Bieberbach, Ludwig. Analytische Fortsetzung. Ergebnisse der Mathematik und ihrer Grenzgebiete, (N.F.), Heft 3, Springer-Verlag, Berlin-Göttingen-Heidelberg, 1955.
2. Cantor, David G. Power series with integral coefficients. Bull. Amer. Math. Soc. 69, 362–366 (1963).
3. Cantor, David G. A simple construction of analytic functions without radial limits. Proc. Amer. Math. Soc. 15, 335–336 (1964).
4. Carlson, F. Über Potenzreihen mit endlich verschiedenen Koeffizienten. Math. Ann. 79, 237–245 (1919).
5. Carlson, F. Über Potenzreihen mit ganzzahligen Koeffizienten. Math. Z., 9, 1–13 (1921).
6. Cassels, J. W. S. An introduction to Diophantine approximation. Cambridge Tracts in Mathematics and Mathematical Physics, No. 45. Cambridge University Press, New York, 1957.
7. Duffin, R. J.; Schaeffer, A. C. Power series with bounded coefficients. Amer. J. Math. 67, 141–154 (1945).
8. Duren, Peter, Theory of H^p spaces. Academic Press, New York and London, 1970.
9. Duren, Peter; Schuster, Alexander Bergman spaces. Mathematical Surveys and Monographs, 100. American Mathematical Society, Providence, RI, 2004.
10. Fatou, P. Séries trigonométriques et séries de Taylor. Acta Math. 30, no. 1, 335–400 (1906).
11. Folland, G. B. Real analysis. Modern techniques and their applications. 2nd Edition. Pure and Applied Mathematics (New York). A Wiley-Interscience Publication. John Wiley & Sons, Inc., New York, 1999.
12. Garnett, John B. Bounded analytic functions. Pure Appl. Math., 96. Academic Press, Inc. [Harcourt Brace Jovanovich, Publishers], New York-London, 1981.
13. Helson, Henry, On a theorem of Szegö. Proceedings of the Amer. Math. Soc., Vol. 6, No. 2, 235–242 (1955).
14. Pisot, Charles. Propriétés arithmétiques des coefficients des séries de Taylor. C. R. Acad. Sci. Paris 224, 438–440 (1947).
15. Salem, Raphaël. Power series with integral coefficients. Duke Math. J. 12, 153–172 (1945).
16. Szegö, G. Über Potenzreihen mit endlich vielen verschiedenen Koeffizienten. Sitzgsber. preuss. Akad. Wiss., Math.-phys. Kl., 88–91 (1922).

Applied and Numerical Harmonic Analysis (112 volumes)

1. A. I. Saichev and W. A. Woyczyński: *Distributions in the Physical and Engineering Sciences* (ISBN: 978-0-8176-3924-2)
2. C. E. D'Attellis and E. M. Fernandez-Berdaguer: *Wavelet Theory and Harmonic Analysis in Applied Sciences* (ISBN: 978-0-8176-3953-2)
3. H. G. Feichtinger and T. Strohmer: *Gabor Analysis and Algorithms* (ISBN: 978-0-8176-3959-4)
4. R. Tolimieri and M. An: *Time-Frequency Representations* (ISBN: 978-0-8176-3918-1)
5. T. M. Peters and J. C. Williams: *The Fourier Transform in Biomedical Engineering* (ISBN: 978-0-8176-3941-9)
6. G. T. Herman: *Geometry of Digital Spaces* (ISBN: 978-0-8176-3897-9)
7. A. Teolis: *Computational Signal Processing with Wavelets* (ISBN: 978-0-8176-3909-9)
8. J. Ramanathan: *Methods of Applied Fourier Analysis* (ISBN: 978-0-8176-3963-1)
9. J. M. Cooper: *Introduction to Partial Differential Equations with MATLAB* (ISBN: 978-0-8176-3967-9)
10. Procházka, N. G. Kingsbury, P. J. Payner, and J. Uhlir: *Signal Analysis and Prediction* (ISBN: 978-0-8176-4042-2)
11. W. Bray and C. Stanojevic: *Analysis of Divergence* (ISBN: 978-1-4612-7467-4)
12. G. T. Herman and A. Kuba: *Discrete Tomography* (ISBN: 978-0-8176-4101-6)
13. K. Gröchenig: *Foundations of Time-Frequency Analysis* (ISBN: 978-0-8176-4022-4)
14. L. Debnath: *Wavelet Transforms and Time-Frequency Signal Analysis* (ISBN: 978-0-8176-4104-7)
15. J. J. Benedetto and P. J. S. G. Ferreira: *Modern Sampling Theory* (ISBN: 978-0-8176-4023-1)

D. Bilyk et al. (eds.), *Recent Advances in Approximation and Potential Theory*, Applied and Numerical Harmonic Analysis,
https://doi.org/10.1007/978-3-031-95551-8

16. D. F. Walnut: *An Introduction to Wavelet Analysis* (ISBN: 978-0-8176-3962-4)
17. A. Abbate, C. DeCusatis, and P. K. Das: *Wavelets and Subbands* (ISBN: 978-0-8176-4136-8)
18. O. Bratteli, P. Jorgensen, and B. Treadway: *Wavelets Through a Looking Glass* (ISBN: 978-0-8176-4280-80)
19. H. G. Feichtinger and T. Strohmer: *Advances in Gabor Analysis* (ISBN: 978-0-8176-4239-6)
20. O. Christensen: *An Introduction to Frames and Riesz Bases* (ISBN: 978-0-8176-4295-2)
21. L. Debnath: *Wavelets and Signal Processing* (ISBN: 978-0-8176-4235-8)
22. G. Bi and Y. Zeng: *Transforms and Fast Algorithms for Signal Analysis and Representations* (ISBN: 978-0-8176-4279-2)
23. J. H. Davis: *Methods of Applied Mathematics with a MATLAB Overview* (ISBN: 978-0-8176-4331-7)
24. J. J. Benedetto and A. I. Zayed: *Sampling, Wavelets, and Tomography* (ISBN: 978-0-8176-4304-1)
25. E. Prestini: *The Evolution of Applied Harmonic Analysis* (ISBN: 978-0-8176-4125-2)
26. L. Brandolini, L. Colzani, A. Iosevich, and G. Travaglini: *Fourier Analysis and Convexity* (ISBN: 978-0-8176-3263-2)
27. W. Freeden and V. Michel: *Multiscale Potential Theory* (ISBN: 978-0-8176-4105-4)
28. O. Christensen and K. L. Christensen: *Approximation Theory* (ISBN: 978-0-8176-3600-5)
29. O. Calin and D.-C. Chang: *Geometric Mechanics on Riemannian Manifolds* (ISBN: 978-0-8176-4354-6)
30. J. A. Hogan: *Time-Frequency and Time-Scale Methods* (ISBN: 978-0-8176-4276-1)
31. C. Heil: *Harmonic Analysis and Applications* (ISBN: 978-0-8176-3778-1)
32. K. Borre, D. M. Akos, N. Bertelsen, P. Rinder, and S. H. Jensen: *A Software-Defined GPS and Galileo Receiver* (ISBN: 978-0-8176-4390-4)
33. T. Qian, M. I. Vai, and Y. Xu: *Wavelet Analysis and Applications* (ISBN: 978-3-7643-7777-9)
34. G. T. Herman and A. Kuba: *Advances in Discrete Tomography and Its Applications* (ISBN: 978-0-8176-3614-2)
35. M. C. Fu, R. A. Jarrow, J.-Y. Yen, and R. J. Elliott: *Advances in Mathematical Finance* (ISBN: 978-0-8176-4544-1)
36. O. Christensen: *Frames and Bases* (ISBN: 978-0-8176-4677-6)
37. P. E. T. Jorgensen, J. D. Merrill, and J. A. Packer: *Representations, Wavelets, and Frames* (ISBN: 978-0-8176-4682-0)
38. M. An, A. K. Brodzik, and R. Tolimieri: *Ideal Sequence Design in Time-Frequency Space* (ISBN: 978-0-8176-4737-7)
39. S. G. Krantz: *Explorations in Harmonic Analysis* (ISBN: 978-0-8176-4668-4)

40. B. Luong: *Fourier Analysis on Finite Abelian Groups* (ISBN: 978-0-8176-4915-9)
41. G. S. Chirikjian: *Stochastic Models, Information Theory, and Lie Groups, Volume 1* (ISBN: 978-0-8176-4802-2)
42. C. Cabrelli and J. L. Torrea: *Recent Developments in Real and Harmonic Analysis* (ISBN: 978-0-8176-4531-1)
43. M. V. Wickerhauser: *Mathematics for Multimedia* (ISBN: 978-0-8176-4879-4)
44. B. Forster, P. Massopust, O. Christensen, K. Gröchenig, D. Labate, P. Vandergheynst, G. Weiss, and Y. Wiaux: *Four Short Courses on Harmonic Analysis* (ISBN: 978-0-8176-4890-9)
45. O. Christensen: *Functions, Spaces, and Expansions* (ISBN: 978-0-8176-4979-1)
46. J. Barral and S. Seuret: *Recent Developments in Fractals and Related Fields* (ISBN: 978-0-8176-4887-9)
47. O. Calin, D.-C. Chang, and K. Furutani, and C. Iwasaki: *Heat Kernels for Elliptic and Sub-elliptic Operators* (ISBN: 978-0-8176-4994-4)
48. C. Heil: *A Basis Theory Primer* (ISBN: 978-0-8176-4686-8)
49. J. R. Klauder: *A Modern Approach to Functional Integration* (ISBN: 978-0-8176-4790-2)
50. J. Cohen and A. I. Zayed: *Wavelets and Multiscale Analysis* (ISBN: 978-0-8176-8094-7)
51. D. Joyner and J.-L. Kim: *Selected Unsolved Problems in Coding Theory* (ISBN: 978-0-8176-8255-2)
52. G. S. Chirikjian: *Stochastic Models, Information Theory, and Lie Groups, Volume 2* (ISBN: 978-0-8176-4943-2)
53. J. A. Hogan and J. D. Lakey: *Duration and Bandwidth Limiting* (ISBN: 978-0-8176-8306-1)
54. G. Kutyniok and D. Labate: *Shearlets* (ISBN: 978-0-8176-8315-3)
55. P. G. Casazza and P. Kutyniok: *Finite Frames* (ISBN: 978-0-8176-8372-6)
56. V. Michel: *Lectures on Constructive Approximation* (ISBN : 978-0-8176-8402-0)
57. D. Mitrea, I. Mitrea, M. Mitrea, and S. Monniaux: *Groupoid Metrization Theory* (ISBN: 978-0-8176-8396-2)
58. T. D. Andrews, R. Balan, J. J. Benedetto, W. Czaja, and K. A. Okoudjou: *Excursions in Harmonic Analysis, Volume 1* (ISBN: 978-0-8176-8375-7)
59. T. D. Andrews, R. Balan, J. J. Benedetto, W. Czaja, and K. A. Okoudjou: *Excursions in Harmonic Analysis, Volume 2* (ISBN: 978-0-8176-8378-8)
60. D. V. Cruz-Uribe and A. Fiorenza: *Variable Lebesgue Spaces* (ISBN: 978-3-0348-0547-6)
61. W. Freeden and M. Gutting: *Special Functions of Mathematical (Geo-)Physics* (ISBN: 978-3-0348-0562-9)
62. A. I. Saichev and W. A. Woyczyński: *Distributions in the Physical and Engineering Sciences, Volume 2: Linear and Nonlinear Dynamics of Continuous Media* (ISBN: 978-0-8176-3942-6)

63. S. Foucart and H. Rauhut: *A Mathematical Introduction to Compressive Sensing* (ISBN: 978-0-8176-4947-0)
64. G. T. Herman and J. Frank: *Computational Methods for Three-Dimensional Microscopy Reconstruction* (ISBN: 978-1-4614-9520-8)
65. A. Paprotny and M. Thess: *Realtime Data Mining: Self-Learning Techniques for Recommendation Engines* (ISBN: 978-3-319-01320-6)
66. A. I. Zayed and G. Schmeisser: *New Perspectives on Approximation and Sampling Theory: Festschrift in Honor of Paul Butzer's 85th Birthday* (ISBN: 978-3-319-08800-6)
67. R. Balan, M. Begue, J. Benedetto, W. Czaja, and K. A. Okoudjou: *Excursions in Harmonic Analysis, Volume 3* (ISBN: 978-3-319-13229-7)
68. H. Boche, R. Calderbank, G. Kutyniok, and J. Vybiral: *Compressed Sensing and its Applications* (ISBN: 978-3-319-16041-2)
69. S. Dahlke, F. De Mari, P. Grohs, and D. Labate: *Harmonic and Applied Analysis: From Groups to Signals* (ISBN: 978-3-319-18862-1)
70. A. Aldroubi: *New Trends in Applied Harmonic Analysis* (ISBN: 978-3-319-27871-1)
71. M. Ruzhansky: *Methods of Fourier Analysis and Approximation Theory* (ISBN: 978-3-319-27465-2)
72. G. Pfander: *Sampling Theory, a Renaissance* (ISBN: 978-3-319-19748-7)
73. R. Balan, M. Begue, J. Benedetto, W. Czaja, and K. A. Okoudjou: *Excursions in Harmonic Analysis, Volume 4* (ISBN: 978-3-319-20187-0)
74. O. Christensen: *An Introduction to Frames and Riesz Bases, Second Edition* (ISBN: 978-3-319-25611-5)
75. E. Prestini: *The Evolution of Applied Harmonic Analysis: Models of the Real World, Second Edition* (ISBN: 978-1-4899-7987-2)
76. J. H. Davis: *Methods of Applied Mathematics with a Software Overview, Second Edition* (ISBN: 978-3-319-43369-1)
77. M. Gilman, E. M. Smith, and S. M. Tsynkov: *Transionospheric Synthetic Aperture Imaging* (ISBN: 978-3-319-52125-1)
78. S. Chanillo, B. Franchi, G. Lu, C. Perez, and E. T. Sawyer: *Harmonic Analysis, Partial Differential Equations and Applications* (ISBN: 978-3-319-52741-3)
79. R. Balan, J. Benedetto, W. Czaja, M. Dellatorre, and K. A. Okoudjou: *Excursions in Harmonic Analysis, Volume 5* (ISBN: 978-3-319-54710-7)
80. I. Pesenson, Q. T. Le Gia, A. Mayeli, H. Mhaskar, and D. X. Zhou: *Frames and Other Bases in Abstract and Function Spaces: Novel Methods in Harmonic Analysis, Volume 1* (ISBN: 978-3-319-55549-2)
81. I. Pesenson, Q. T. Le Gia, A. Mayeli, H. Mhaskar, and D. X. Zhou: *Recent Applications of Harmonic Analysis to Function Spaces, Differential Equations, and Data Science: Novel Methods in Harmonic Analysis, Volume 2* (ISBN: 978-3-319-55555-3)
82. F. Weisz: *Convergence and Summability of Fourier Transforms and Hardy Spaces* (ISBN: 978-3-319-56813-3)
83. C. Heil: *Metrics, Norms, Inner Products, and Operator Theory* (ISBN: 978-3-319-65321-1)

84. S. Waldron: *An Introduction to Finite Tight Frames: Theory and Applications.* (ISBN: 978-0-8176-4814-5)
85. D. Joyner and C. G. Melles: *Adventures in Graph Theory: A Bridge to Advanced Mathematics.* (ISBN: 978-3-319-68381-2)
86. B. Han: *Framelets and Wavelets: Algorithms, Analysis, and Applications* (ISBN: 978-3-319-68529-8)
87. H. Boche, G. Caire, R. Calderbank, M. März, G. Kutyniok, and R. Mathar: *Compressed Sensing and Its Applications* (ISBN: 978-3-319-69801-4)
88. A. I. Saichev and W. A. Woyczyński: *Distributions in the Physical and Engineering Sciences, Volume 3: Random and Fractal Signals and Fields* (ISBN: 978-3-319-92584-4)
89. G. Plonka, D. Potts, G. Steidl, and M. Tasche: *Numerical Fourier Analysis* (978-3-030-04305-6)
90. K. Bredies and D. Lorenz: *Mathematical Image Processing* (ISBN: 978-3-030-01457-5)
91. H. G. Feichtinger, P. Boggiatto, E. Cordero, M. de Gosson, F. Nicola, A. Oliaro, and A. Tabacco: *Landscapes of Time-Frequency Analysis* (ISBN: 978-3-030-05209-6)
92. E. Liflyand: *Functions of Bounded Variation and Their Fourier Transforms* (ISBN: 978-3-030-04428-2)
93. R. Campos: *The XFT Quadrature in Discrete Fourier Analysis* (ISBN: 978-3-030-13422-8)
94. M. Abell, E. Iacob, A. Stokolos, S. Taylor, S. Tikhonov, J. Zhu: *Topics in Classical and Modern Analysis: In Memory of Yingkang Hu* (ISBN: 978-3-030-12276-8)
95. H. Boche, G. Caire, R. Calderbank, G. Kutyniok, R. Mathar, P. Petersen: *Compressed Sensing and its Applications: Third International MATHEON Conference 2017* (ISBN: 978-3-319-73073-8)
96. A. Aldroubi, C. Cabrelli, S. Jaffard, U. Molter: *New Trends in Applied Harmonic Analysis, Volume II: Harmonic Analysis, Geometric Measure Theory, and Applications* (ISBN: 978-3-030-32352-3)
97. S. Dos Santos, M. Maslouhi, K. Okoudjou: *Recent Advances in Mathematics and Technology: Proceedings of the First International Conference on Technology, Engineering, and Mathematics, Kenitra, Morocco, March 26-27, 2018* (ISBN: 978-3-030-35201-1)
98. Á. Bényi, K. Okoudjou: *Modulation Spaces: With Applications to Pseudodifferential Operators and Nonlinear Schrödinger Equations* (ISBN: 978-1-0716-0330-7)
99. P. Boggiato, M. Cappiello, E. Cordero, S. Coriasco, G. Garello, A. Oliaro, J. Seiler: *Advances in Microlocal and Time-Frequency Analysis* (ISBN: 978-3-030-36137-2)
100. S. Casey, K. Okoudjou, M. Robinson, B. Sadler: *Sampling: Theory and Applications* (ISBN: 978-3-030-36290-4)

101. P. Boggiatto, T. Bruno, E. Cordero, H. G. Feichtinger, F. Nicola, A. Oliaro, A. Tabacco, M. Vallarino: *Landscapes of Time-Frequency Analysis: ATFA 2019* (ISBN: 978-3-030-56004-1)
102. M. Hirn, S. Li, K. Okoudjou, S. Saliana, Ö. Yilmaz: *Excursions in Harmonic Analysis, Volume 6: In Honor of John Benedetto's 80th Birthday* (ISBN: 978-3-030-69636-8)
103. F. De Mari, E. De Vito: *Harmonic and Applied Analysis: From Radon Transforms to Machine Learning* (ISBN: 978-3-030-86663-1)
104. G. Kutyniok, H. Rauhut, R. J. Kunsch, *Compressed Sensing in Information Processing* (ISBN: 978-3-031-09744-7)
105. P. Flandrin, S. Jaffard, T. Paul, B. Torresani, *Theoretic Physics, Wavelets, Analysis, Genomics: An Indisciplinary Tribute to Alex Grossmann* (ISBN: 978-3-030-45846-1)
106. G. Plonka-Hoch, D. Potts, G. Steidl, M. Tasche, *Numerical Fourier Analysis, Second Edition* (ISBN: 978-3-031-35004-7)
107. P. Alonso Ruiz, M. Hinz, K. Okoudjou, L. Rogers, A. Teplyaev, *From Classical Analysis to Analysis on Fractals: A Tribute to Robert Strichartz, Volume 1* (ISBN: 978-3-031-37799-0)
108. S. Casey, M. Dodson, P. Ferreira, A. Zayed, *Sampling, Approximation, and Signal Analysis: Harmonic Analysis in the Spirit of J. Rowland Higgins* (ISBN: 978-3-031-41129-8)
109. J. Feuto, B. A. Kpata, *Harmonic Analysis and Partial Differential Equations: Proceedings of the Workshop in Abidjan, Côte de'Ivoire, May 22-26, 2023* (ISBN: 978-3-031-66374-1)
110. S. Foucart, S. Wojtowytsch, *Explorations in the Mathematics of Data Science: The Inaugural Volume of the Center for Approximation and Mathematical Data Analytics* (ISBN: 978-3-031-66496-0)
111. E. Hernández, M. Peloso, F. Ricci, F. Soria, A. Tabaco, *The Mathematical Heritage of Guido Weiss* (ISBN: 978-3-031-76792-0)
112. D. Bilyk, E. Iacob, A. Martinez-Finkelshtein, A. Stokolos, *Recent Advances in Approximation and Potential Theory* (ISBN: 978-3-031-95550-1)

For an up-to-date list of ANHA titles, please visit http://www.springer.com/series/4968

MIX
Papier aus verantwortungsvollen Quellen
Paper from responsible sources
FSC® C105338

If you have any concerns about our products,
you can contact us on
ProductSafety@springernature.com

In case Publisher is established outside the EU,
the EU authorized representative is:
Springer Nature Customer Service Center GmbH
Europaplatz 3, 69115 Heidelberg, Germany

Printed by Libri Plureos GmbH
in Hamburg, Germany